THE LIVING PLANET IN CRISIS

THE LIVING PLANET IN CRISIS

BIODIVERSITY SCIENCE AND POLICY

Joel Cracraft and Francesca T. Grifo

COLUMBIA UNIVERSITY PRESS ▟▙ NEW YORK

Columbia University Press

Publishers Since 1893

New York   Chichester, West Sussex

Library of Congress Cataloging-in-Publication Data

The living planet in crisis : biodiversity science and policy /

 [edited by] Joel Cracraft and Francesca T. Grifo.

   p.   cm.

  Based on a conference held at the American Museum of Natural
History in 1995.

  Includes bibliographical references and index.

  ISBN 0–231–10864–8 (cloth : alk. paper). — ISBN 0–231–10865–6
(pbk. : alk. paper)

  1. Biological diversity conservation—Government policy—
Congresses. 2. Endangered species—Social aspects—Congresses.
3. Biological diversity—Congresses.   I. Cracraft, Joel.
II. Grifo, Francesca T.

QH75.A1L58   1999

333.95'16—dc21   98–31465

Casebound editions of Columbia University Press books
are printed on permanent and durable acid-free paper.

Printed in the United States of America

c 10 9 8 7 6 5 4 3 2 1

p 10 9 8 7 6 5 4 3 2 1

The artwork for the figures in chapter 4 was
provided by Steven S. Kometani.

CONTENTS

Edward O. Wilson

The Living Planet in Crisis: Biodiversity Science and Policy addresses a very important subject as viewed by some of its foremost experts. The writing is sufficiently detailed and up-to-date to be used by other professionals, yet clear enough to be understood by a general audience. In short, this book is what we expect from the much admired and emulated American Museum of Natural History. Among its strongest features are dispatches from the research front by leading experts on systematics and biogeography, telling us what is known and what is yet to be learned about the status of global biodiversity.

The authors bring into focus what I consider to be the two defining images of the biodiversity crisis: the Malthusian wall, which threatens the welfare of humanity and the very existence of biodiversity, and the bottleneck we must pass through to avoid it. The wall now stares us in the face: the human population is already precariously large, and will become critical before peaking sometime in the second half of the coming century. In 1997, as I write, the world is crowded with 5.8 billion people, up from 2 billion in 1940. The per capita global birthrate is dropping, fortunately, having declined from 4.1 children per woman in 1963 to 2.6 children now. Eventual zero population growth can be attained at 2.1 children. But even if the 2.1 level were reached immediately, there would be about 7.7 billion people in 2060. In fact, it will not be reached. We are almost certain to climb to 8 billion by 2020, and then grow still more. Most demographers believe the human population will peak at 8 to 10 billion, then begin slowly to subside.

How many people can Earth actually support? If everyone agreed to become vegetarian, leaving nothing for livestock, the 3.5 billion acres of land under cultivation today could feed about 10 billion at present levels of productivity. But the standard of living of the industrialized countries cannot be attained by the rest of the world. At present, each person in the United States is supported by an average of 12 acres of productive land, by which is meant land commandeered somewhere in the world for food, wood, fuel, and waste disposal. In most developing countries this ecological footprint is 1 acre. To raise the standard of living of the whole world to the American level with existing technology and resources would require two more planet Earths.

It will not happen. We are drawing down Earth's capital too rapidly. Here are some of the grim facts:

- From 1950 to the mid-1990s the area of cropland per person fell by half, to less than the size of a soccer field. The green revolution, by boosting fertilizer and high-yield varieties of rice and other crops, kept production rising, but it is now approaching its limit. By 1985 the growth in agricultural yield began to slow. That subsidence, combined with population growth, initiated an absolute decline in per capita production. By 1996 the world grain carryover stocks, humanity's emergency food supply, had fallen 50 percent from the all-time peak reached in 1987.

- Too many people already compete for too little water. Water tables are dropping in many parts of the world—120 feet beneath Beijing between 1965 and 1995, for example, and 10 feet through a fifth of the Ogallala aquifer, a major water source of the central United States, during the 1980s alone. On a global scale, humanity now uses a quarter of the accessible water released to the atmosphere by evaporation and plant transpiration and somewhat more than half that available in rivers and other runoff channels.

- All 17 of the world's oceanic fisheries are now being harvested beyond their capacity, and the world food catch has leveled off at about 100 million tons.

- The remnant forests and savannas could be cleared and planted, but at the cost of most of Earth's remaining fauna and flora and with no more than minor agricultural gain. Nearly half of the forests and savannas in tropical countries are underlain by soils of low natural fertility: 42 percent of the untapped area of sub-Saharan Africa, for example, and 46 percent of that in Latin America.

In summary, the human population is approaching the limit at which an acceptable quality of life for its entirety can be attained. Further population growth will soon hit the wall, constrained not by energy or minerals, but by a shortage of food and water and the land to produce them. This brings us to the bottleneck through which our species must now squeeze itself. In the decades immediately ahead, as the global population continues to soar to increase production, we will need every technological fix that genius can provide to increase production. It will also be necessary for the developed countries to shrink the size of their ecological footprints, by technology directed at higher sustainable production and by a new ethic of self-restraint based on a decent concern for the rest of the world and for future generations everywhere.

We hope—surely we have to believe—that our species will emerge from the environmental bottleneck, and safely past the wall, in better condition than we entered. But implicit in this odyssey of the twenty-first century is the great responsibility that cannot be ignored: preserving Creation by taking as much of life with us as possible. That is the clear and urgent message conveyed by the authors of this book.

The chapters in this volume grew out of a conference of leading experts on biodiversity held at the American Museum of Natural History in 1995. The goal of the conference was to assess the status of global biodiversity and discuss how scientists and policymakers might work together on conservation solutions. The conference, attended by over 600 people, was the first to the sponsored by the AMNH Center for Biodiversity and Conservation.

We would like to thank all the staff members of the AMNH, too numerous to mention by name, for their hard work in making the conference possible. Valeda Slade contributed significantly to the editorial process, for which we are appreciative. We are especially grateful for the input of Walter V. Reid and the World Resources Institute for the help they gave in preparing for the conference. We also thank the numerous scientists who participated in discussion panels but did not contribute to this volume: Walter V. Reid, Elliott A. Norse, Robert Repetto, Robert T. Watson, Michael Balick, Julie M. Feinsilver, Nazli Choucri, Eduardo Fuentes, Jonathan Coddington, Rodrigo Gámez, Kenton Miller, Kathryn A. Saterson, Mary Pearl, Niles Eldredge, Brian Boom, Don Melnick, John Robinson, Peter Thomas, and Kathryn Fuller. Finally, we gratefully acknowledge the support and encouragement of Ellen V. Futter, president of the AMNH, and Michael J. Novacek, senior vice president and provost.

The Living Planet in Crisis conference was made possible through the generous support of the following sponsors and contributors to the Center for Biodiversity and Conservation: American Express Publishing Corporation, The Bay Foundation, Condé Nast Publications Inc., The Geraldine R. Dodge Foundation, The Starr Foundation, Mr. and Mrs. Andrew L. Farkas, Mrs. Karen Lauder, Origins Natural Resources Inc.–Estée Lauder Companies, Howard Phipps Foundation, and the Rockefeller Brothers Fund Inc.

THE LIVING PLANET IN CRISIS:
BIODIVERSITY SCIENCE AND POLICY

Joel Cracraft and Francesca T. Grifo

The transformation of Earth's habitats and ecosystems, along with an inevitable loss of biological diversity, involves a complicated set of social, economic, and political issues. Whereas most public discourse describes the problem from a first-world perspective of the loss of endearing species or the wanton destruction of natural beauty, the real story of biodiversity loss is a complex melange of mind-numbing concerns such as economic inequalities and pernicious incentives, poverty, agricultural policies and food security, political justice, and many other societal manifestations of the way in which people relate to the land and the resources it houses.

What this implies is that the guaranteed way of keeping the status quo—of ensuring that habitats and ecosystems will continue to degrade and that species loss will continue unabated—is to keep visualizing the problem from an overly simplified, developed-world perspective. Yet because the problem is so vast, the causal interconnections and consequences so complex and often indeterminate (at least before the fact), it is often easy for people to deny there is a problem or to claim it is intractable and thus fail to act. Those having a vested interest in maintaining the status quo, particularly many policymakers, are loath to look beyond the economic and political here-and-now and take a more global, long-term perspective. That stance is made easier by scientific knowledge that appears, or *is made to appear*, indeterminate and conflicting—for example, whether there is global warming—or by uncertainties in the societal consequences of scientific knowledge, even when there is general consensus about the latter.

The scholars in this volume share a common view: despite scientific uncertainties about Earth's biological diversity and its rate of loss, what we do know is that species loss is accelerating at an alarming rate; despite uncertainties about the details of the long-term societal impact of this loss, those impacts have been and will continue to be profound; finally, despite the certainties about how humans have adversely interacted with Earth's biosphere and what current trends portend for the future, policy responses remain inadequate. As a consequence, societies everywhere will be at increasing risk if current trends continue.

This book does not pretend to provide insight into all the disciplines and nuances of biodiversity science, nor does it explore all the conservation and policy implications of the biodiversity crisis. Some important topics are missing, in several cases because promised contributions to this volume did not materialize; also, from the outset it was recognized that it is impossible to cover all aspects of biodiversity science and policy in a single compendium. Thus important topics such as the benefits of biodiversity, including ecosystem services, marine biodiversity, and a comprehensive summary of the factors leading to the loss of biodiversity, are not covered here, but detailed treatments of them can be found in many publications (Norse 1993; National Research Council 1995; Heywood 1995; Daily 1997). The task here is different. The last decade in particular has seen a steadily growing literature on the biodiversity crisis. Much of that literature finds its home within the academic community, but an increasing component is in the public domain. Few volumes attempt to have a foot in both communities, and the chapters included here have been chosen to bridge that gap in a way not found in similar efforts. The first section contains six scientific viewpoints that ask what we really know about the magnitude of biodiversity and extinction, how we might improve our knowledge, and what this information says is happening to Earth's species. The second section contains a series of chapters and perspectives that attempt to construct a framework for seeing biodiversity loss within a societal context and predicting what it means for the future of societies. The third section takes up some of the policy implications of biodiversity loss, primarily at an international level and especially through the Convention on Biological Diversity, and introduces the dialogue that must take place between scientists and policymakers. Then, in a short final section, that dialogue takes center stage in two chapters that explore how scientific knowledge and uncertainties about the loss of biological diversity and its consequences are not eliciting adequate responses by policy-making institutions.

THE SCIENCE OF DIVERSITY AND EXTINCTION

What are the undeniable scientific conclusions we can state about diversity and extinction? There are many, of course, but three have implications above all others. The first circumscribes our knowledge about the abundance of the natural world: we know that about 1.75 million species have been discovered and described (Heywood 1995) but that many millions remain unknown to science. The second undeniable conclusion has two parts and constitutes the evidence of our misuse of the natural world and of its slipping away: we know that, for at least the last five centuries the global rate of species extinction has been far higher than the geological background rate and that this spike in extinction is caused by human activities (Lawton and May 1995). Moreover, we know that the rate of local extirpation of species—the first signal that local habitats and ecosystems are functionally unraveling—is accelerating rapidly, essentially at an immeasurable pace. Finally, the third undeniable scientific conclusion is that the functional integrity of habitats and ecosystems is unraveling across the globe as local extinction tears apart the building blocks of the ecological nexus that provides the ecosystem services supporting all human endeavors (Heywood 1995; Daily 1997).

These facts are the foundation of the biodiversity crisis. They are not the only things we know. Indeed, we need to know much more. How many species make up Earth's ecosystems? Nigel Stork (chapter 1) considers this question, one of the more controversial in the scientific study of biodiversity. Estimates run from 4 or 5 million to over 100 million. Most opinion settles in the range of 15 to 60 million, and in these discussions the answer one arrives at depends on how insect numbers are estimated. Stork settles on a "working figure" of a little more than 13 million species, with 8 million being insects. Others believe that Earth's habitats and ecosystems may harbor another 30 million species of insects. The debate will not be settled until more precise methods and more extensive studies are undertaken to sample the world's biota. One suggestion by Norman Platnick (chapter 2) is to concentrate on a sampling regime for some of the "megadiverse" groups, especially those with a worldwide or nearly worldwide distribution. He argues that in so doing we will come to have a broader understanding of global patterns of diversity than if we just focused our attention on a few localities around the globe.

For many people it may not be important how many species there actually are, yet such a view overlooks the benefits that currently known biological diversity provides societies everywhere. If at most 5 to 10 percent of the world's species yield the benefits we now enjoy—food, fiber, trade, medicines, industrial products—it is not difficult to imagine what knowledge of the other 90 percent (or 99 percent if the global species number is many tens of millions) could mean for future generations. In addition to these economic benefits, the more we understand about species diversity, the more we understand about the structure and function of ecosystems. Such knowledge is the linchpin of maintaining and efficiently managing the ecological services provided by the various habitats and ecosystems.

Without a clear understanding of how many species there are, we cannot come to grips with extinction. The question of concern to everyone, of course, is how quickly species are being driven to extinction. Given a certain regime of habitat fragmentation or landscape disturbance, how many species are likely to be driven to extinction, locally or globally, in the short term or in the long term? Such questions cannot be answered satisfactorily without a better estimate of species numbers.

Both Stork (chapter 1) and Ross MacPhee and Clare Flemming (chapter 4) examine our knowledge of extinction rates, but from two different perspectives. Stork reviews the various approaches used to estimate global extinction rates, all of which, despite their differences and margin of error, show that a catastrophe is at hand. MacPhee and Flemming, on the other hand, take a single "well-known"

xvi CRACRAFT AND GRIFO

group—mammals—and in the most comprehensive analysis of how-do-we-know-what-we-think-we-know about extinction, they demonstrate that getting the science right when counting extinction events is often not straightforward and that efforts to put together lists of extinctions are fraught with difficulties.

This is not to say such lists are not informative. Indeed, the general finding of such exercises is that we have lost innumerable species through human activities. The real focus, however, should be on the loss of habitats and ecosystems; at the rate these have been lost over the last 100 years, it may be academic whether this species or that species is or is not extinct. Melanie Stiassny (chapter 3) argues that because freshwater ecosystems are particularly vulnerable and disappearing at a rapid rate from increasing human demand for freshwater and from pollution, fully 25 percent of global vertebrate diversity, in the form of freshwater fish, is at risk of extinction. She points out that as ecosystems such as freshwater environments are progressively fragmented and become habitat islands, an increase in extinction rate is inevitable.

The commentaries by Diana H. Wall and G. Carleton Ray pick up a thread running throughout all these chapters: despite many years of scientific research we still know so little about the natural world that precise statements about diversity and its distribution and loss are difficult to make, especially in the developing, species-rich nations of the world. This is especially true, as they point out, for all the poorly known habitats and ecosystems—including soils, the deep sea, most freshwater ecosystems, forest canopies, and many more. Thus we may have accumulated a substantial body of knowledge in the biodiversity sciences, but if we are to understand what is happening to Earth's diversity and how to manage it more intelligently and cost-effectively, we need to know much more and we need to see beyond the obvious and seek out the groups and habitats that are poorly known.

THE EXTINCTION CRISIS AND THE FUTURE OF SOCIETIES

If the loss of biodiversity had no measurable effect on the well-being of human populations, then aside from the aesthetic or ethical values we see in biodiversity there might not be great cause for alarm. But the fact that the international community, national governments, nongovernment organizations, and countless millions of people are alarmed manifests the undeniable conclusion that biodiversity does matter. And obviously it does. The benefits have been enumerated by many.

Yet more often than not, discussions about the linkage between biodiversity loss and societal well-being focus on the ways in which humans are causing the destruction of biological diversity, and much too little attention is being paid to the consequences of biodiversity loss. Common sense leads one to conclude that those consequences are and will be profound. If 70 or 80 percent of the world's people depend on the direct appropriation of diversity to meet their daily needs—food, medicines, firewood for cooking, shelter, and trade and commerce—then the loss of diversity means that over time societies will become increasingly disrupted. The fact that more and more wildlands are being converted to human uses, and then those same lands are being abandoned at an ever-increasing rate, is evidence that the land as currently managed is increasingly unable to meet human needs.

It is clear that the loss and disruption of local ecosystems can have direct adverse affects on people's livelihoods. There are many examples of this dysfunctional relationship at a local level. Loss of arable land and loss of sources of firewood are two. But what about at national and regional levels? As biodiversity continues to diminish, what will be the consequences for nations as a whole? There is surprisingly little analysis of this question. The chapters in the second section provide various perspectives.

Joel Cracraft (chapter 5) looks at a large series of indicators of species diversity, threats to diversity, and the capacity to respond to those threats for 77 developing countries. About a third of these countries, most of which are in Africa, have moderate to high threat to their biodiversity but at the same time have low capacity (economic, educational, scientific) to address those threats. The obvious conclusion is that some countries and regions are likely to feel the consequences of biodiversity loss sooner than others. But what might these effects be and what are the implications for their neighbors?

The most obvious effect is on the capacity of a country or region to feed itself. In various areas of the world, particularly parts of Africa, the population growth rate exceeds the rate of growth in food production, a fact that has many implications for national and regional security. John Burnett (chapter 6) reviews the historical and ongoing story of biodiversity loss from expanding agriculture, which carries with it the loss of soils and increased need for a dwindling supply of freshwater. He also raises the specter of the loss of genetic variability accompanying the rise of high-yield cultivated varieties. Although these domesticates have increased food production, they carry with them many environmental costs and social impacts and they place constraints on the future adaptability of our food crops to environmental change. As Burnett notes, the loss of wild genetic diversity in food plants has many implications for the future of sustainable agriculture. Much the same theme underlies David Pimentel's perspective in this section. He also raises other linkages between biodiversity loss and the future of agriculture, namely that without wild biodiversity the difficulties for realizing sustainable agriculture or forestry are enormous because of the loss of natural biological pest control, pollinators, and agents of nitrogen fixation.

Functioning societies need food security. They also need health security, and that depends on the maintenance of wild biodiversity, as discussed by Francesca Grifo and Eric Chivian (chapter 7). The derivation of most of our critical pharmaceuticals is to be found in wild biodiversity and the indigenous knowledge that has developed about those species. Modern medicine has spread throughout the world, yet it is clear that this form of medicine is economically inaccessible to most of the world's people, who still rely on traditional medicines for most of their daily health needs. For many people, their pharmacy is in the local forest or wildlands. It does not take much imagination to see that if this pharmacy is lost, the level of health for people dependent on medicinals from these areas will suffer, with all the societal costs that entails. Intergovernment and government health agencies see world health primarily in terms of the spread of modern developed-world medicine, yet much more research must be undertaken that investigates the consequences of the loss of traditional medicines on national and regional health security in many areas of the world.

If the loss of biodiversity has an adverse influence on a country's food supply or level of health, then it can be said to have an effect on the country's internal security. Arthur Westing (chapter 8) goes further and notes that using resources in an unsustainable manner, though perhaps contributing to national wealth in the short term, eventually erodes national security in the long term. He argues that it is extremely difficult to establish a direct causal link between environmental degradation and human conflict in any particular instance, but as he notes, that may not be the point: if one sees a nation's security as being dependent on the well-being of its citizenry, then the loss of biodiversity will have an obvious linkage to long-term security. It is clear that much more research and analysis are needed to develop a better understanding of environmental degradation and its role in shaping national and regional security issues.

Losing (converting) biodiversity may appear to generate wealth over the short term, but it is a complex issue to determine whether that gain is offset by short- and long-term costs because many values can be assigned to intact habitats and ecosystems. In recent decades the field of ecological or environmental economics has burgeoned, and Dominic Moran and David Pearce (chapter 9) guide us through some of the complexities. The use of biodiversity improves human welfare and thus has an economic value, but biodiversity also has intrinsic values that cannot be assigned a cost. "Getting the price right" on biodiversity is crucial for intelligent management and use, but Moran and Pearce raise a little-appreciated issue: because our scientific knowledge of biodiversity is limited, this uncertainty constrains the accuracy of any economic values we might want to place on it. Therefore, they stress, improved scientific understanding of biodiversity is crucial. In the meantime, prudence dictates the application of the precautionary principle to how we value our biological resources.

GETTING SOMEONE TO LISTEN: EFFECTING POLICY CHANGE

There are probably few, if any, environmentalists who think society is doing enough, that governments' policy responses are sufficient. By and large, policy-making institutions, whether local or even intergovernmental, respond to their constituencies. Unfortunately, the problem is that biological diversity does not have much of a political constituency. Nevertheless, everyone would probably agree that some progress is being made—the Convention on Biological Diversity is a shining example, despite its difficulties of implementation.

The chapters in earlier sections of this volume circumscribe some of the reasons why having effective policies toward biological diversity meets so many roadblocks. First, there are large gaps in our knowledge of biological systems. It takes a sophisticated scientific understanding of a region to manage its biological diversity effectively, yet scientific knowledge is insufficient by itself; one must also integrate information about how those resources are being used, how people and biological diversity meld into a whole.

Second, the context for conserving biological diversity is interleaved with societal needs, thereby making the constituency for policy decisions about how biodiversity is to be managed large indeed, and often conflicting. As Norman Myers (chapter 10) remarks, we must expand the policy arena for biodiversity conservation because the older approaches—seeing protected areas, especially national parks, as the locus of conservation effort—are not getting the job done. Myers sees the problem precisely as policymakers should see it: "There is much complementarity between our policy responses to the biodiversity problem and our responses to other problems of the biosphere." Save the world, you save biodiversity. Because of the linkages between biodiversity and societal well-being, policy responses must be broader than the current focus of many conservationists and governments.

As noted earlier, however, policymakers respond to their constituents, and people tend to be more concerned with other issues than with biodiversity loss. Why? Because they do not have the information they need to see biodiversity as an issue that is important to them. Thomas Lovejoy (chapter 11) makes this point when he argues we need to find a better way to get people to see the benefits provided by biodiversity. Self-interest is the great motivator; at present people do not see the conservation of biodiversity as a matter of self-interest. Nor do they understand what biodiversity loss might mean for their children. If they do not get these fundamental facts and see the implications of biodiversity loss for humanity, how can they be expected to call for policy change? Lovejoy makes a further point, echoed throughout the volume: lack of knowledge in the biological, economic, and social sciences is hindering our efforts to create sustainable societies through informed decision-making.

Perhaps the most fundamental change in the collective policy psyche of the world's nations toward biological diversity is the Convention on Biological Diversity (CBD). Stripped of its droll legalese, it is to a large extent a manifesto to save the world and save biodiversity together. It reflects the linkages that Myers refers to. Mulongoy, Bragdon, and Ingrassia (chapter 12), who have worked within UNEP and the CBD Secretariat, lead us through the objectives of the convention. Their description of the convention demonstrates that it is a remarkable blueprint for progress, even leaving aside all the hedging and compromising language that international documents of this type are burdened with.

BRINGING TOGETHER SCIENCE AND POLICY

Policy instruments such as the CBD are constructed by politicians, who (we hope) are aware, or reflect government policy that is aware, of the realities of the biodiversity crisis and its complex linkages to society. Biodiversity loss, like the loss of global ozone, is first and foremost a knowledge issue. Biodiversity scientists, on the ground, have led the way in documenting the loss of diversity, its many ecological values, and the consequences of that loss. At the same time, the linkages between degrading environments and societal well-being have been explicated by other sciences. Scientists have been involved directly because policymakers attempting to deal with biodiversity loss would be hopelessly in the dark without the information provided by accurate science. Yet as Jeffrey McNeely (chapter 13) describes, science and policy are often at odds with one another. Policy change is usually designed to provide benefits to constituents, not constraints, and at least in the developed world these constituents gobble up the benefits that flow from biodiversity depletion. They cannot easily identify one or two obvious culprits causing the problem, and they do not see biodiversity loss as affecting them directly. Causally biodiversity loss is complex—in this way it is not at all like atmospheric ozone depletion—and therefore it does not lend itself to easy policy responses. However, scientists must learn to function within a policy environment and must become more involved.

Thus a meaningful policy framework for saving biodiversity depends on knowledge, and the preceding discussion stresses the role of science in improving management of biological resources. But knowledge can also be used to shift values, and all policy decisions are founded on a particular set of values. Peter Raven and Joel Cracraft (chapter 14) argue that our value systems in the developed world are out of step with any policies that might have lasting, positive influence on biodiversity. Citizens of developed nations generally misconstrue how the remainder of the world—the developing world—is operating. That world contains the vast majority of the world's poor, undereducated, and sick, only about 6 percent of the world's scientists, but probably 80 to 90 percent of its biodiversity. People in that world—the world of the poor—use biodiversity in fundamentally different ways, and at fundamentally different rates of consumption, than we do. Collectively, we do not appear to understand that difference or its implications for the future and do not seem terribly interested in formulating policy to address the situation. The results of science must be digested and interpreted within a broader setting, one that moves our value systems from those anchored squarely in maintaining or expanding our cultural hegemony over the economically disadvantaged to one that recognizes and accepts a shared destiny for the planet.

A haunting coda to these concerns, and to the book, is provided by Strachan Donnelley in his perspective. As a philosopher who sees himself as part of the public outside the scientific mainstream, he laments that scientists have largely failed to create a public informed on the issue of biodiversity loss: "In the main, the scientific community has fed our economic and technological boosterism and left us bulls in the China shop of nature. . . . The gauntlet of public education is thrown and ought not be ignored. Our citizen ignorance is an integral part of the living planet in crisis."

REFERENCES

Daily, G. C. 1997. *Nature's services: Societal dependence on natural ecosystems.* Washington, D.C.: Island Press.

Heywood, V., ed. 1995. *Global biodiversity assessment* (UNEP). Cambridge, U.K.: Cambridge University Press.

Lawton, J. H. and R. M. May, eds., 1995. *Extinction rates.* Oxford, UK: Oxford University Press.

National Research Council. 1995. *Understanding marine biodiversity.* Washington, D.C.: National Academy Press.

Norse, E. A., ed. 1993. *Global marine biological diversity: A strategy for building conservation into decision making.* Washington, D.C.: Island Press.

ABTI	All-Biota Taxon Inventory
AMNH	American Museum of Natural History, New York
ANIC	Australian National Insect Collection, Canberra
ATBI	All Taxa Biodiversity Inventory
BMNH	British Museum (Natural History), London
BTI	biodiversity threat index
CBD	Convention on Biological Diversity
CEPII	cost-effective priority investment index
CITES	Convention on International Trade in Endangered Species
COP	Conference of the Parties
CPTI	conservation potential/threat index
CRI	capacity response index
CV	contingent valuation; cultivated variety
EED	effective extinction date
EEE	eastern equine encephalitis
ERIN	Environmental Resource Information Network
FAO	Food and Agriculture Organization
GEF	Global Environment Facility
GIS	geographic information systems
HDI	Human Development Index
HPR	host-plant resistance
HYCV	high-yield cultivated variety
IBP	International Biological Programme
ICBP	International Council for Bird Preservation
ICCBD	Intergovernmental Committee on the Convention on Biological Diversity
IOC	Intergovernmental Oceanographic Commission

IPCC	Intergovernmental Panel on Climate Change
IPGRI	International Plant Genetic Resources Institute
IPM	integrated pest management
ISNB	Institut Royal des Sciences Naturelles de Belgique
ISRIC	International Soil Reference and Information Centre
IUBS	International Union of Biological Sciences
IUCN	International Union for the Conservation of Nature/World Conservation Union
LEO	list of extant organisms
MNHN	Muséum National d'Histoire Naturelle, Paris
MRAC-ORSTOM	Musée Royal de l'Afrique Central-Office de la Recherche Scientifique et Technique Outre-Mer
NCC	Nature Conservation Council
NHRS	Naturhistoriska Riksmusset, Stockholm
NIPAS	National Integrated Protected Areas System
NRA	natural resource accounting
OXUM	University Museum, Oxford (UMO)
PSP	paralytic shellfish poisoning
RCYRBP	radiocarbon years before present
RDB	*Red Data Book*
SBSTTA	Subsidiary Body on Scientific, Technical and Technological Advice
SDI	species diversity index
SMFD	Forschungsinstitut und Naturmuseum, Frankfurt (Senkenberg Museum)
UNDP	United Nations Development Programme
UNEP	United Nations Environment Programme
USNM	U.S. National Museum (National Museum of Natural History, Smithsonian Institution)
WCMC	World Conservation Monitoring Centre
WI	West Indies
WMO	World Meteorological Organisation
WRI	World Resources Institute
WWF	World Wildlife Fund
YRBP	years before present (i.e., the radiocarbon datum, 1950)
ZFMK	Zoologische Forschungsinstitut und Museum "Alexander Koenig," Bonn
ZIL	Universitetskaya (ZMAS), Leningrad
ZMHB	Museum für Naturkunde der Humboldt Universität (MNHU), Berlin
ZSBS	Zoologische Sammlung des Bayerischen Staates (ZSMC), Munich

THE LIVING PLANET IN CRISIS

PART I

SCIENCE OF DIVERSITY AND EXTINCTION

CHAPTER 1

THE MAGNITUDE OF GLOBAL BIODIVERSITY AND ITS DECLINE

Nigel E. Stork

We know a remarkable amount about the physical nature of the world around us. For example, the moon is an average of 384,400 km from the earth and it takes about 1.25 sec for a beam of light to travel this distance; there are 106 elements on the periodic table, which was largely completed by the middle of the nineteenth century, and so on. Yet, surprisingly, we cannot say how many species we share this planet with, how many have so far been named, and how many have become extinct or may become extinct in the near future. This chapter concentrates on the species level of biodiversity and examines how much we know about the magnitude of biodiversity on Earth and its decline in recent years. What we know about the genetic level of biodiversity or about the phylogenetic relationships of organisms is not discussed.

First, I provide a historical context for our understanding of biodiversity because this has resulted in considerable geographic and biological biases. Second, I examine how much we know at the species level of biodiversity. In particular, I examine the latest attempts to count how many valid species names exist and how little we know about most of these species. Third, I review the recently well-discussed topic of how many species there are on Earth and suggest where new information may help narrow future estimates. Finally, I discuss what information is currently available on how many species are known to have become extinct in recent years and I review estimates of future extinction levels.

HISTORICAL CONTEXT

The diversity of life has been recognized and used by humans on all continents for millennia, but it was not until 1775, when Carolus Linnaeus published his *Systema Naturae* , that a common language for identifying life forms came into being. His hierarchical system in Latin, the language of the learned, provided the means to

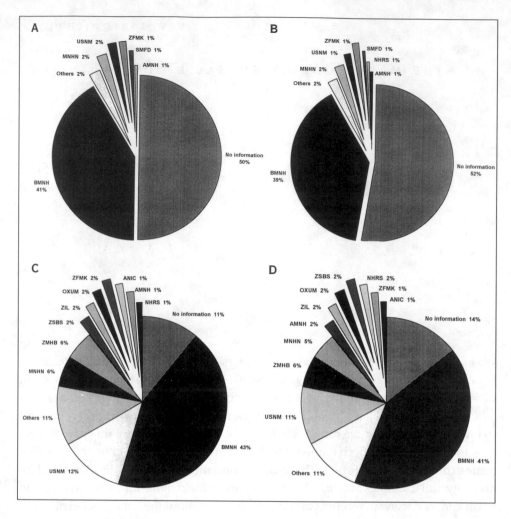

FIGURE 1.1

Pie charts showing the distribution of primary type specimens of **A,** 11,900 valid species and sub-species of Ennominae, **B ,** 15,384 valid species, subspecies, and synonyms of Ennominae, **C,** 25,103 valid species and subspecies of Noctuidae, and **D,** 35,026 valid species, subspecies, and synonyms of Noctuidae in major collections throughout the world holding more than 1% of primary types for these groups.

Abbreviations: AMNH, American Museum of Natural History, New York; ANIC, Australian National Insect Collection, Canberra; BMNH, British Museum (Natural History), London; MNHN, Museum National d'Histoire Naturelle, Paris; NHRS, Naturhistoriska Riksmusset, Stockholm; OXUM, University Museum, Oxford (UMO); SMFD, Forschungsinstitut und Naturmuseum, Frankfurt (Senkenberg Museum); USNM, Smithsonian Institution, Washington, DC; ZFMK, Zoologische Forschungsinstitut und Museum "Alexander Koenig," Bonn; ZIL, Universitetskaya (ZMAS), Leningrad; ZMHB, Museum für Naturkunde der Humboldt Universitat (MNHU), Berlin; ZSBS, Zoologische Sammlung des Bayerischen Staates (ZSMC), Munich.

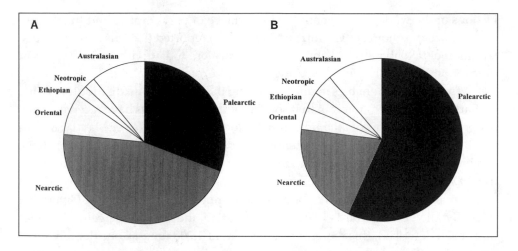

FIGURE 1.2

A, Distribution among Wallace's biogeographic realms of about 16,000 authors of papers listed under "ecology" in *Biological Abstracts* 1982–1983 (after Golley 1984 and Gaston and May 1992). **B,** Biogeographic distribution of about 1,200 taxonomists borrowing specimens from the insect collections at the Natural History Museum in London between January 1986 and June 1991. *Palearctic:* Europe and Siberia; *Nearctic:* North America; *Oriental:* Near East to Far East and Malaysia; *Ethiopian:* sub-Saharan Africa; *Neotropic:* Central and South America; *Australasian:* New Zealand, Papua New Guinea, and the Pacific Islands (after Gaston and May 1992).

catalogue all organisms and for scientists to communicate about them. There followed an extraordinary frenzy of activity, almost entirely based in western Europe, to describe the world's fauna and flora. This process of chaotic and competitive activity produced the new breed of biologists—taxonomists—vying to describe new species. As more species were described, competing hypotheses about the hierarchical relationships of these organisms were devised. Despite the immense diversity of taxonomists from different cultural backgrounds and the lack of an evolutionary view of life (Darwin's *Origin of Species* did not appear until 1859 and was not widely accepted until much later), the Linnaean system for naming and describing species has been closely adhered to. Much of the activity in describing species took place in Europe, with many museums and herbaria amassing very large and important collections. Collecting was largely through expeditions, with the geographic basis of these collections usually determined by the distribution of the empires of different European countries. The pie charts in figure 1.1, for example, show the distribution in museums of the 'type' specimens (when describing and naming a species, a single specimen is usually selected as typical of that species and is called the holotype or type) of two of the most species-rich groups of insects, Noctuidae and Ennominae (Geometridae). It is clear that most types of these groups are in European and North American museums even though the vast majority of these species are not European, but are probably from tropical countries. This European and North American bias in the distribution of the major col-

lections of the world reflects both the importance of past empires and the distribution of past taxonomists. Gaston and May (1992) reported that this bias in the distribution of taxonomists still appears to be prevalent today, with about 80 percent
of taxonomists residing in North America and Europe (figure 1.2).

A further important bias in understanding the diversity and distribution of life
on Earth has been taxonomic in that some groups of organisms have received a dis-
proportionate amount of attention from biologists. Table 1.1 (after May 1988) provides a rough indication of the attention that different groups of organisms have
received, as measured by the average number of papers listed for those groups in
the *Zoological Record*. Note, for example, that there are 100 times as many papers
written per described species of bird as there are per species of beetle. Given that
virtually all species of birds have been identified and only 10 percent of beetles
have been described, this represents a real bias of 1,000 times. Considerably more
descriptive effort on the part of taxonomists has been expended on some groups
than others. Only a few species of birds are described each year, compared to the
many thousands of species of insects (see tables 4.5 and 4.6 in Hammond 1992).
Table 1.2 (after Simon 1983; May 1990) shows that most descriptive effort for
groups such as birds and mammals occurred soon after Linnaeus, whereas for
groups such as Crustacea, most species have been described only in recent years.
For groups such as insects, fungi, and microorganisms we have yet to see a tailing
off of descriptive effort.

HOW MUCH DO WE KNOW?

How far have taxonomists progressed in the task of describing all species? Recent
estimates, based on careful study of checklists and other sources of data for a wide
range of groups of organisms, suggest that this total is 1.75 million species, as
shown in figure 1.3 (data from Stork 1988; Hammond 1995). The precise figure is
uncertain for several reasons, the main one being that there is no recognized central register of names for described species (although such registers do exist for a
few groups), so some species have been described many times. For example, someone describing a species in India may be unaware that the species has already been
described in Pakistan. In other cases the natural variation of a species is unknown
and different forms of the same species are given different names. The common
European 10-spot ladybird, *Adalia decempunctata* L., for example, has at least 40 different synonyms, many of these having been used for the color morphs. Gaston
and Mound (1993) noted that although only about 4,000 species of mammal are
currently recognized (Corbet and Hill 1980), the collections at the Natural History
Museum (London) contain types for 9,000 names. They also suggest that for insects
the level of synonymy is about 20 percent (table 1.3). Therefore, although only
about 1.75 million species have been described, many more names are being used
by taxonomists. Of course, for biologists the definition of a species is still a funda-

TABLE 1.1

Rough Indication of the Relative Effort Devoted to Animals from Different Taxonomic Groups

PHYLUM SUBPHYLUM CLASS ORDER	AVERAGE NUMBER OF PUBLICATIONS PER YEAR (COEFFICIENT OF VARIATION IN PERCENT)[a]	APPROXIMATE NUMBER OF RECORDED SPECIES	PAPERS PER SPECIES PER YEAR
Protozoa	3,900 (10)	260,000	0.15
Porifera	190 (22)	10,000	0.02
Coelenterata	740 (12)	10,000	0.07
Echinoderma	710 (15)	6,000	0.12
Nematoda	1,900 (1)	15,000	0.13
Annelida	840 (9)	15,000	0.06
Brachiopoda	220 (14)	350	0.63
Bryozoa	160 (15)	4,000	0.04
Entoproctra	7 (53)	150	0.04
Mollusca	1,000 (8)	100,000	0.04
Arthropoda			
Crustacea	3,300 (9)	39,000	0.09
Chelicerata			
Arachnida	2,000 (6)	63,000	0.03
Uniramia			
Insecta	17,000 (7)	1,000,000 (?)	0.02
Coleoptera	2,900 (6)	300,000	0.01
Diptera	3,200 (7)	85,000	0.04
Lepidoptera	3,500 (9)	110,000	0.03
Hymenoptera	2,200 (9)	110,000	0.02
Hemiptera	1,700 (7)	40,000	0.04
Chordata			
Vertebrata			
Pisces	7,000 (13)	19,000	0.37
Amphibia	1,300 (12)	2,800	0.47
Reptilia	2,400 (7)	6,000	0.41
Aves	9,000 (10)	9,000	1.00
Mammalia	8,100 (12)	4,500	1.80

(continued on next page)

TABLE 1.1 *(continued from previous page)*

Rough Indication of the Relative Effort Devoted to Animals from Different Taxonomic Groups

PHYLUM SUBPHYLUM CLASS ORDER	AVERAGE NUMBER OF PUBLICATIONS PER YEAR (COEFFICIENT OF VARIATION IN PERCENT)[a]	APPROXIMATE NUMBER OF RECORDED SPECIES	PAPERS PER SPECIES PER YEAR
Mammalian orders			
Monotremata	20.00	3	6.80
Marsupialia	269.00	266	1.00
Insectivora	270.00	345	0.80
Dermoptera	2.20	2	1.10
Chiroptera	402.00	951	0.40
Primates	956.00	181	5.30
Edentata	38.00	29	1.30
Pholidota	5.00	7	0.70
Lagomorpha	173.00	58	3.00
Rodentia	1,538.00	1,702	0.90
Cetacea	360.00	76	4.80
Carnivora	1,157.00	231	5.00
Tubulidentata	2.70	1	2.70
Proboscidea	94.00	2	47.00
Hyracoidea	12.00	11	1.00
Sirenia	43.00	4	10.80
Perissodactyla	142.00	16	8.90
Artiodactyla	1,124.00	187	6.00
Pinnipedia	218.00	33	6.60

After May (1988). Note that May incorrectly listed Nematoda as having 1,000,000 recorded species. The figure used here is from Hammond (1995).

[a]Average number of papers listed in the *Zoological Record,* 1978–1987.

mental problem and it is difficult, if not impossible, to compare viruses with, say, flowering plants or birds.

It seems incredible that taxonomists and other biologists have managed to carry out their work for hundreds of years without a single master list of described species. Bisby (1994) lists the few groups of organisms for which there are good master species databases, but for the groups with the greatest number of described species, such as most insect groups, no such lists are available. Stork (1993) and Hammond (1995) have both called for a simple but comprehensive list of all named species. There is an urgent need for such a global list of described species (variously called Global Species Archive, Lifelist, and List of Extant Organisms or LEO).

TABLE 1.2

Taxonomic Activity from 1758 to 1970 for Different Animal Groups, as Revealed in Patterns of Recording New Species

ANIMAL GROUP	ESTIMATED NUMBER OF SPECIES RECORDED UP TO 1970	LENGTH OF TIME (YR) BEFORE 1970 TO RECORD SECOND HALF OF TOTAL IN PREVIOUS COLUMN	PERIOD OF MAXIMUM RATE OF DISCOVERY OF NEW SPECIES
Protozoa	32,000	21	1897–1911
"Vermes"	41,000	28	1859–1929
Arthropoda (excluding insects)	96,000	10	1956–1970
Arthropoda (insects only)	790,000	55	1859–1929
Coelenterata	9,600	58	1899–1928
Mollusca	45,000	71	1887–1899
Echinodermata	6,000	63	1859–1911
Tunicata	1,600	68	1900–1911
Chordata			
Pisces	21,000	62	1887–1929
Amphibia	2,500	60	1930–1970
Reptilia	6,300	79	1859–1929
Aves	8,600	125	1859–1882
Mammalia	4,500	118	1859–1898

After Simon (1983) and May (1990).

Such a list will help to stabilize the names of many species and may stimulate further taxonomic work. It should also show critical areas where future taxonomic effort should be expended. Although questions about the true dimensions of global biodiversity have had a central role in discussions of biodiversity (namely, "How many species are there?" and "How many species have been described?"), one related and important issue seems to have been largely neglected. Some nontaxonomists may assume that once a species has been named and described, the task of taxonomists and related biologists is virtually complete. This is far from true. As Lawton (1993) suggested, "Intriguingly, I have never seen anybody discuss what we actually know about these 1.7 million species that do have names. Overwhelmingly, the answer will be nothing, except where they were collected and what they look like."

Description of a species is only the first step in understanding the biology of an organism. At least three other aspects are also important. First, its genealogi-

FIGURE 1.3

Proportions of described valid species for
the major taxa (1.7 million total).
Reproduced from Stork (1997).

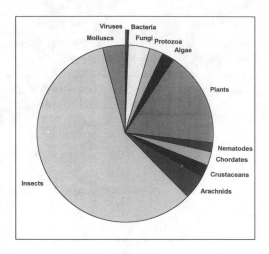

cal relationship with respect to other species and other major taxa is important in
determining its *phylogenetic* significance. Second, what the organism does for a
living and its ecological relationship with other organisms is important for determining its *ecological* significance. Third, its distribution is important in conjunction with the previous two to determine its overall *biological* and *conservation*
significance. To many nonbiologists it may seem that we know a great deal about
all three factors for all described species. In reality this far from true. Many
species are named and described on the basis of single specimens or a series of
specimens from a single location. Furthermore, many if not most species are still
known only from their original descriptions, which may be in old, poorly prepared, or inaccessible publications. For example, in one recent study Stork (1997)
and Stork and Hine (unpublished) found that in a random sample of 46 taxonomic papers from the *Zoological Record* for 1987, 53 percent of the 186 beetle
species in the sample are known from a single locality and 90 percent from fewer
than six localities. Furthermore, 13 percent of the species are known from single
specimens. Given that half of these 186 species are ones that previously had been
described and that beetles represent a quarter of all described species on Earth,
for many species we know little of their distribution. The laudable missions of
Systematics Agenda 2000 (Systematics Agenda 2000 1994a, 1994b) are "to describe all species, determine their phylogenetic relationships and their biology
and distributions and to make this information available for the benefit of humans." It would seem that we still have an enormous amount of work to do if
such goals are to be achieved.

HOW MANY SPECIES ARE THERE?

The biodiversity debate centers on the age-old question, "How many species
are there?" It is a simple enough question, but the answer is extraordinarily im-

TABLE 1.3

Present Levels of Synonymy in Insects

	NAMES		% SYNONYMY	SOURCE
	TOTAL	ACCEPTED CURRENTLY		
Odonat[a]	7,694	5,667	26	Bridges (1991)
Isoptera[b]	2,000	1,600	20	Snyder (1946)
Thysanoptera	6,479	5,062	22	L. A. Mound (unpublished results)
Homoptera				
Aleyrodidae	1,267	1,156	9	Mound and Halsey (1978)
Aphididae	5,900	3,825	35	Eastop and Hille Ris Lambers (1976)
Siphonaptera	2,692	2,516	7	D. Lewis (unpublished results)
Diptera				
Simuliidae	1,800	1,460	19	Crosskey (1978)
Lepidoptera				
Noctuidae	35,473	28,175	21	Poole (1989)
Papilionidae and Pieridae[a]	9,075	1,792	80	Bridges (1988a)
Lycaenidae and Riodinidae[a]	13,108	5,757	56	Bridges (1988b)
Hesperiidae[a]	8,445	3,589	58	Bridges (1988c)
Hymenoptera				
Chalcidoidea	22,533	18,601	18	Noyes (1990)

From Gaston and Mound (1993).
[a]Subspecific names treated as synonyms.
[b]These figures are estimates, due to problems in data interpretation.

portant because it has implications for fields such as biology, systematics, ecology, and conservation biology. Recent concerns over global extinction rates have fueled a renewed interest in this question, because until we have an accurate assessment of global species richness we cannot determine the frequency and magnitude of species losses.

The question of the number of species on the planet has been a source of fascination for biologists for centuries. In 1833 Westwood quoted Ray, a fellow British entomologist, and suggested that global insect richness might be as high as 20,000 species. Advances in species identification and collection over the last 150 years indicate that global species richness is much greater than that, with more than 20,000 species described for Britain alone. Until recently, however, the question of global species richness has received little attention in biological texts, with many suggesting that total species richness is just a little larger than our current list of described species (2 to 3 million). Furthermore, it was believed by many that we would one day know all species, simply through advances in taxonomic description and species collection. The 1980's heralded a dramatic change in these views, as a variety of new approaches were developed to estimate global species richness. These methods, developed from both ecological and taxonomic patterns and extrapolations from empirical data, are discussed in this chapter.

Ratios of Known to Unknown Faunas

One simple method of global species number estimation is based on extrapolation of unknown:known species ratios for particular taxonomic groups. For well-known taxa (primarily mammals and birds) evidence suggests that there are about twice as many tropical species as temperate species (Raven 1985). Raven argued that if this pattern is consistent across all groups, then, with a total number of described species around 1.5 million—and this dominated by temperate species—the global species richness might be as high as 3 million. Gaston (1994), however, has shown that most species described in the 1980's are from nontropical regions and nonmegadiverse countries.

Further evidence (table 1.4) disproving Raven's hypothesis is provided by a survey of the countries represented by species in every hundredth drawer in the beetle collection of the Natural History Museum (Stork 1997). A total of 1,753 species were examined, suggesting that the total collection is approximately 175,300 species, or about half of all of the beetle species described in the world. Of these, 625 were from temperate countries, 946 from tropical countries, and 182 from countries with temperate and tropical regions (e.g., Australia).

Another approach used to estimate global species richness uses the number of species from a taxonomic group as a ratio of the total fauna or flora. Stork and Gaston (1990) developed this approach using the ratio of 67:22,000 for butterfly species to all insect species from the well-known British fauna. If this ratio is representative of a worldwide pattern, then with 15,000–20,000 species of butterfly in the world, the global insect species richness may be between 4.9 and 6.6 million.

Similar extrapolations, based on known ratios of temperate fungi species to vascular plant species of 1:1.4 and 1:6, have been made to suggest that there may be 1.5 million fungi species worldwide (Hawksworth 1991). This estimate may be conservative because it is believed that the ratio of vascular plants to fungi may be even higher in tropical regions.

These approaches are attractively simple, but there is little evidence to suggest that any of the assumed relationships between ratios of taxa really exist. For example, the relative proportions of the four most species-rich insect groups (Coleoptera, Diptera, Lepidoptera, and Hymenoptera) vary markedly in different parts of the world (Gaston 1991).

TABLE 1.4

Number of Beetle Species and Predicted Total Number of Beetle Species for the Top 15 Countries Represented in the Natural History Museum in London

NUMBER OF SPECIES SAMPLED	PREDICTED NUMBER OF SPECIES IN COLLECTION	COUNTRY
143	14,300	**Australia**
142	14,200	South Africa
139	13,900	U.S.A.
134	13,900	**Brazil**
118	11,800	**India**
86	8,600	**Mexico**
65	6,500	Philippines
61	6,100	Japan
60	6,000	**Malaysia**
58	5,800	Former USSR
49	4,900	**Indonesia**
44	4,400	**Colombia**
44	4,400	Guatemala
39	3,900	Papua New Guinea
37	3,700	**Madagascar**

From Stork (1997). Data were produced by examining the species in every hundredth drawer of the 11,500-drawer beetle collection. In total, 1,753 species were recorded, suggesting that the total holdings of identified beetle species is 175,300. This does not include unidentified species in the accessions and several small separate collections of beetles. "Megadiversity" countries (see Gaston 1994) are in bold.

Extrapolations from Samples

ERWIN'S 30 MILLION SPECIES

A radical new approach was developed in 1982 by Erwin, based on knock-down insecticide samples of beetles he collected from 19 individuals of *Luehea see-mannii*, a tropical rainforest tree in Panama. Sampling over three seasons, he collected more than 955 beetle species, excluding weevils. On previous inventories in Brazil, Erwin had collected as many weevils as leaf beetles, so he added 206 species to his count before rounding to 1,200 for convenience. He then took an average of 70 genus-group trees where he thought host specificity might play a role with regard to arthropods, based on there being between 40 and 100 tree species or genera in 1 hectare of forest. Because no data were available for the level of host specificity for trophic groups, Erwin used the following estimates: 20 percent of the herbivorous beetles (i.e., must use this tree species in some way for successful reproduction), 5 percent of the predators (i.e., are tied to one or more of the host-specific herbivores), 10 percent of the fungivores (i.e., are tied to fungus associated

TABLE 1.5

Numbers of Host-Specific Beetle Species per Trophic Group in Samples Collected by Knockdown Insecticide Fogging from *Luehea Seemannii*

TROPHIC GROUP	NUMBER OF SPECIES	% HOST-SPECIFIC SPECIES (ESTIMATED)	NUMBER OF HOST-SPECIFIC SPECIES (ESTIMATED)
Herbivores	682	20	136.4
Predators	296	5	14.8
Fungivores	69	10	6.9
Scavengers	96	5	4.8
	1,200[a]		162.9

From Erwin (1982).
[a]Note that Erwin rounded the species total to 1,200.

only with this tree), and 5 percent of the scavengers (i.e., are associated in some way with only the tree or with the other three trophic groups) (table 1.5).

Using these levels of specificity, or an average of 13.5 percent host specificity, Erwin estimated that *Luehea* had 163 host-specific beetles. Thus in 1 ha with 70 generic-group tree species, there would be 11,410 host-specific beetle species, plus an additional 1,038 transient species, for a total of 12,448 species per hectare of tropical forest canopy. Extrapolating further, Erwin estimated that 40 percent of all arthropod species were beetles, so there would be as many as 31,120 arthropod species per hectare of tropical forest canopy.

Erwin suggested that the canopy fauna is twice as rich as that of the forest floor and thus added an additional one-third to his canopy estimates to arrive at a total of 41,389 species per hectare. Erwin estimated that there are 50,000 tropical tree species, so he generated a tropical arthropod richness estimate of 30,000,000 species.

Erwin's calculations raised global species estimates by a factor of 10 and, as a consequence, heightened awareness of global diversity. They also highlighted how little we know about the fauna and flora of the planet and, when combined with estimates of deforestation and habitat loss, suggested to many biologists that more species are threatened with extinction than previously thought (Myers 1989). Inevitably, Erwin's global species richness estimate has become a biopolitical tool.

But how accurate was Erwin's estimate of global species richness? The assumptions he used to generate his estimate have become the target of a rigorous research agenda (May 1988, 1990; Stork 1988, 1993, 1997) and analyses of each of Erwin's assumptions follow.

Host Specificity   Erwin's first assumption related to the level of host specificity exhibited by beetles. Recent evidence suggests that rather than being a conservative estimate, the level of host specificity assumed in his study (13.5 percent) is higher than that observed in the field.

Gaston (1992) found that there were typically 10, and no more than 25, insects specifically associated with a plant species in a variety of temperate regions. Although Gaston used an indirect measure of specificity (ratio of insect:plant species numbers), this result indicates that host specificity is lower than suggested by Erwin. Upon further extrapolation (assuming that beetles represent 40 percent of all arthropods), Gaston's results indicate that just 4 of the 10 host-specific insects on each plant would be beetles. Although this is based on temperate rather than tropical plant and insect numbers, it is markedly different from the 163 host-specific beetles per tree suggested by Erwin.

It is ultimately difficult to determine the host specificity of insects because many herbivorous species feed on more than one species, genus, or family of plant. Furthermore, most canopy samples are characterized by a large number of singletons, which presumably do not feed on the sample tree (Stork 1991). Typical rank abundance curves of beetles and chalcid wasps from insecticide-fogged samples from 10 trees in Borneo are presented in table 1.6. Note that singletons represent more than half of the species collected for both of these groups. Given the remarkable botanical diversity of tropical forests, the presence of so many singletons, with no apparent host associations, is hardly surprising.

Thomas (1990) found that regional extrapolations of species numbers based on collections at single locations can overestimate regional insect species numbers. He found an average of 7.2 Passifloraceae species and 9.7 Heliconiinae butterfly species across the 12 central American sites he studied. Given that there are 360 Neotropical species of Passifloraceae, the scaled-up estimate of Heliconiinae species is 485 ($9.7/7.2 \times 360$). In reality there are only 66 species. These butterflies use different host plants in different regions of their distributions. Despite the fact that only 100–150 Passifloraceae species occur below 1,500 m (the upper limit for Heliconiinae), the principle remains the same.

Further analysis has been conducted by May (1990), who adopted a theoretical distribution function $p_k(i)$ (that is, the fraction of canopy insects found on tree species k that use a total of i different tree species) to determine the specificity of beetles associated with oak trees in the United Kingdom. May also calculated f, which is the proportion of species effectively specialized to each tree species. Given the well-known British beetle fauna, he was able to predict that about 10 percent of the herbivorous species were specifically associated with trees of the genus *Quercus*.

Erwin's estimated level of host specificity has been further challenged by more recent data. Hammond and Owen (in press) collected 983 species of beetles in a long-term study conducted in Richmond Park, an oak woodland of about 20 square miles in the United Kingdom. Their dataset represents 25 percent of all known beetle species in Britain. A part of this study involved the fogging of 40 oak and several other species of trees over 2 years, during which time 198 species were

TABLE 1.6

Diversity of Coleoptera (859 species, 3,919 individuals) and Chalcidoidea (739 species, 1,455 individuals) Collected from 10 Trees in Borneo Using Knockdown Insecticide Fogging

COLEOPTERA		COLEOPTERA *(continued)*		CHALCIDOIDEA	
INDIVIDUALS	SPECIES	INDIVIDUALS	SPECIES	INDIVIDUALS	SPECIES
1	499	26	1	1	437
2	133	30	1	2	160
3	62	31	2	3	54
4	24	32	1	4	31
5	35	35	1	5	18
6	21	36	1	6	10
7	8	39	1	7	8
8	13	40	2	8	4
9	4	45	1	9	6
10	4	49	1	10	3
11	4	53	1	11	2
12	5	66	2	12	1
13	4	77	1	13	1
14	2	81	1	17	1
15	2	90	1	19	1
17	4	112	1		
18	2	129	1		
19	1	137	1		
22	3	140	1		
23	2	194	1		
24	3	235	1		

From Stork (1991, 1993, 1995). Table shows the number of species represented by each number of individuals. For example, 499 species of Coleoptera were each represented by only one individual, 133 species were each represented by two individuals, and so on.

collected (Hammond 1994). Of these, 18 percent were found to be specifically associated with trees, although as few as 3 percent of the total number were associated only with oak trees, the dominant tree species. Mawdsley and Stork (1997) also used May's function to determine a specificity level of less than 5 percent for beetle species collected in canopy fogs of 10 trees in Borneo.

Beetles as a Proportion of Canopy Species   Erwin's assumption that 40 percent of the tropical arthropods are beetles also has been questioned. My former colleagues from the Natural History Museum in London and I made almost a complete sort of the 25,000 arthropods I collected by canopy fogging in Brunei. More than 4,000 species were sorted. This was the first time a complete sort to species had been made for tropical forest samples and beetles were found to make up between 8.7 and 26.7 percent of the arthropod fauna collected from each tree, or an average of 16.9 percent (Stork 1991). Similarly, beetles represent only about 18 percent (4,000 species) of the total British arthropod fauna. Gaston (1991) investigated differences in the relative numbers of species for the four largest insect orders (Coleoptera, Diptera, Lepidoptera, and Hymenoptera) in mostly temperate faunas, only to conclude that there may be strong latitudinal differences for these groups. Despite this conclusion, there appears to be a remarkable similarity in the relative proportions of species of different guilds of insects in canopy samples from Brunei, the United Kingdom, and South Africa (Stork 1987) and the relative numbers of species for different families of beetles in canopy samples from Brunei and Panama (Stork 1993). Irrespective of the consistency of these trends, we are still a long way from understanding of the relative contributions of these key groups to global species richness.

Canopy: Ground Ratios for Arthropods   Recent comparisons of canopy and ground arthropod faunas in Borneo and Sulawesi suggest that Erwin's 2:1 ratio should be reversed. According to Hammond (1994) and Hammond et al. (1997), only 10 percent of more than 4,000 species collected in a hectare of lowland rainforest in Sulawesi are "canopy specialists." Furthermore, as much as 20 percent of the fauna was found to be comprised of ground specialists. These findings are supported by Mawdsley's (1995) analysis of 3,000 beetle species collected in Borneo.

In summary, it seems likely that Erwin grossly overestimated the number of host-specific species for a given tree species but underestimated the relative contributions of arthropods other than beetles and the number of species associated with the ground.

ESTIMATES FROM INTENSIVE SAMPLING IN SULAWESI

As part of the Project Wallace expedition (an entomological survey by the Royal Entomological Society and the Indonesian Department of Science in the Dumoga-Bone area of North Sulawesi conducted in 1985), the Natural History Museum in London organized a highly intensive insect sampling program. Many thousands of species were collected from a small area and sorted to species. One of the aims of this study was to examine regional and global species richness.

Hodkinson and Casson (1991) examined the collection of 1,690 species of Hemiptera from north Sulawesi and, in consultation with other experts, established that 62.5 percent were previously unknown. Providing some statistical support for the assertion that this proportion of new species would be found worldwide, they estimated global species richness for Hemiptera to be 184–193,000 species. They suggested that 7.5 to 10 percent of all insects belong to the Hemiptera and that therefore there are 1.84 to 2.57 million species of insects in the world. Using similar arguments to Erwin, they predicted that there should be as many as 105,600 new species of Hemiptera worldwide (based on their collection of 1,056 new species from an estimated 500 tree species in the region and Erwin's hypothesized 50,000 tree species in the world). Combined with the number of described species, this yields a total of 187,300 Hemiptera species, producing a global insect diversity of 1.84 to 2.57 million species.

The estimates calculated by Hodkinson and Casson are not without problems (Stork 1993). First, despite their assumption to the contrary, it is unlikely that every species of Hemiptera within the Dumoga-Bone area was collected, as cumulative species number curves for the more rigorously sampled beetles showed no signs of flattening out (figure 2 in Stork 1993). Indeed, many of the Hemiptera species they examined were from the same canopy fogging samples as the beetles and similar numbers of singletons were found. Second, it is not clearly demonstrated that the percentage of new species collected in the Dumoga-Bone area is representative of other areas in the world. Third, because the Hemiptera include numerous economically important pest species, they may have been subjected to a greater level of taxonomic description than have other orders. If this is the case, then Hemiptera is likely to represent much less than the 7.5 to 10 percent of all arthropod species that they suggest.

Some 6,000 beetle species were also sorted from the samples (Hammond 1990) and a minimum estimate of 10,000 species was projected for the region (Hammond 1990; Stork 1993). Hodkinson and Casson's hypotheses can be tested using these two figures. Using their calculations and assuming that beetles represent 15 to 20 percent of all insect species results in global estimates of 3 to 4 million insect species and 5 to 6.7 million insect species. This highlights the difficulties with extrapolations from incomplete datasets: if samples represent just 50 percent of the total fauna, then the consequent global estimate will be half the true amount. Only much more intensive sampling, or even complete inventories, of tropical forests will generate datasets without the deficiencies described here.

Other Models for Species Estimation

BODY SIZE AND SPECIES NUMBER

It is an obvious feature of biological communities that there are many fewer large-bodied organisms than smaller-bodied organisms. May (1978, 1988) found that there is an inverse relationship between numbers of species and body size, at least for organisms larger than about 1 mm. According to fractal arguments, $S \approx L^{-x}$ (where S is the number of species and L is body length), the factor x should be between 1.5 and 3.0. May used this relationship to estimate global species richness of 10 to 50 million, simply by using these figures and extrapolating down to 1 mm (figure 2 in May 1988). Subsequently, he extended his analysis to include organisms down to 0.2 mm in length and, based on the assumption that species richness increases 100-fold for a 10-fold decrease in size, he added a further 10 million species to his global estimate (May 1990). André et al. (1994), through their work on communities of mites and collembolans in the soil, suggested that May's extrapolation should continue down to at least 136 to 165 μm, adding a further 10 million species to the global estimate. Further additions to the global estimate are likely if smaller organisms such as the Protozoa, Nematoda, Enchytraeidae, and Tardigrada are included in the analysis.

According to André et al. (1994) there may be 10 million or more species within the Collembola, Acarina, and Nematoda. To date, there is little evidence in support of this assertion. Lawton et al. (1996), for example, sampled the nematode fauna of soil and leaf litter in a lowland forest in Cameroon and found just 483 morphospecies. Furthermore, their analysis indicates that the number of new species collected was dropping off with additional sampling, which suggests that they had sampled most of the species present. Theoretical considerations further suggest that small organisms are not as diverse as hypothesized by André et al. Because smaller organisms are easily dispersed in air or water, it is feasible that conditions promoting allopatric speciation (i.e., separation of populations) may be rarely, if ever, met. This increased mobility or dispersal ability may therefore lead to a reduction in diversity with body size rather than an increase.

Perhaps then, despite the high abundance of the soil invertebrate mesofauna, the level of species diversity in those groups is unlikely to be as high as it is in groups such as Coleoptera, Diptera, and Hymenoptera.

SPECIES TURNOVER

Species accumulation with time and distance is another method that has been used (primarily in marine systems, but see Hammond 1990 and Mawdsley 1995 for similar terrestrial work) to estimate global species richness. Due to spatial and temporal heterogeneity, more species will continue to be collected in samples at increasing distance from a given starting point. Beta diversity, or species turnover, relates to the differences in diversity between samples collected from different sites; this measure can be used to estimate species numbers. For example, Grassle and Maciolek (1992) studied species data for a range of deep-sea macroinvertebrates collected over a 176-km transect of 14 stations at a depth of 2,100 m. From a total of 90,677 individuals, they found 798 species of annelids, molluscs, and arthropods, with 480 of these being new to science. They suggested that a straight-line prediction from the upper end of the species accumulation curve would indicate a linear rate of increase of one new species per kilometer. By extrapolating from kilometers to square kilometers and incorporating an oceanic area of 3×10^8 km^2 deeper than 1,000 m, they hypothesized that there are several hundred million new species yet to be sampled. Because these samples were collected from the continental shelf (where densities are much greater than those on the bottom of the deep sea) they finally rescaled their estimate of benthic marine species to 10^7.

According to May (1992), Grassle and Maciolek falsely assumed a straight line for the upper part of their rarefaction curve. A more appropriate extrapolation would have been a simple doubling of the number of benthic species on the basis of the number of new to known species they collected.

Erwin also used evidence from beta diversity to support his figures. He found that only 1 percent of 1,080 beetle species collected by canopy fogging in the Amazon were collected at all four sites in the same area (Erwin 1988). Furthermore, collections of moths from sites in Bolivia and Peru (which are 500 km apart) that totaled 933 species (1,748 individuals) and 1,006 species (1,731 individuals), respectively, were found to have just 60 species (3.2 percent) shared between the two (Erwin 1991). Both of these results are impressive, but the fact that most insect species are extremely rare in canopy (and other) samples (table 1.6) suggests that the chance of catching the same species in two samples even at the same site is extremely low. These results are consequently misleading and do not support high levels of endemism or reduced distributions of the collected species. More realistic analyses of endemism require comparisons of the percentage shared between subsamples at the same site and the percentage shared between different sites. Alternatively, analysis of this problem is facilitated through estimation of the regional species pool. If we assume, for the sake of argument, a modest global insect richness figure of 5.3 million (see table 3 in Stork 1993), of which, say, 1.2 million occur in tropical South America, that 300,000 of these are beetles, and that half of these are widespread (but probably mostly not very abundant), then the regional beetle fauna where Erwin collected his canopy samples might be as high as 100,000 species. Based on this estimate, a collection of 1,080 species from four forest types is very small and this undersampling of the fauna explains the 1 percent value for species shared among all four sites. Irrespective of the accuracy of the assumptions used here, the problem of sample size appears to be a real phenomenon. Future methods of global species estimation may require immense datasets and advanced statistical analyses.

TAXONOMISTS' VIEWS

The taxonomic identification of species is likely to vary markedly according to the taxa studied, their biology, the techniques used to distinguish species, and the level of clarity chosen by individual taxonomists. For example, much of the preceding discussion has focused on beetles and other insect groups for which a morphospecies concept has been adopted. This is a rather crude level of taxonomic clarity, and recent advances in molecular systematics have shown that single species that appear to be valid on normal morphological characteristics are sometimes found to comprise several genetically distinct species. This lack of taxonomic clarity has the potential to further cloud our view of the number of species inhabiting the planet.

How Many Species Are There?: Conclusions

The preceding sections demonstrate how difficult it has been for scientists to agree on just how many species there are on Earth. Ultimately our confidence in any of these estimates will depend on their scientific basis and on the degree to which numbers have been extrapolated upward. Table 1.7, for example, gives a rough guide to levels of extrapolations required for a working figure of about 13.4 million species. Extrapolations of more than a factor of 10 are hard to accept without good reasons.

What the various methods of estimating global numbers of species indicate is that some fundamental questions in biology, particularly in tropical rainforest biology, are unanswered. For example, on the broader scale we know little of alpha, beta, and gamma diversity for most organisms. Similarly, we know little about species ranges in the tropics other than for vertebrates and large plants. At the community level there also seem to be some basic questions that need tackling and widely held assumptions that need testing, such as the level of specificity of insects to particular plants, or whether more species are associated with ground habitats than live in the canopy of trees.

At this point it seems reasonable to opt for a rough working figure of about 13 million species of organisms on Earth. Critical data for some of the more fundamental aspects of tropical biology could easily see this figure doubled or halved.

TABLE 1.7

Numbers of Described Species and Estimated Numbers of All Species on Earth Showing the Extrapolation Factor for the Latter Estimates

	NUMBER OF SPECIES DESCRIBED (1)	WORKING FIGURE FOR TOTAL SPECIES ON EARTH (2)	EXTRAPOLATION FACTOR ((2) ÷ (1))
Viruses	4,000	400,000	100
Bacteria	4,000	1,000,000	250
Fungi	70,000	1,500,000	20
Protozoa	40,000	200,000	5
Plants	40,000	400,000	10
Algae	300,000	350,000	1.2
Nematodes	25,000	400,000	8
Chordates	45,000	50,000	1.1
Crustaceans	40,000	150,000	3.7
Arachnids	75,000	750,000	10
Insects	1,000,000	8,000,000	8
Molluscs	70,000	200,000	2.8
Total	1,713,000	13,400,000	7.8

RECENT RECORD OF EXTINCTION

In this section I examine empirical evidence of recent extinctions of species and the problems associated with finding such evidence and I examine current methods of estimating present and future extinction rates. I also call for a more scientific approach to the subject of extinction and extinction rates and the increased use of existing but largely untapped data in many museums and herbaria.

New Methods of Estimating Species Extinction Rates

Initial fears in the early 1980s concerning rising extinction rates were based on the inferred relationship between the loss of tropical forest through timber removal and loss of associated tropical species (Myers 1979; Lovejoy 1980; Raven 1987, 1988a, 1988b; Reid 1992; table 1.8). Models of species–area relationships developed in the 1960's suggested that just as the number of species on islands increased with

TABLE 1.8

Number of Recorded Extinctions Since 1600

	NO. OF SPECIES CERTIFIED EXTINCT SINCE 1600	APPROXIMATE TOTAL OF RECORDED EXTANT SPECIES	PERCENTAGE EXTINCT
ANIMALS			
Molluscs	286	10^5	0.29
Crustaceans	4	4×10^3	0.10
Insects	73	10^6	0.01
Vertebrates	257	4.7×10^4	0.55
Fishes	36	2.4×10^4	0.15
Amphibians	4	3×10^3	0.13
Reptiles	20	6×10^3	0.33
Birds	114	9.5×10^3	1.20
Mammals	83	4.5×10^3	1.84
Total	620	1.4×10^6	0.04
PLANTS			
Gymnosperms	2	758	0.26
Dicotyledons	120	1.9×10^5	0.06
Monocotyledons	462	5.2×10^4	0.89
Palms	4	2,820	0.14
Total	588	2.4×10^5	0.24

Plants: after WCMC (1992), IUCN (1986, 1988, 1990), and Smith et al. (1993a). Animals: After Groombridge (1993).

increasing size of the islands, so the number of tropical species should decrease as the areas of tropical forest decrease. In essence, species–area arguments suggested that a 90 percent reduction in size of an area should result in loss of half of its species. Concern about rates of global extinctions were further heightened by increased estimates of the number of species in tropical forests (Erwin 1982).

A novel method of estimating extinction rates for some well-known groups was recently introduced by Smith et al. (1993a, 1993b). They used the IUCN *Red Data Book* status of animals and plants in two ways. First they summed the number of species added to the list of extinctions (as recorded on the World Conservation Monitoring Centre's database) for a 4-year period (1986–1990) for animals, and 2-year period (1990–1992) for plants. With these data and the estimated number of species worldwide for different taxa, they were able to calculate the time required for 50 percent of the species to go extinct.

For birds and mammals the figures were 1,500 and 6,500 years, respectively. Second, they estimated the net changes in the *Red Data Book* status of species (from rare, vulnerable, endangered, to probably extinct) and again estimated 50 percent extinction values. The resulting values of 350, 250, and 70 years for birds, mammals, and palms, respectively, were much lower than those from their first method.

Smith et al. were unable to produce estimates for 50 percent extinction for less well-known groups such as insects and other invertebrates because of the lack of adequate information on their threatened status. However, Mawdsley and Stork (1995) suggested that one such source of information was the *Red Data Book* status of British invertebrates. They noted that insects, molluscs, and spiders were on average 7.1 more threatened than birds in Britain (table 1.9) and suggested that because 1 percent of birds and mammals have become extinct since 1600 (WCMC 1992), this "relative extinction rate" would indicate that 0.14 percent of insects also may have become extinct in the same period of time, equivalent to 11,200 insect species, assuming a figure of 8 million insect species on Earth. Using the estimated 12- to 55-fold increase in extinction rates of birds over the next 300 years (Smith et al. 1993a, 1993b) a further 100–500,000 insect species would also become extinct. Mawdsley and Stork (1995) also demonstrated that their relative extinction rate could be tested by back-prediction as follows. Shirt (1987) lists 99 or 0.7 percent of the 13,746 British insect species included in the *Red Data Book* as having become extinct since 1600. A relative extinction rate of 7.1 would suggest that 5 percent or 11 British breeding birds should also have been lost. This is confirmed by Sharrock (1974), who states that 11 species of bird have become extinct in Britain and Ireland, with two later recolonizing.

A third method of estimating extinction rates has been presented by Mace (1995). She also used IUCN *Red List* categories but ones that have been recently refined and she made species-by-species assessments of extinction probabilities over time (Mace 1994a, 1994b; Seal et al. 1993). Her calculations, based on expected extinction times for half of 10 vertebrate taxa, would suggest average extinction rates of 100 to 1,000 years or characteristically 300–400 years for mammals and birds.

TABLE 1.9

Estimated Rates of Extinction

ESTIMATE	% GLOBAL LOSS PER DECADE	METHOD OF ESTIMATION	REFERENCE
One million species between 1975 and 2000	4	Extrapolation of past exponentially increasing trend	Myers (1979)
15–20% of species between 1980 and 2000	8–11	Estimated species–area curve; forest loss based on Global 2000 projections	Lovejoy (1980)
50% of species by 2000 or soon after; 100% by 2010–2025	20–30	Various assumptions	Ehrlich and Ehrlich (1981)
9% extinction by 2000	7–8	Estimates based on Lovejoy's calculations using Lanly's (1982) estimates of forest loss	Lugo (1988)
12% of plant species in Neotropics, 15% of bird species in Amazon basin	—	Species—area curve ($z = 0.25$)	Simberloff (1986)
2000 plant species per year in tropics and subtropics	8	Loss of half the species in area likely to be deforested by 2015	Raven (1987)
25% of species between 1985 and 2015	9	As above	Raven (1988a, 1988b)
At least 7% of plant species	7	Half of species lost over next decade in 10 hot spots covering 3.5% of forest area	Myers (1988)
0.2–0.3% per year	2–6	Half of rainforest species assumed lost in tropical rain forests to local endemics and becoming extinct with forest loss	Wilson (1988, 1989, 1993)
2–13% loss between 1990 and 2015	1–5	Species–area curve ($0.15 < z < 0.35$); range includes current rate of forest loss and 50% increase	Reid (1992)
Red Data Books: Selected taxa: 50% extinct in 50–100 yrs. (palms), 300–400 yrs. (birds and mammals)	1–10	Extrapolating current recorded extinction rates and the dynamics of threatened status	Smith et al. (1993a, 1993b)
Selected vertebrate taxa	0.6–5	Fitting of exponential extinction functions based on IUCN categories of threat	Mace (1995)

After Reid (1992), Smith et al. (1993a), and Mace (1994a, 1994b). Note that the estimated rates include species "committed" to extinction; see text and Heywood et al. (1994) (after Reid 1992; Mawdsley and Stork 1995).

Empirical Evidence of Recent Extinctions

The most recent published list of species that are officially recognized by the International Union of Conservation of Nature and Natural Resources (IUCN) as having become extinct since 1600 has just 1,104 species names on it (WCMC 1992; IUCN 1994; table 1.10). To many this number must seem remarkably small considering the enormous amount of publicity on increasing extinction rates of species. Indeed, if some of the predicted extinction rates of the early 1980s (Myers 1979, 1988; Lovejoy 1980; Ehrlich and Ehrlich 1981; Raven 1987, 1988a, 1988b) were accurate we should have witnessed the loss of many tens of thousands if not hundreds of thousands of species in the last 20 years. Why is this not the case?

The first and most obvious reason is that it is very difficult to determine when the last individual of a species has died (Diamond 1987; Ehrlich and Wilson 1991; Mawdsley and Stork 1995) and figures for recent extinctions are probably woefully inaccurate. Biologists are still discovering many large and obvious species of birds and mammals that have never been seen before, such as a new species of bovid in Vietnam (Mackinnon 1993) and several species of primate in one year in Brazil (see Mawdsley and Stork 1995), so it is not surprising that the last remaining individuals of many threatened species of these vertebrates are often impossible to observe and monitor. For small vertebrates and for invertebrates it is virtually impossible to monitor the fate of more than a handful of such threatened species. The species losses recorded in table 1.10 must be seen as just the tip of the iceberg.

It is also important to recognize that the 1994 IUCN list of extinct species is in reality a list for 1944 because a species is officially recognized as extinct only when no living individual has been seen for 50 years. There have been remarkable changes in the global environment over the last 50 years, with large increases in rates of desertification or the loss of dry and moist tropical forests, and increases in pollution.

In practice, there is little direct evidence of widespread mass global extinction of species on continental landmasses through loss of forest, as some authors have already noted (Heywood and Stuart 1992, but see Raven and McNeely in press). For example, although many believed that the massive forest loss in the Atlantic forests in northeast Brazil would result in many extinctions of the endemic species, there is little evidence so far to show that this has happened. Brown and Brown (1992) point out that although the Atlantic rainforests have been reduced to about 12 percent of their original extent, zoologists there could not find a single animal species that could be declared as extinct. Simple species–area predictions would suggest that 50 percent of the animal species there are threatened with extinction. Some of the reasons for this lack of extinction in the Atlantic forests are obvious, such as the fragmented nature of the area. However, Brown and Brown suggest that the heterogeneous nature of the environment itself may have given rise to a fauna with a great deal of genetic plasticity. One other explanation is that it can take many years for factors inducing extinction to take effect (Heywood et al. 1994; Janzen 1987). Thus some species may be committed to extinction by external fac-

TABLE 1.10

Relative Rates of Extinction Estimated Using Data on the Threatened Status of British Animals

	1 NUMBER OF BRITISH SPECIES	2 NUMBER OF THREATENED BRITISH SPECIES	3 NUMBER OF ENDANGERED BRITISH SPECIES	4 % OF BRITISH SPECIES THREATENED	5 % OF WORLD SPECIES THREATENED	6 % OF BRITISH SPECIES ENDANGERED	7 RELATIVE EXTINCTION RATE: THREATENED (FROM [4])	8 RELATIVE EXTINCTION RATE: ENDANGERED (FROM [6])	9 MEAN RELATIVE EXTINCTION RATE (FROM [8] AND [9])
Mammals	55	16	No data	29.10	11	—	2.3	—	2.30
Birds	210	117	76	55.70	11	36.20	4.3	9.80	7.10
Insects	13,746[a]	1,786	506	12.90	0.07	3.70	1.00	1.00	1.00
Araneae	622	86	22	13.80	—	3.50	1.1	0.90	1.00
Mollusca	210[b]	30	10	14.20	0.40	4.80	1.1	0.80	1.00

Data in columns 1–6 from Shirt (1987), Bratton (1991), Batten (1990), NCC (1989), Knox (1992), and Kierney (1994). Table from Mawdsley and Stork (1995). "Threatened" includes all *Red Data Book* categories; "endangered" is RDB category 1. Columns 7 and 8 are derived from comparing the relative percentages between taxa in columns 4 and 6, respectively. Note that the percentage of British birds threatened represents an overestimate on account of the data used to calculate this figure; see text for further details.
[a]There are more than 22,000 species of insects in Britain but Shirt (1987) discusses the RDB status of only some of these.
[b]Only terrestrial, freshwater, and brackish Mollusca considered.

tors some tens or hundreds of years before they actually go extinct. Britain was al-most totally forested 5,000 years ago, yet by Roman times much of the forests had been cut down through human settlement. Some of the British species extinctions that have occurred in the last 4–500 years, and particularly more recently, may be the consequence of the forest loss more than 1,000 years before. Janzen (1987) has called such species committed to extinction the "living dead."

Despite such delayed extinctions, there appear to be considerably fewer record-ed species extinctions arising from forest loss than would be expected from species–area models. It has been suggested that this is because of the nature of forest loss and fragmentation of forests. Perhaps the critical factor here is in the way the area of "lost forest" has been calculated. In many instances it is assumed that logged forests no longer contribute to biodiversity counts. Biologists still have a poor un-derstanding of what happens to biodiversity (insects, fungi, and other microorgan-isms, as well as birds and mammals) when areas are selectively logged or how bio-diversity changes as forests regenerate after logging (but see Lawton *et al.* 1998).

Studies to date have focused largely on changes to birds, mammals, and flow-ering plants. Organisms such as insects, fungi, and other microorganisms general-ly have been ignored. Interpretation of the results of existing studies has also been complicated because too often the different or compounding effects of aspects of forest fragmentation (e.g., edge effects, size of fragments, degree of isolation, or length of time of isolation) or forest disturbance (e.g., amount of loss of vegetation cover, degree of soil compaction, degree of soil erosion, or hydrologic effects) are not recognized.

In contrast to the situation on continental areas, there is a vast amount of evi-dence of extinction on islands. Pimm et al. (1995) suggest that 50 percent or more of bird species on many of the Pacific islands including the Hawaiian islands, southeast Polynesia, the Marianas, and New Caledonia are missing, endangered, extinct, or known from bones only. For the Hawaiian islands this figure is closer to 90 percent. Although there is less information available, the widespread loss of species of insects on isolated islands is also true. Of the 61 insect species listed as extinct by WCMC (1992), 51 are from islands, including 42 from the Hawaiian is-lands. In general there is little evidence of extinction of aves on continental land masses, with most recent extinctions having taken place on islands. It also seems that not all island faunas and floras are equally extinction prone. For example, as a broad generalization, Pacific islands have experienced much greater extinction rates than Atlantic islands. The reasons for this difference are unclear.

IMPROVING THE CHANCES OF SURVIVAL OF ENDANGERED SPECIES

Much of what has been written about extinction and extinction rates predicts a very bleak future for many organisms, yet there is a body of literature that provides some hope. If, as evidence from the Brazilian Atlantic forests and elsewhere suggests, there is considerable delay in extinction of species due to loss of habitat, then it seems that there may be some hope for many endangered species through species survival programs and through habitat restoration. Perhaps much greater efforts should be placed on the latter because this will lead to the long-term survival of species and whole communities.

From our understanding of the ecology of many groups of organisms we should be able to make some predictions about which ecological qualities make a species more or less prone to extinction. Two sets of factors appear to be important in this respect. The first are the intrinsic qualities of a given species or group of species. The second are external or extrinsic factors that act on species and reduce their populations. For both sets of factors it seems that all species are not equal and that some are both intrinsically prone to extinction and more affected by extrinsic factors (e.g., species that are high on the trophic scale such as predators and parasitoids), have low fecundity and small population sizes, and are narrowly distributed in rare ecosystems. For tropical forest birds, for example, the relative abundance in the forest and the habitat specificity and use of disturbed habitats are important determinants of local extinction (Newmark 1991), whereas for nonflying mammals, Laurance (1991) found that natural rarity does not appear to be important, leaving abundance in disturbed habitats as the best predictor of the probability of extinction.

Many of the chapters in the recent book on extinction rates by Lawton and May (1995) highlight apparent differences in extinction rates among taxonomic groups and places. Indeed, there is much theoretical and empirical evidence that not all species are equal in terms of their likelihood of extinction. In simple terms it seems that large vertebrates are more threatened than nonvertebrates. As Mawdsley and Stork (1995) indicated, gross extrapolations of extinction rates for all organisms based on data for large vertebrates may be misleading. We need to refine our understanding of these differences if we are to develop more meaningful models for extinction rates. Future research should focus on the effects of habitat fragmentation and loss on nonvertebrates, such as invertebrates, fungi, and other microorganisms, because these organisms are often most responsible for critical ecosystem processes such as pollination and decomposition (see Didham et al. 1996) and on the intrinsic qualities of species that make them more or less prone to extinction (e.g., Newmark 1991; Laurance 1991; Mawdsley and Stork 1995).

Museums, botanical gardens, and other repositories of natural history organisms have a vital role to play in identifying the species and geographic locations that are most threatened. A rapid survey of the largest museums and botanical gardens to identify the species that are most likely to be threatened or are already extinct would be easy to achieve and yet would advance our knowledge of the threatened status of life on Earth enormously. As biologists we should be ashamed to present such poor data as in table 1.10 when we know that many more species are either extinct or are well on the way to extinction. How can we expect to be taken seriously by other scientists and by politicians if we do not do a better job of presenting good supporting data for our arguments?

ACKNOWLEDGMENTS

I thank Joel Cracraft and Michael Novacek for inviting me to prepare this material for the New York Conference. I thank Wade Hadwen and Harriet Eeley for assistance in preparing the manuscript and Martin Honey and David Goodger for information on the types of Noctuidae and Geometridae.

REFERENCES

André, H. M., M.-I. Noti, and P. Lebrun. 1994. The soil fauna: the other last biotic frontier. *Biodiversity and Conservation* 3: 45–46.

Batten, L. A. et al. 1990. *Red data birds in Britain: Action for rare, threatened and important species.* London: NCC and RSPB, T. and A.D. Poyser.

Bisby, F. A. 1994. Global master species databases and biodiversity. *Biology International* 29: 33–38.

Bratton, J. H. 1991. *British red data books. 3. Invertebrates other than insects.* Peterborough, U.K.: Joint Nature Conservation Committee.

Brown Jr., K. S. and G. G. Brown. 1992. Habitat alteration and species loss in Brazilian forests. In T. C. Whitmore and J. A. Sayer, eds., *Tropical deforestation and species extinction,* 119–142. London: Chapman & Hall.

Corbet, G. B. and J. E. Hill. 1980. *A world list of mammalian species.* Ithaca, N.Y.: Cornell University Press.

Diamond, J. M. 1987. Extant unless proven extinct? Or extinct unless proven extant? *Conservation Biology* 1: 77–79.

Didham, R. K., J. Ghazoul, N. E. Stork, and A. J. Davis. 1996. Insects in fragmented forests: A functional approach. *Trends in Ecology and Evolution* 11: 255–260.

Ehrlich, P. R. and A. H. Ehrlich. 1981. *Extinction. The causes of the disappearance of species.* New York: Random House.

Ehrlich, P. R. and E. O. Wilson. 1991. Biodiversity studies: Science and policy. *Science* 253: 758–762.

Erwin, T. L. 1982. Tropical forests: Their richness in Coleoptera and other arthropod species. *Coleoptera Bulletin* 36: 74–75.

Erwin, T. L. 1988. The tropical forest canopy: The heart of biotic diversity. In E. O. Wilson and F. M. Peters, eds., *Biodiversity,* 123–129. Washington, D.C.: National Academy Press.

Erwin, T. L. 1991. How many species are there? Revisited. *Conservation Biology* 5: 330–333.

Gaston, K. J. 1991. The magnitude of global insect species richness. *Conservation Biology* 5: 283–296.

Gaston, K. J. 1992. Regional numbers of insects and plants. *Functional Ecology* 6: 243–247.

Gaston, K. J. 1994. Spatial patterns of species description: How is our knowledge of the global insect fauna growing? *Biological Conservation* 67: 37–40.

Gaston, K. J. and R. M. May. 1992. Taxonomy of taxonomists. *Nature* 356: 281–282.

Gaston, K. J. and L. A. Mound. 1993. Taxonomy, hypothesis testing and the biodiversity crisis. *Proceedings of the Royal Society of London, Series B* 251: 139–142.

Golley, F. B. 1984. Introduction. In J. H. Cooley and F. B. Golley, eds., *Trends in ecological research in the 1980s ,* 1–4. New York: Plenum.

Grassle, J. F. and N. J. Maciolek. 1992. Deep-sea species richness: Regional and local diversity estimates from quantitative bottom samples. *American Naturalist* 139: 313–341.

Groombridge, B., ed. 1993. *1994 IUCN red list of threatened animals*. Gland, Switzerland: IUCN.

Hammond, P. M. 1990. Insect abundance and diversity in the Dumoga-Bone National Park, N. Sulawesi, with special reference to the beetle fauna of lowland rain forest in the Toraut region. In W. J. Knight and J. D. Holloway, eds., *Insects and the rain forests of South East Asia (Wallacea)* , 197–254. London: Royal Entomological Society of London.

Hammond, P. M. 1992. Species inventory. In B. Groombridge, ed., *Global biodiversity, status of the earth's living resources* , 17–39. London: Chapman & Hall.

Hammond, P. M. 1994. Practical approaches to the estimation of the extent of biodiversity in speciose groups. *Philosophical Transactions of the Royal Society of London, Series B* 345: 119–136.

Hammond, P. M. 1995. Described and estimated species numbers: An objective assessment of current knowledge. In D. Allsopp, R. R. Colwell, and D. L. Hawksworth, eds., *Microbial diversity and ecosystem function*, 29–71. Wallingford, U.K.: CAB International.

Hammond, P. M. and Owen, J. In press. *The beetles of Richmond Park SSSI: A case study*. Peterborough, U.K.: English Nature.

Hammond, P. M., N. E. Stork, and M. J. D. Brendell. 1997. Comparison of the composition of the beetle faunas of the canopy and other ecotones of lowland rainforest in Indonesia. In N. E. Stork and J. Adis, eds., *Canopy arthropods*, 184–223. London: Chapman & Hall.

Hawksworth, D. L. 1991. The fungal dimension of biodiversity: Magnitude, significance and conservation. *Mycological Research* 95: 641–655.

Heywood, V. H., G. M. Mace, R. M. May, and S. N. Stuart. 1994. Uncertainties in extinction rates. *Nature* 368: 105.

Heywood, V. H. and S. N. Stuart. 1992. Species extinctions in tropical forests. In T. C. Whitmore and J. Sayer, eds., *Tropical deforestation and species extinctions* , 91–117. London: Chapman & Hall.

Hodkinson, I. D. and D. Casson. 1991. A lesser predilection for bugs: Hemiptera (Insecta) diversity in tropical forests. *Biological Journal of the Linnaean Society*, 43: 101–109.

International Union for the Conservation of Nature. 1986, 1988, 1990, 1994. Red list of threatened animals. Gland, Switzerland: IUCN.

Janzen, D. H. 1987. Insect diversity of a Costa Rican dry forest: Why keep it, and how? *Biological Journal of the Linnaean Society* 30: 343–356.

Knox, A. G. 1992. *Checklist of birds in Britain and Ireland*. Tring: British Ornithologists Union.

Laurance, W. F. 1991. Ecological correlates of extinction proneness in Australian tropical rain forest mammals. *Conservation Biology* 5: 79–89.

Lawton, J. 1993. On the behaviour of autecologists and the crisis of extinction. *Oikos* 67: 3–5.

Lawton, J. H., D. E. Bignell, G. F. Bloemers, P. Eggleton, and M. E. Hodda. 1996. Carbon flux and diversity of nematodes and termites in Cameroon forest soils. *Biodiversity and Conservation* 5(2): 261–273.

Lawton, J. H., D. E. Bignell, R. Bolton, G. F. Bloemers, P. Eggleton, P. M. Hammond, M. Hodda, R. D. Holt, T. B. Larsen, N. A. Mawdsley, N. E. Stork, D. S. Srivastava, and A. D. Watt. 1998. Biodiversity inventories, indicator taxa and effects of habitat modification in tropical forests. *Nature* 391: 72–76.

Lawton, J. H. and R. M. May, eds., 1995. *Extinction rates*. Oxford, U.K.: Oxford University Press.

Lovejoy, T. E. 1980. A projection of species extinctions. In *Council on Environmental Quality (CEQ) : The Global 2000 Report to the President*, Vol. 2: CEQ, 328–331. Washington, D.C.: U.S. Government Printing Office.

Lugo, A. E. 1988. Estimating reductions in the diversity of tropical forest species. In E. O. Wilson and F. M. Peter, eds., *Biodiversity* , 58–70. Washington, D.C.: National Academic Press.

Mace, G. M. 1994a. An investigation into methods for categorising the conservation status of species. In P. J. Edwards, R. M. May, and N. R. Webb, eds., *Large scale ecology and conservation biology,* 295–314. Oxford, U.K.: Blackwell.

Mace, G. M. 1994b. Classifying threatened species: Means and ends. *Philosophical Transactions of the Royal Society of London, Series B* 344: 91–97.

Mace, G. M. 1995. Classification of threatened species and its role in conservation. In J. H. Lawton and R. M. May, eds., *Extinction rates,* 197–213. Oxford, U.K.: Oxford University Press.

Mackinnon, J. 1993. A new species of living bovid from Vietnam. *Nature* 363: 443–445.

Mawdsley, N. A. 1995. *Community structure of Coleoptera in a Bornean lowland forest.* Ph.D. Thesis, Imperial College, University of London.

Mawdsley, N. A. and N. E. Stork. 1995. Species extinctions in insects: Ecological and biogeographical considerations. In R. Harrington and N. E. Stork, eds., *Insects in a changing environment* , 322–369. London: Academic Press.

Mawdsley, N. A. and N. E. Stork. 1997. Host-specificity and the effective specialisation of tropical canopy beetles. In N. E. Stork, J. Adis, and R. K. Didham, eds., *Canopy arthropods* , 104–130. London: Chapman & Hall.

May, R. M. 1978. The dynamics and diversity of insect faunas. In L. A. Mound and N. Waloff, eds., *Diversity of insect faunas* , 188–204. Oxford, U.K.: Blackwell.

May, R. M. 1988. How many species are there on Earth? *Science* 241: 1441–1449.

May, R. M. 1990. How many species? *Philosophical Transactions of the Royal Society of London, Series B* 330: 293–304.

May, R. M. 1992. Bottoms up for the oceans. *Nature* 357: 278–279.

Myers, N. 1979. *The sinking ark. A new look at the problem of disappearing species.* New York: Pergamon.

Myers, N. 1988. Threatened biotas: "Hotspots" in tropical forests. *Environ* 8: 1–20.

Myers, N. 1989. *Deforestation rates in tropical forests and their climatic implications.* London: Friends of the Earth.

Nature Conservation Council. 1989. *Guidelines for the selection of biological sites of special scientific interest: Rationale, operational approach and criteria.* Peterborough, U.K.: NCC.

Newmark, W. D. 1991. Tropical forest fragmentation and the local extinction of understory birds in the eastern Usambara mountains, Tanzania. *Conservation Biology* 5: 67–68.

Pimm, S. L., M. P. Moulton, and L. J. Justice. 1995. Bird extinctions in the central Pacific. In J. H. Lawton and R. M. May, eds., *Extinction rates* , 75–97. Oxford, U.K.: Oxford University Press.

Raven, P. H. 1985. Disappearing species: A global tragedy. *Futurist* 19: 8–14.

Raven, P. H. 1987. The scope of the plant conservation problem world-wide. In D. Bramwell, O. Hamann, V. Heywood, and H. Synge, eds., *Botanic gardens and the world conservation strategy,* 19–29. London: Academic Press.

Raven, P. H. 1988a. Biological resources and global stability. In S. Kawano, J. H. Connell, and H. Hidaka, eds., *Evolution and coadaptation in biotic communities* , 3–27. Tokyo: University of Tokyo Press.

Raven, P. H. 1988b. Our diminishing tropical forests. In E. O. Wilson and F. M. Peter, eds., *Biodiversity,* 119–122. Washington, D.C.: National Academy Press.

Raven, P. H. and J. A. McNeely. In press. Biological extinctions: Its scope and meaning for us. In L. Guruswamy and J. A. McNeely, eds., *Protection of global diversity: Converging strategies* . Durham, N.C.: Duke University Press.

Reid, W. V. 1992. How many species will there be?In T. C. Whitmore and J. A. Sayer, eds., *Tropical deforestation and species extinction*, 55–73. London: Chapman & Hall.

Seal, U. S., T. J. Foose, and S. Ellis-Joseph. 1993. Conservation assessment and management plans (CAMPs) and global captive action plans (GCAPs). In G. M. Mace, P. J. Only, and A. T. C. Feistner, eds., *Creative conservation*, 312–25. London: Chapman & Hall.

Sharrock, J. T. R. 1974. The changing status of breeding birds in Britain and Ireland. In D. L. Hawksworth, ed., *The changing flora and fauna of Britain* , 203–220. London: Academic Press for the Systematics Association.

Shirt, D. B., ed. 1987. *British red data books. 2. Insects*. Peterborough, U.K.: Nature Conservation Council.

Simberloff, D. 1986. Are we on the verge of a mass extinction in tropical rain forests? In D. J. Elliot, ed., *Dynamics of extinction*, 165–180. New York: Wiley.

Simon, H. R. 1983. *Research and publication trends in systematic zoology*. Ph.D. thesis, The City University, London.

Smith, F. D. M., R. M. May, R. Pellew, T. H. Johnson, and K. S. Walter. 1993a. Estimating extinction rates. *Nature* 364: 494–496.

Smith, F. D. M., R. M. May, R. Pellew, T. H. Johnson, and K. S. Walter. 1993b. How much do we know about the current extinction rate. *Trends in Ecology and Evolution* 8: 375–378.

Stork, N. E. 1987. Guild structure of arthropods from Bornean rain forest trees. *Ecological Entomology* 12: 69–80.

Stork, N. E. 1988. Insect diversity: Facts, fiction and speculation. *Biological Journal of the Linnaean Society* 35: 321–337.

Stork, N. E. 1991. The composition of the arthropod fauna of Bornean lowland rain forest trees. *Journal of Tropical Ecology* 7: 161–180.

Stork, N. E. 1993. The biodiversity crisis: An agenda for global research. *Biology International Special Issue* 29: 59–64.

Stork, N. E. 1995. Measuring and inventorying arthropod diversity in temperate and tropical forests. In T. Boyle and B. Boontawee, eds., *Measuring and monitoring biodiversity in tropical and temperate forests* , 257–270. Bogor, Indonesia: Center for International Forestry.

Stork, N. E. 1997. Measuring global biodiversity and its decline. In E. O. Wilson, M. L. Reaka-Kudla, and D. E. Wilson, eds., *Biodiversity II: Understanding and protecting our natural resources* , 41–68. Washington, D.C.: Joseph Henry / National Academy Press.

Stork, N. E. and K. G. Gaston. 1990. Counting species one by one. *New Scientist* 1729: 43–47.

Systematics Agenda 2000. 1994a. Charting the biosphere. New York: American Museum of Natural History.

Systematics Agenda 2000. 1994b. Charting the biosphere. Technical Report. New York: American Museum of Natural History.

Thomas, C. D. 1990. Fewer species. *Nature* 347: 237.

Westwood, J. O. 1833. On the probable number of species in the Creation. *Magazine of Natural History* 6: 116–123.

Wilson, E. O. 1988. The current state of biological diversity.In E. O. Wilson and F. M. Peter, eds., *Biodiversity* , 3–18. Washington, D.C.: National Academy Press.

Wilson, E. O. 1989. Threats to biodiversity. *Scientific American* September: 108–116.

Wilson, E. O. 1993. *The diversity of life*. Cambridge, Mass.: Belknap Press.

World Conservation Monitoring Centre (WCMC). 1992. *Global biodiversity: Status of the earth's living resources*. London: Chapman & Hall.

CHAPTER 2

DIMENSIONS OF BIODIVERSITY: TARGETING MEGADIVERSE GROUPS

Norman I. Platnick

The ratio of taxonomists to species is an order-of-magnitude greater for vertebrates than for plants, and two orders-of-magnitude greater for vertebrates than for invertebrates. These disparities are mirrored in publications per species (May 1988). This is no way to run a business.

—R. M. May (1994:18)

My brief here is to review what we know and what we still need to learn about taxonomic, rather than genetic or ecological, aspects of biodiversity. A superb general account of the species-level aspects of biodiversity has already been provided by Hammond (1992; see also Hammond 1995), who summarized what is known about the numbers of species already described for all major groups of living organisms and evaluated estimates of what the actual sizes of those groups might be. With regard to determining the numbers of known species, Hammond pointed out the difficulties in distinguishing, among available estimates, those that merely summarize the total number of specific names ever established in a group from those that tackle the much more difficult questions of how many of the available specific names are currently regarded as valid, or how many of the available specific names are likely to prove valid when fully evaluated by modern revisionary studies. For some groups (even discounting those with significant questions involving the status of putative subspecies), the number of valid species may be one-third smaller than the number of available names (Gaston and Mound 1993: table 1).

Hammond's results are splendidly summarized in two figures (1992:figures 4.5 and 4.6) showing the distribution, among groups, of his current estimates of about 1.75 million known species and about 12.75 million probably extant species. I have no quarrel with Hammond's figures, but by necessity they refer to higher taxa, such as the class Arachnida, that are so large and varied as to be opaque to detailed analysis. Among the orders of arachnids, for example, the problems in estimating the global diversity of scorpions or mites are of two different orders of magnitude. I will use the arachnid order I happen to know best (spiders) as an example of a megadiverse taxon and will ask two main questions: how many species of spiders are there?, and does it matter, for spiders or any other group?

HOW MANY SPECIES OF SPIDERS ARE THERE?

Spiders are among the most diverse groups on the planet. Of the other taxa ranked as orders, only five animal groups include a larger number of described species. They are the five largest insect orders: Coleoptera, Hymenoptera, Lepidoptera, Diptera, and "Hemiptera" (Parker 1982; Coddington and Levi 1991). There are probably many more extant species of mites than of spiders, but fewer acarine species have been described to date (Johnston 1982). The literature already published on spiders is enormous. Although existing catalogs of that literature currently total 14 volumes, the 3 most recent of which I have produced, we do not yet have an accurate total of the number of spider species already described or currently considered valid. My cataloging activities began in 1986, but a complete world catalog is still 5 years of work away. The best estimate I can offer at the moment is that the order includes some 36,000 currently valid species, placed in about 3,150 genera and 106 families.

It would be comforting to think that at high levels of the taxonomic hierarchy, such as the family level for spiders, our task has largely been completed. Although the number of spider families increased significantly in the years after 1967, the number of families recognized was 105 at the end of 1987 and remained at that number at the end of 1991 (Platnick 1989, 1993). However, even that limited stability is only coincidental; over the intervening 4 years, five groups earlier treated as families (the Aphantochilidae, Dolomedidae, Hadrotarsidae, Loxoscelidae, and Platoridae) were downgraded to lower rank and five other groups (the Lamponidae, Prodidomidae, Synotaxidae, Trechaleidae, and Zoropsidae) were elevated to family status. More recently, the monogeneric groups Bradystichidae and Argyronetidae have been synonymized with larger families (Platnick and Forster 1993; Grothendieck and Kraus 1994), the genera *Pimoa* and *Periegops* were elevated to the family level (Hormiga 1993; Forster 1995), and a new family (apparently with no previously described members) was described from Madagascar (Jocqué 1994).

At the family level, there is still a final frontier for spider studies; it is not in tropical rainforest canopies, but rather in Australia, where many known genera cannot readily be placed into the current familial classification, which for historical reasons was based largely on north temperate faunas. I would not be surprised if an additional 20 families eventually must be recognized for Australian spider taxa.

In a survey of spider systematics, Coddington and Levi admitted that "Because spiders are not thoroughly studied, estimates of total species diversity are difficult" (1991:566). The question of how many spider species might exist was of only tangential interest to their task of summarizing what we have learned to date about the interrelationships among spider families, and they devoted only one paragraph to the problem, but Coddington and Levi nevertheless published what is, to my knowledge, the only modern estimate of the total diversity of the group. They accepted the estimate by Raven (1988, after Monteith and Davies 1984) that only about 20 percent of the Australian fauna had been described at that time, suggested that perhaps 60 to 70 percent of the New Zealand fauna had been described and cited Coddington, Larcher, and Cokendolpher (1990) to the effect that "The Nearctic fauna is perhaps 80% described" (Coddington and Levi 1991:566). They indicated that "other areas, especially Latin America, Africa, and the Pacific region are much more poorly known" and concluded that "if the above statistics suggest that 20% of the world fauna is described, then about 170,000 species of spiders are extant."

It is important, then, to see what available data might suggest about the proportion of the total fauna that has been described to date. Coddington and Levi illustrated one possible approach to this problem, which involves comparing estimates of the size of particular groups before and after modern taxonomic revisions have been conducted. For example, they cited four large revisions of orb-weavers by Levi, in which "60–70% of the species in available collections were new." For those groups, of course, not all of the newly described species represented a net gain in known diversity because, as they indicated, "for each 50 previously known species about 75 names exist, as common species had been given different names in different countries." Thus their figures suggest that for every 75 described species, 50 would have been valid and somewhere between 75 and 116 new species would have been described, for a total of 125–166 actual species. The previously described 75 species thus represented a diversity estimate that was somewhere between 45 and 60 percent of the newly estimated total faunal size.

The question, then, is whether that result is typical of spider groups in general. As Coddington and Levi pointed out, "available collections are biased toward medium- and larger-sized species from easily accessible habitats." Orb-weavers are indeed medium- and larger-sized spiders and might have better than average representation in available collections. As evidence that the Neotropical orb-weavers might have been disproportionately well known, Coddington and Levi cited a monograph by Forster and Platnick (1985) "on the poorly known south temperate family Orsolobidae," in which "85% of the species were new." The actual figures from that paper are 41 previously known species, 129 new species, and no synonymies, so that the total proportion of previously known species was 24 percent, not 15 percent. For South American areas only, however, the three orsolobid species previously described from Chile and Argentina were joined by 29 new species, so that only 9 percent of the American fauna had previously been described. Of course, it is precisely in groups such as orsolobids, which require specialized collecting techniques and are scarcely represented at all in classic collections, that the number of existing names that are actually synonyms will be lowest, so that net gains are much easier to achieve.

The problem of synonymies and net gains can be serious. As aptly pointed out by Gaston and Mound (1993), synonymies represent falsifications of earlier hypotheses about species limits and can severely affect estimates of the amount of work necessary to complete systematic projects. For example, Alderweireldt and

Jocqué (1994) recently compared the diversity of spiders in Africa and Europe, estimating that almost 3,500 species have been described from Europe and almost 6,000 species from Africa. Some of their results were surprising. For example, they compiled data on the rates of species description, by decade, for each region, but the two species accumulation curves scarcely differ. In neither case do the curves appear to have leveled off. They also applied the method used by Coddington and Levi, listing the results of a dozen recent taxonomic revisions of African spiders. For those papers, before the revisions, a total of 297 species were recognized. After the revisions, a total of 268 species were recognized, representing a net loss of 29 species. The net loss, of course, means that the number of older hypotheses about species identities that were falsified during these studies was greater than the number of new hypotheses about species identities that were put forward.

No one would contend that this is a representative result, or that the actual size of the African spider fauna is smaller than the currently described 6,000 species. Indeed, the figures were biased in part by one revision of an orb-weaver genus in which the number of species decreased from 95 to 20. If anything, that result tends to support the idea that the Neotropical orb-weaver results show too low a net gain to be representative of the Neotropical fauna in general.

I can provide a few examples of modern revisions of south temperate groups that could help support the view that only 20 percent of the fauna has been described to date. Combining the results of revisions of the families Anapidae and Synotaxidae (Platnick and Forster 1989; Forster et al. 1990), 1 previously known species in Chile expanded to 20, 7 previously known species in Australia expanded to 43, and 8 previously known species in New Zealand and New Caledonia expanded to 68. For these two families in those areas, only 12 percent of the total species number was previously known.

However, there is little reason to judge these groups typical. One can easily select revisions from the other end of the spectrum; combining the results of recent revisions of the Mimetidae and tracheline Corinnidae of Chile and adjacent Argentina (Platnick and Shadab 1993; Platnick and Ewing 1995), for example, 14 new species were described but 16 older names were sunk, so that a total of 30 previously existing names represents 103 percent of the actual known diversity.

Both the Alderweireldt and Jocqué results and these few southern examples show dramatically that using just a small sample of modern revisionary studies to estimate the real number of spider species is unlikely to provide an accurate estimate of anything. The variance in those results is enormous. The African examples range from −82 to +700 percent (in a group that increased from 2 to 14 species). The question, then, is what a larger sample of revisions might show. To investigate that, I have used as a sample some 54 revisions of various Latin American spider groups that I and my coauthors have published in the *American Museum Novitates* and *Bulletin* over the past two decades (table 2.1). Producing such tallies is not unproblematic because some species have been named (often more than once) only from the United States, even though they are now known to occur in Mexico or farther

TABLE 2.1

Changes in Group Size During Revisions by Platnick et al. (1974–1995) of Spiders from Mexico, the West Indies, Central America, and South America

FAMILY	SPECIES # BEFORE	SPECIES # AFTER	% KNOWN BEFORE
Actinopodidae	1	1	100
Anapidae	11	48	23
Anyphaenidae	13	26	50
Atypidae	0	1	0
Austrochilidae	1	7	14
Caponiidae	3	8	37
Corinnidae	62	80	77
Ctenizidae	1	1	100
Diguetidae	1	3	33
Gnaphosidae	163	269	61
Idiopidae	2	9	22
Liocranidae	0	1	0
Malkaridae	0	1	0
Mecicobothriidae	1	2	50
Mecysmaucheniidae	3	20	15
Micropholcommatidae	1	2	50
Microstigmatidae	1	4	25
Migidae	2	3	67
Mimetidae	10	8	125
Miturgidae	12	3	400
Mysmenidae	8	16	50
Orsolobidae	3	32	9
Palpimanidae	29	34	85
Prodidomidae	6	23	26
Symphytognathidae	4	13	31
Synotaxidae	1	5	20
Zodariidae	2	5	40
Total	341	625	55

south. For this table, I counted a species as previously known if it had even a single name from anywhere in its range, but I deducted newly synonymized names from the totals of previously known species only if the type localities of two or more synonyms were from Latin America. For each family listed, the numbers refer only to genera or species groups that happen to have been revised, not necessarily to the entire known Neotropical fauna.

The families included in this list range from large, burrowing mygalomorphs through medium-sized ground-dwelling hunting spiders to tiny litter-dwelling and orb-weaving species. As expected, they show a large range of values, with the previously existing number of names representing from 0 to 400 percent of the actual fauna. Also as expected, the medium-sized hunters that are better represented in classic collections show less extensive expansions in species numbers than do the tiny litter-dwellers or the large, burrowing mygalomorphs, each of which requires special collecting techniques. But most relevant are the figures on the bottom line, which show that for this sample of groups, the total of earlier names is fully 55 percent of the newly estimated number of species. In other words, the expansion over this entire list of revisions was less than one new species per previously existing name.

I do not have similar figures for all modern revisions in all parts of the world, but can provide a worldwide summary of the total numbers of species described or synonymized each year from 1955 through 1993 (table 2.2). Similar figures have been provided by Hammond (1992) for various (much larger) groups and by Bonnet (1979 and earlier papers) for spiders. In both cases the previously published figures represent numbers counted from the *Zoological Record*, and my figures (based on a far more complete cataloging of the literature) indicate that numbers taken from the *Zoological Record*, at least for spiders, are underestimates (for 1959 through 1972, 13.6 percent of the new species were missed, if Bonnet's and my counts are accurate).

Since 1954, arachnologists have described more than 12,000 new spider species, but they also synonymized more than 4,000 older names (including 625 of the newer names), resulting in a net gain of 8,182 species.

A few observations can be made about these figures. Only 2 years show a net loss of species, where the number of older names sunk exceeded the number of newly described species. Overall, the rate of synonymy of newer names is acceptably low. Only 2 years (1955 and 1963) show alarmingly high numbers, and in each case those numbers reflect a single large, unfortunate publication. In fact, one of those papers significantly biased Alderweireldt and Jocqué's African results, for it provided the grist for many of the synonymies made in a revision that reduced a genus from 80 to 14 species. On the more positive side, the 4 most recent years show a significant increase in net gains, adding 1,854 species. Few of those names are likely to be synonymized in the future, mostly from cases in which increasingly active work in Russia, China, Korea, and Japan sometimes results in nearly simultaneous descriptions of the same new taxa in two or more of those neighboring countries.

These figures cannot be related directly to those in table 2.1 because they include many isolated species descriptions that involved little or no revisionary work. But the 4,047 synonyms that were found refer to 2,351 other names. At the very least, then, 6,398 previously described taxa were examined seriously during the four decades of worldwide work sampled here, which means that no more than two new species were described for each previously described species that was seriously examined. In rough terms, 6,000 or more earlier names produced some 14,000 currently valid species, of which the total of earlier names thus constituted, at the very least, 42.8 percent of the resulting total of names now considered valid. Of course, that figure does not include the many older names that were reexamined during those decades and found to be valid but without new synonyms. Here again, the real figure probably lies somewhere between 45 and 60 percent.

So what have we learned from examining statistics about species description rates? As mentioned earlier, Raven estimated that only 20 percent of the Australian spider fauna is described. Based on my knowledge of Australian collections, that estimate seems reasonably accurate, but I also believe that there is no other major geographic region where the figure is actually that low. A more realistic global estimate would consider the Neotropical percentages typical, not the Australian ones. Thus the 36,000 currently valid names probably represent somewhere between 45 and 60 percent of the total number of extant species, leading to an estimated range of 60,000–80,000 spider species in the world.

Of course, this result merely raises the question of whether there is an alternative approach to the problem that might extrapolate from something other than changes published in modern revisions or counts of modern descriptions and synonymies in general. In short, is their some other approach that might help us check the accuracy of these estimates? One such approach extrapolates from geographic patterns in the diversity of groups of organisms that are well known (or at least are much better known than are spiders).

As Colwell and Coddington (1994) indicated, these kinds of extrapolations compare some reference site, where ratios are calibrated using some putatively well-known indicator taxon, with one or more comparative sites at which only one factor in the ratio is estimated directly. Reference sites can be as small as one tree species in Panama (Erwin 1982); the extensive criticisms of Erwin's widely publicized estimate of 30 million species of tropical arthropods are ably summarized by Stork (1993; see also Gaston 1991a, 1991b; Erwin 1991). More reasonable applications generally use a sizable study site, such as the Project Wallace efforts in Sulawesi (Noyes 1989; Hammond 1990; Stork and Brendell 1990; Hodkinson and Casson 1991) and similar projects elsewhere (Stork 1987, 1988, 1991, 1994).

In perhaps its simplest realistic form, this approach can be applied by concentrating on the part of the world that is best known and comparing only two groups. For spiders, that geographic area would be northwestern Europe, particularly Great Britain. Because butterflies are probably the best-known group of arthro-

TABLE 2.2
Numbers of Spider Species Described or Synonymized

YEAR	VALID, IN SAME GENUS	VALID, IN ANOTHER GENUS	SUNK BEFORE 1992	TOTAL DESCRIBED	OLDER NAMES SUNK	NET GAIN
1955	300	71	83	454	23	431
1956	79	26	10	115	23	92
1957	71	16	13	100	101	-1
1958	112	35	16	163	42	121
1959	236	44	24	304	64	240
1960	234	66	30	330	23	307
1961	91	13	12	116	27	89
1962	169	20	18	207	103	104
1963	339	33	96	468	67	401
1964	173	31	24	228	64	164
1965	194	16	32	242	51	191
1966	96	21	14	131	48	83
1967	118	14	17	149	231	-82
1968	208	63	17	288	80	208
1969	122	7	12	141	68	73
1970	253	33	16	302	94	208
1971	160	28	16	204	101	103
1972	156	22	6	184	83	101
1973	342	17	12	371	135	236
1974	156	48	8	212	91	121

TABLE 2.2 *(continued from previous page)*

Numbers of Spider Species Described or Synonymized

YEAR	VALID, IN SAME GENUS	VALID, IN ANOTHER GENUS	SUNK BEFORE 1992	TOTAL DESCRIBED	OLDER NAMES SUNK	NET GAIN
1975	155	8	22	185	87	98
1976	192	23	7	222	66	156
1977	198	19	14	231	147	84
1978	312	37	9	358	113	245
1979	360	31	9	400	108	292
1980	291	26	18	335	128	207
1981	304	20	8	332	136	196
1982	256	13	5	274	75	199
1983	325	10	20	355	185	170
1984	317	22	12	351	138	213
1985	409	25	12	446	225	221
1986	394	27	2	423	186	237
1987	529	11	6	546	208	338
1988	429	5	3	437	198	239
1989	316	5	1	322	79	243
1990	470	12	1	483	72	411
1991	775	1	0	776	102	674
1992	671	11	—	682	191	491
1993	362	0	—	362	84	278
Total	10,674	930	625	12,229	4,047	8,182

pods, we could apply what might be called the British butterfly test (Stork and Gaston 1990). Butterflies are not ideal for this use; as in some other groups heavily studied by amateurs, variation in species concepts may significantly affect the results. As pointed out by Vane-Wright (1992:15), butterflies "are thought to comprise about 17,500 full species, but the number of currently recognised subspecies approaches 100,000. Many of these subspecific taxa (particularly those from small islands or isolated mountains) are fully diagnosable. . . . Such subspecies would qualify as species under a phylogenetic species concept."

Ignoring this potentially serious problem in counting species for the moment, Britain contains 67 of the world's butterfly species; thus the world butterfly fauna, which is generally considered to be as close to completely known as is the world tetrapod fauna, comprises about 261 times as many species as does the British fauna (Hammond 1992). The spider fauna of the British isles includes some 626 species (Merrett et al. 1985; Merrett and Millidge 1992). If spider species are distributed in a manner roughly comparable to butterflies, there should thus be some 163,500 species of spiders in the world.

Note, however, that if even half of the described butterfly subspecies proved to represent phylogenetic species, this estimate would produce the absurd figure of 625,000 species of spiders. Large numbers are unsettlingly easy to come up with. For example, Hammond (1992, 1994) estimated that there are 4,000 described species of British beetles and that somewhere between 0.8 and 3.1 million beetles species might exist in the world, so by the British beetle test, there might be as many as 485,150 spider species in the world. Similarly, Hammond (1992, 1994) estimated 22,000 total British insects and perhaps somewhere between 5.7 and 8.0 million total world insects; even by the British insect test, there might be over 227,600 spider species in the world.

Given a choice among only these estimates, I suspect that most arachnologists would opt for the British butterfly results based on "full species" counts only. But the disparity in these results, none of which are founded on totally unreasonable assumptions, surely demonstrates that basing estimates on any single well-known group, and any single well-studied place, is an extremely risky proposition.

However, the general approach can be applied in a much more comprehensive fashion, as admirably illustrated in a recent paper on insect diversity by Gaston and Hudson (1994). Those authors tabulated data, for eight well-known groups, on the percentages of the total number of known species that are found in each of nine biogeographic regions that subdivide the globe. The groups they examined were higher plants, amphibians, birds, mammals, and four groups of insects (dragonflies, tiger beetles, dynastine scarab beetles, and swallowtail butterflies). Although they were not able to obtain data for each group in every region, their results show, for example, that for these groups, 4.8 to 8.5 percent of the known species occur in the United States and Canada, 28.9 to 53.2 percent occur in the Neotropics, 2.1 to 9.2 percent occur in Europe, 13.2 to 20.9 percent occur in sub-Saharan Africa, and 3.0 to 12.6 percent occur in Australia.

Gaston and Hudson then used various estimates of insect diversity in the Nearctic and Australia, in comparison with each of the well-studied groups, to generate estimates of actual global diversity. For example, if insect diversity is distributed like that of higher plants and there are 150,000 species of Nearctic insects, then there should be a total of about 2.3 million species of insects in the world. In no case did any of their more realistic estimates of the numbers of Nearctic or Australian insects produce a global insect diversity estimate of more than 6.6 million species.

In table 2.3, a similar set of comparisons is presented for spiders. For the United States and Canada, the most recent estimate of spider diversity is by Roth (1994), who reported 3,500 described species and an additional 350 species known in collections but not yet described. Roth's figures accord well with the earlier estimates by Coddington, Larcher, and Cokendolpher (1990) of 3,400 described and 300–700 undescribed species and are used here. Similarly, Raven's (1988) estimate for Australia of 9,380 species (1,876 described and about four times that number undescribed) is also accepted and used here.

As indicated in table 2.3, the resulting estimates of total spider species diversity are all smaller and exhibit a much smaller range (varying from 76,000 to 157,500) than the extrapolations from Britain alone. Might there be reasons to prefer some of those estimates over others? In other words, given that some of the well-known groups will more closely model the global biodiversity patterns of spiders than will the others, are there reasons to suggest which model group (or groups) are most similar to spiders in this regard? At least two factors seem

TABLE 2.3

Regional Patterns of the Distribution of Biodiversity, Applied to Spiders

	NORTH AMERICA	AUSTRALIA	BOTH	NUMBER OF SPIDER SPECIES
Higher plants	6.5	5.7	12.2	108,400
Amphibians	4.9	4.5	9.4	140,700
Birds	6.1	6.0	12.1	109,300
Mammals	8.4	6.5	14.9	88,800
Dragonflies	8.5	5.5	14.0	94,500
Tiger beetles	5.5	3.8	9.3	143,000
Dynastines	4.8	12.6	17.4	76,000
Swallowtails	5.4	3.0	8.4	157,500

Patterns from Gaston and Hudson (1994). The first column provides a list of well-known groups, the second column the percentage of the known species of those groups found in the United States and Canada, the third column the percentage found in Australia, the fourth column the combination of those two figures, and the fifth column the estimated number of spider species worldwide if spiders resemble the better-known groups in their global distribution patterns.

promising. First, the better-known arthropod groups used for comparison are not very speciose, so one might reasonably ask whether their patterns are typical of larger arthropod groups. It has been estimated, both for the much larger and well-cataloged group Diptera and for the vastly larger insect fauna as a whole, that fully 5 percent of the world's species diversity occurs in Australia (Colless and McAlpine 1991). It would seem, then, that the global distribution patterns of spiders are more likely to resemble those of the well-known groups in which at least 5 percent of the extant fauna occurs in Australia. The other better-known groups (amphibians, tiger beetles, and swallowtails) are probably less accurate indicators of spider diversity than are the remainder; if so, the range of estimates narrows to 76,000–109,300.

Indeed, we already have reason to believe that for spiders, the Australian fauna has more than 2.4 times the number of species that are found in the United States and Canada. Of the better-known groups in table 2.3, only the dynastine beetles resemble spiders in that regard; indeed, the dynastines are the only one of the better-known groups in which the Australian fauna is not smaller than the Nearctic one. One might therefore reasonably regard the estimate provided by the dynastines (76,000) as the best-founded in the group, even though it is farthest from the British butterfly analogy.

At this point, some readers may be chafing at all these estimates based on faunal sizes outside the tropics. After all, is it not true that the tropics are both much more diverse and much less studied than the temperate areas? As I have pointed out before (Platnick 1991), most such claims reflect severe boreal and megafaunal biases. They refer mostly to differences between just the north temperate areas and the tropics and to the species-poor groups that happen to be well known, such as vertebrates (estimated in Hammond's working figures to constitute less than 0.5 percent of the world's biota) or even the higher plants (estimated in Hammond's figures at less than 2.5 percent of the world's total biota). I have argued that for spiders and many other groups, we know only that the north temperate areas—North America and Eurasia—are depauperate relative to both the tropics and the south temperate areas, and Eggleton (1994) has since found that termites, like spiders, live in a pear-shaped world. Similar patterns have even been found for some marine crustacean groups (Barnard 1991; Reid 1994).

But the point I want to make here is a different one. In suggesting that spider distributions more closely resemble those of dynastine scarabs than those of the other seven well-known groups examined by Gaston and Hudson (1994), I have not discriminated against the tropics. Indeed, if anything the contrary is true. Based on a list I supplied, Coddington and Levi (1991:566) indicated that "about one third of all [spider] genera (1,090 in 83 families) occur in the Neotropics." But no less than 53.2 percent of the world's known dynastine species occur in the Neotropics. That is the highest percentage shown by any of the eight indicator groups.

I suspect that, in comparison to spiders, the parts of the world that are under-represented by dynastines are temperate, not tropical. Among dynastines, 4.8 percent of the species are Nearctic but only 2.1 percent occur in Europe; at least for known spider species, Europe has almost exactly the same total as the United States and Canada. Only 1 percent of the world's known dynastines occur in Asia (minus India and the southeast), but the figures for spiders suggest that Asia has more species than Europe, not fewer. The total for Japan, for example, is over 1,017 (Yaginuma 1984), compared to 626 in the British isles. Finally, New Zealand has only 0.4 percent of the known dynastine species, but its spider fauna may turn out to be fully as large as the Nearctic one.

The dynastine percentage for sub-Saharan Africa is 13.2 percent, with 12.7 percent for India, southeast Asia, and the Indo-Pacific and 4.9 percent for Madagascar and nearby islands. Compared to totals of currently known spider species, the African percentage is low (for known spider species, the figure is about 16.6 percent). However, in the comparative, quantitative sampling of sites in the Neotropics and in Africa that has been carried out to date by Coddington and his associates (personal communication; see Coddington et al. 1991, for details on the sampling methods), the African sites show far lower numbers of species, making the dynastine figures believable for spiders also.

In short, the two different approaches taken here together suggest estimates of total spider diversity that fall within the range of 76,000–80,000 species. These figures provide an analog, for spiders, of the conservative working estimate chosen for larger groups by Hammond (1992).

DOES IT MATTER?

In a nutshell, yes. If there are actually 170,000 species of spiders in the world and systematic work continues at the pace it has exhibited since 1955, it will take arachnologists another 638 years to finish describing the world spider fauna. In other words, for spiders and other megadiverse groups we can abandon all hope of learning enough to do anything more than guess at worldwide diversity patterns in our lifetimes. But that means that we will have to base all our attempts to cope with the biodiversity crisis on the few groups that are already well known. Because one of the few things we do know is that those particular groups are not generally representative of biodiversity (Gaston 1992; Prendergast et al. 1993; Cornutt et al. 1994), such limitations are simply unacceptable.

One approach that has been suggested to cope with this dilemma is the All Taxa Biodiversity Inventory (or ATBI). It starts with the entirely reasonable desire to have at least one place on Earth where we actually know the entire biota and where we can actually calibrate all the sampling methods we could then apply far more extensively. Such an ATBI has been very conservatively estimated to take 7 years and cost $100 million.

I support that endeavor, but also have to point out that, in many respects, an ATBI represents an ecologist's view of the world. Inventorying the entire biota at one site has obvious benefits to the country housing that site, but those benefits diminish rapidly with increasing distance from the site. The results of a Costa Rican ATBI, for example, are of significantly diminished utility by the time one studies organisms in southern Colombia, and knowledge of the Costa Rican biota is of very limited usefulness to biodiversity prospectors or land managers in Cameroon or Indonesia.

It has also been claimed that a large percentage of the world's biota can be covered by conducting an ATBI at a small number of sites. I do not support that claim. For example, ecologist Dan Janzen is quoted (in Langreth 1994:81) as stating, "If scientists performed ATBI surveys in 10 or 12 carefully chosen locations, you'd cover maybe 40 to 50 percent of the world's biodiversity." That is, at best, wishful thinking. Consider a group such as the metaltelline spiders, which have about 55 species (mostly still undescribed) in Chile. One of those species is widespread and would probably be collected at any ATBI site in central Chile. The remaining species seem to have very narrow ranges and it is unlikely that more than two of them would be found at any single ATBI site. Thus one would need five ATBI sites in Chile alone to capture 40 percent of the Chilean metaltelline species alone. At $100 million per ATBI, this would be an extremely inefficient way to survey the diversity of metaltellines.

No one would claim that Chilean metaltellines are representative even of the Chilean biota in general, but consider the results of spider sampling in Bolivia (Coddington, personal communication; Colwell and Coddington 1994). In a transect of three sites, separated by about 110 km and covering a range of elevations of 100–1,900 meters, the number of species found in comparable samples from any pair of sites was never higher than 3 percent. The total number of species found in samples from all three sites was zero. Here again, no one would claim that these results are necessarily representative; for example, the sampling methods were probably biased against some hunting spider groups that probably do have species common to all three sites. Moreover, the high proportion of species that are represented by single specimens suggests that rare but widespread species are probably underrepresented in these samples. Nevertheless, the degree of species overlap with similar samples taken in Peru, just a few hundred kilometers to the north of the Bolivian sites, was again virtually nil. In short, there are sufficiently high numbers of species in the megadiverse arthropod groups, that have sufficiently small range sizes, to make it certain that focusing on just a dozen sites scattered around the world would overlook the vast majority of the world's biodiversity.

Wheeler (1995) has presented a complementary approach: an All-Biota Taxon Inventory (ABTI). This initiative represents a systematist's view of the world and advocates inventorying and classifying the entire world's biota of one or more higher taxa, with obvious benefits to every nation. The parallels are striking: in choosing the first ATBI site, one country wins and the others lose out, whereas in choosing the first ABTI subject, one taxonomic group wins and the others lose out. How might we reasonably make the choice?

Some of the criteria that have already been suggested include the following:

- The group should be megadiverse, certainly with over 5,000 described species, preferably with at least 20,000 described species.
- The group should be well known, in the sense that there are at least one or two places on Earth (probably northern Europe or the northeastern United States) where almost all the species are described and identifiable.
- The group should nevertheless be very incompletely known, in the sense that at least half of the world fauna is probably still undescribed. The few groups that are already better known are simply not diverse enough to tell us about worldwide biodiversity patterns, which are by definition set primarily by the megadiverse groups.
- Species in the group should have the smallest possible distribution ranges, to maximize the value of the ABTI products for making local conservation and land management decisions within every country. The group should also have demonstrable economic potential, at scales ranging from chemical prospecting through biological control to ecotourism.
- A high percentage of the species in the group should be collectable by mass collecting techniques (such as Malaise traps, Berlese sampling, and pitfall traps) and easily and durably maintained in collections (e.g., groups such as mites, in which every specimen must be individually mounted on a slide, are unlikely to be chosen).
- The available taxonomic expertise on the group must be sufficient to handle the human resource question. In other words, there must be enough people working on the group today to train, within a 5-year period, enough new taxonomists to complete the project over the following 10 or so years.
- Available knowledge about the group should be readily accessible through printed or electronic taxonomic catalogs.

Although this list of possible criteria is not at all definitive, it is easy to show, for example, that of the 11 orders of arachnids, only spiders are a suitable candidate for an ABTI. Among other megadiverse groups, ants and bees are also obvious candidates. Workers in those areas will have to decide how feasible the idea is, but for spiders, I suggest that it would be possible to conduct an ABTI, over a 15-year period, at a cost roughly equal to that of a single ATBI conducted at a single site.

Let us make the (probably overgenerous) assumption that there are 90,000 spider species in the world. At the rate of progress shown since 1955, it would take some 257 years to complete our knowledge of the group. But imagine a working group of four individuals, including one Ph.D.-level systematist, aided by a full-time artist and two full time-technicians to help collect, process, sort, database, describe, and classify specimens. Setting up such a working group might cost perhaps $225,000 a year in salaries and expenses for field and laboratory work. Over a 15-year period, the cost per group would be some $3,375,000, and 30 such groups would cost just over $100 million. To cover the world's fauna, each such group would together have to study 3,000 species over the 15-year period, an average of 200 species per year.

No present-day arachnologist routinely covers 200 species a year, but no present-day arachnologist has a support staff of three full-time people, and that productivity goal for such a team is not unreasonable. Given the present state of human resources in arachnology, the required Ph.D.-level systematists could probably be supplied immediately, but the leeway provided by the overly large estimate for the total fauna to be studied would allow some years of training for group members to be included in the budget.

What I am suggesting is that for less than the probable cost of a complete inventory of the biota of a single site, we can bring the systematics of a megadiverse group such as spiders to a worldwide level approaching that which it now has in Great Britain or the northeastern United States, and we can do it in 15 years. I suggest further that the value of having global data on the diversity patterns of such a group will vastly exceed that cost. Many spider species, for example, have small enough distribution ranges that thorough revisions backed up by complete specimen-level databases will supply information for worldwide land management on a scale far more detailed than anything currently available.

Consider, for example, a recent study on central Chile by Morrone, Katinas, and Crisci (1997). Those authors compared distributions and relationships of several genera of angiosperms (Asteraceae), beetles (Buprestidae and Curculionidae), and spiders (Gnaphosidae) that have speciated extensively within Chile. Those authors recognized four areas of endemism within central Chile (from Coquimbo south to Ñuble), but the gnaphosid spider genus *Echemoides* alone suggests that there are at least two additional areas of endemism within that region. For a land manager who has to choose which Chilean forests to lumber and which to preserve, the smaller the areas of endemism we can identify are, the more precisely we can make those choices.

In stressing the potential value of knowledge about the distribution patterns of megadiverse groups of species with small ranges, I do not mean to neglect other important criteria. Both spider venoms and spider silks, for example, have potentially enormous economic impact. Research in neurobiology is increasingly taking advantage of the ability of some spider venoms to inhibit transmission of nerve impulses across synapses in vertebrates, and that kind of research could lead eventually to cures for diseases such as epilepsy. The U.S. military is investing heavily in research on spider silks, in hopes of synthesizing materials (for use in parachutes, bulletproof vests, and the like) that might duplicate the extraordinary physical properties of spider silks, which (despite their incredibly light weight) have a tensile strength greater than steel strands of the same diameter. By the same token, I do not mean to neglect the enormous value of the ecosystem services provided by the spiders that help keep us alive by eating the insects that would otherwise devour all our crops, or the potential of spiders as biological control agents.

Similar ABTI initiatives for other megadiverse groups could no doubt yield a similarly great benefit:cost ratio. The sheer biomass and chemical complexity of ants and the pollination services of bees make them obvious candidates. Such projects would require international funding and cooperation on a scale not yet seen in systematics, but they are a necessity if we are to target the biodiversity crisis with anything more effective than a cap pistol or a pop gun.

In short, Hammond's eminently reasonable figures paint a very disturbing scenario. The only groups we know very much about are vertebrates and higher plants, but together those groups probably represent just 3 percent of the world's biota, quite possibly the least representative 3 percent at that. If we are to achieve even the most cursory understanding of global biodiversity patterns in our lifetimes, we must target some megadiverse groups for intensive study so that we can bring the light they have to shed to bear on the problems we confront. It is imperative that we do so before that light is extinguished.

ACKNOWLEDGMENTS

I thank Drs. Jonathan Coddington, Valerie Davies, Ray Forster, Pablo Goloboff, Charles Griswold, Rudy Jocqué, Peter Hammond, and Robert Raven for their helpful comments on the manuscript.

REFERENCES

Alderweireldt, M. and R. Jocqué. 1994. Biodiversity in Africa and Europe: The case of spiders (Araneae). *Biologisch Jaarboek Dodonaea* 61: 57–67.

Barnard, J. L. 1991. Amphipodological agreement with Platnick. *Journal of Natural History* 25: 1675–1676.

Bonnet, P. 1979. Troisième note sur le nombre des espèces nouvelles, d'araignées décrités chaque année. *Revue Arachnologique* 2: 273–274.

Coddington, J. A., C. E. Griswold, D. Silva, E. Pearanda, and S. F. Larcher. 1991. Designing and testing sampling protocols to estimate biodiversity in tropical ecosystems. In E. C. Dudley, ed., *The unity of evolutionary biology: Proceedings of the Fourth International Congress of Systematic and Evolutionary Biology*, 44–60. Portland: Dioscorides Press.

Coddington, J. A., S. F. Larcher, and J. C. Cokendolpher. 1990. The systematic status of Arachnida, exclusive of Acari, in North America north of Mexico. In M. Kosztarab and C. W. Schaefer, eds., *Systematics of the North American insects and arachnids: Status and needs*, 5–20. Blacksburg: Virginia Polytechnic Institute and State University.

Coddington, J. A. and H. W. Levi. 1991. Systematics and evolution of spiders (Araneae). *Annual Review of Ecology and Systematics* 22: 565–592.

Colless, D. H. and D. K. McAlpine. 1991 (2d ed.). Diptera (flies). In *The insects of Australia*, 717–786. Carlton: Melbourne University Press.

Colwell, R. K. and J. A. Coddington. 1994. Estimating terrestrial biodiversity through extrapolation. *Philosophical Transactions of the Royal Society of London, Series B* 345: 101–118.

Cornutt, J., J. Lockwood, H.-K. Luh, P. Nott, and G. Russell. 1994. Hotspots and species diversity. *Nature* 367: 326–327.

Eggleton, P. 1994. Termites live in a pear-shaped world: A response to Platnick. *Journal of Natural History* 28: 1209–1212.

Erwin, T. L. 1982. Tropical forests: Their richness in Coleoptera and other arthropod species. *Coleoptera Bulletin* 36: 74–75.

Erwin, T. L. 1991. How many species are there? Revisited. *Conservation Biology* 5: 330–333.

Forster, R. R. 1995. The Australasian spider family Periegopidae Simon, 1893 (Araneae: Sicarioidea). *Records of the Western Australian Museum, Supplement* 52: 91–105.

Forster, R. R. and N. I. Platnick. 1985. A review of the austral spider family Orsolobidae (Arachnida, Araneae), with notes on the superfamily Dysderoidea. *Bulletin of the American Museum of Natural History* 181: 1–229.

Forster, R. R., N. I. Platnick, and J. A. Coddington. 1990. A proposal and review of the spider family Synotaxidae (Araneae, Araneoidea), with notes on theridiid interrelationships. *Bulletin of the American Museum of Natural History* 193: 1–116.

Gaston, K. J. 1991a. The magnitude of global insect species richness. *Conservation Biology* 5: 283–296.

Gaston, K. J. 1991b. Estimates of the near-imponderable: A reply to Erwin. *Conservation Biology* 5: 564–566.

Gaston, K. J. 1992. Regional numbers of insect and plant species. *Functional Ecology* 6: 243–247.

Gaston, K. J. and E. Hudson. 1994. Regional patterns of diversity and estimates of global insect species richness. *Biodiversity and Conservation* 3: 493–500.

Gaston, K. J. and L. A. Mound. 1993. Taxonomy, hypothesis testing, and the biodiversity crisis. *Proceedings of the Royal Society of London, Series B* 251: 139–142.

Grothendieck, K. and O. Kraus. 1994. Die Wasserspinne *Argyroneta aquatica*: Verwandtschaft und Spezialisation (Arachnida, Araneae, Agelenidae).*Verhandlung des Naturwissenschaftlichen Verein im Hamburg (N.F.)* 34: 259–273.

Hammond, P. M. 1990. Insect abundance and diversity in the Dumoga-Bone National Park, N. Sulawesi, with special reference to the beetle fauna of lowland rain forest in the Toraut region. In W. J. Knight and J. D. Holloway eds., *Insects and the rain forests of South East Asia (Wallacea)*, 197–254. London: Royal Entomological Society of London.

Hammond, P. M. 1992. Species inventory. In B. Groombridge, ed., *Global biodiversity: Status of the earth's living resources*, 17–39. London: Chapman & Hall.

Hammond, P. M. 1994. Practical approaches to the estimation of the extent of biodiversity in speciose groups. *Philosophical Transactions of the Royal Society of London, Series B* 345: 119–136.

Hammond, P. M. 1995. Described and estimated species numbers: An objective assessment of current knowledge. In D. Allsopp, R. R. Colwell, and D. L. Hawksworth, eds., *Microbial diversity and ecosystem function*. Wallingford, U.K.: Cabi.

Hodkinson, I. D. and D. Casson. 1991. A lesser predilection for bugs: Hemiptera (Insecta) diversity in tropical rain forests. *Biological Journal of the Linnaean Society* 43: 101–109.

Hormiga, G. 1993. Implications of the phylogeny of Pimoidae for the systematics of linyphiid spiders (Araneae, Araneoidea, Linyphiidae). *Memoirs of the Queensland Museum* 33: 533–542.

Jocqué, R. 1994. Halidae, a new spider family from Madagascar (Araneae). *Bulletin of the British Arachnological Society* 9: 281–289.

Johnston, D. E. 1982. Acari. In S. B. Parker, ed. *1982 Synopsis and classification of living organisms,* Vol. 2, 111–117. New York: McGraw-Hill.

Langreth, R. 1994. The world according to Dan Janzen. *Popular Science* December 1994: 79–82, 112–113.

May, R. M. 1988. How many species are there on Earth? *Science* 241: 1441–1449.

May, R. M. 1994. Conceptual aspects of the quantification of the extent of biological diversity. *Philosophical Transactions of the Royal Society of London, Series B* 345: 13–20.

Merrett, P., G. H. Locket, and A. F. Millidge. 1985. A check list of British spiders. *Bulletin of the British Arachnological Society* 6: 381–403.

Merrett, P. and A. F. Millidge. 1992. Amendments to the check list of British spiders. *Bulletin of the British Arachnological Society* 9: 4–9.

Monteith, G. B. and V. T. Davies. 1984. Preliminary account of a survey of arthropods (insects and spiders) along an altitudinal rainforest transect in tropical Queensland. In G. L. Werren and A. P. Kershaw, eds., *Australian National Rainforest Study report. Vol. I: Proceedings of a workshop on the past, present and future of Australian rainforests, Griffith University, December 1983,* 402–412. World Wildlife Fund (Australia) Project 44. Melbourne: Geography department, Monash University.

Morrone, J. J., Katinas, L., and Crisci, J. V. 1997. A cladistic biogeographic analysis of central Chile. *Journal of Comparative Biology* 2: 25–42.

Noyes, J. S. 1989. The diversity of Hymenoptera in the tropics with special reference to Parasitica in Sulawesi. *Ecological Entomology* 14: 197–207.

Parker, S. B. ed. 1982. *Synopsis and classification of living organisms,* Vol. 2. New York: McGraw-Hill.

Platnick, N. I. 1989. Advances in spider taxonomy 1981–1987: A supplement to Brignoli's *A catalogue of the Araneae described between 1940 and 1980.* Manchester: Manchester University Press.

Platnick, N. I. 1991. Patterns of biodiversity. In N. Eldredge, ed., *Systematics, ecology, and the biodiversity crisis,* 15–24. New York: Columbia University Press.

Platnick, N. I. 1993. *Advances in spider taxonomy 1988–1991, with synonymies and transfers 1940–1980.* New York: New York Entomological Society.

Platnick, N. I. and C. Ewing. 1995. A revision of the tracheline spiders (Araneae, Corinnidae) of southern South America. *American Museum Novitates* 3128: 1–41.

Platnick, N. I. and R. R. Forster. 1989. A revision of the temperate South American and Australasian spiders of the family Anapidae (Araneae, Araneoidea). *Bulletin of the American Museum of Natural History* 190: 1–139.

Platnick, N. I. and R. R. Forster. 1993. A revision of the New Caledonian spider genus *Bradystichus* (Araneae, Lycosoidea). *American Museum Novitates* 3075: 1–14.

Platnick, N. I. and M. U. Shadab. 1993. A review of the pirate spiders (Araneae, Mimetidae) of Chile. *American Museum Novitates* 3074: 1–30.

Prendergast, J. R., R. M. Quinn, J. H. Lawton, B. C. Eversham, and D. W. Gibbons. 1993. Rare species, the coincidence of diversity hotspots and conservation strategies. *Nature* 365: 335–337.

Raven, R. J. 1988. The current status of Australian spider systematics. *Australian Entomological Society Miscellaneous Publication* 5: 37–47.

Reid, J. W. 1994. Latitudinal diversity patterns of continental benthic copepod species assemblages in the Americas. *Hydrobiology* 292/293: 341–349.

Roth, V. D. 1994 (3d ed.). *Spider genera of North America.* Gainesville, Fla.: American Arachnological Society.

Stork, N. E. 1987. Guild structure of arthropods from Bornean rain forest trees. *Ecological Entomology* 12: 69–80.

Stork, N. E. 1988. Insect diversity: Facts, fiction and speculation. *Biological Journal of the Linnaean Society* 35: 321–337.

Stork, N. E. 1991. The composition of the arthropod fauna of Bornean lowland rain forest trees. *Journal of Tropical Ecology* 7: 161–180.

Stork, N. E. 1993. How many species are there? *Biodiversity and Conservation* 2: 215–232.

Stork, N. E. 1994. Inventories of biodiversity: More than a question of numbers. In P. I. Forey, C. J. Humphries, and R. I. Vane-Wright, eds., *Systematics and conservation evaluation,* 81–100. Oxford, U.K.: Oxford University Press.

Stork, N. E. and M. J. D. Brendell. 1990. Variation in the insect fauna of Sulawesi trees with season, altitude and forest type. In W. J. Knight and J. D. Holloway, eds., *Insects and the rain forests of South East Asia (Wallacea),* 173–190. London: Royal Entomological Society London.

Stork, N. E. and K. J. Gaston. 1990. Counting species one by one. *New Scientist* 1729: 43–47.

Vane-Wright, R. I. 1992. Species concepts. In B. Groombridge, ed., *Global biodiversity: status of the earth's living resources,* 13–16. London: Chapman & Hall.

Wheeler, Q. 1995. Systematics, the scientific basis for inventories of biodiversity. *Biodiversity and Conservation* 4: 476–489.

Yaginuma, T. 1984. The sequel to "A list of Japanese spiders" (revised in 1977). *Faculty of Letters Review, Otemon Gakuin University* 18: 249–260.

CHAPTER 3

THE MEDIUM IS THE MESSAGE:
FRESHWATER BIODIVERSITY IN PERIL

Melanie L. J. Stiassny

Freshwater, so fundamentally important for all life processes, is quite unlike any other of the earth's natural resources. "Difficult to purify, expensive to transport, and impossible to substitute" (Engelman and LeRoy 1993), water is essential to human welfare, economic and social development, and the maintenance and integrity of life itself. As I have claimed elsewhere, the truth of the edict "the medium is the message" is underscored in considerations of the conservation of aquatic biodiversity (Stiassny 1996). The message from freshwater worldwide is as loud and clear as it is unambiguous. The planet's freshwater habitats are being undermined at an unprecedented rate as freshwater resources around the globe are consumed and degraded. There is little question that we are in serious trouble, yet until recently this message has gone essentially unheeded. A strange type of "water blindness" pervades much of the discussion of conservation and the biodiversity crisis. But today this is no longer possible and water issues, once considered subsidiary to other major environmental problems of this century such as chronic food shortage and disease,[1] economic decline, and energy crises, have moved into the global arena (figure 3.1). More and more the availability and quality of freshwater are recognized as the primary limiting factors for much of human development. As we approach the new millennium, the crisis of freshwater will increasingly set the agenda and frame discussion for future development. Of course, this is nowhere more starkly evident than in the arid and semiarid regions of the globe, where water scarcity and stress are already familiar (figure 3.2), but it is also becoming increasingly true worldwide (Gleick 1993; Clarke 1993; Engelman and LeRoy 1993; Brown et al. 1996). For example, even in Great Britain, a nation with a well-earned reputation for its drizzle and damp, demand for water is increasing at a rate of about 0.5 to 1.0 percent a year and water conservation measures are increasingly necessary (Hamer 1990).

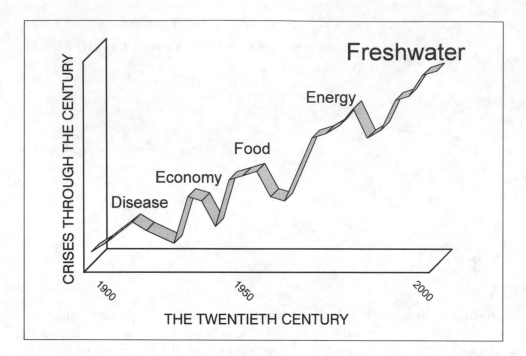

FIGURE 3.1

Crises of the twentieth century.
Modified after Wetzel (1992).

Before going on to consider the fragile limits of the planet's freshwater resources, I want briefly to outline one example of the type of loss we have already sustained. This is the example of the Aral Sea, aptly described as one of the planet's greatest environmental tragedies. The fate of the Aral illustrates both the enormity of our disruptive power and is a stark omen to keep in mind as the litany of loss proceeds.

THE ARAL: A SEA OF TEARS

Until very recently the Aral Sea, once the fourth largest inland water body on the planet, was home to a complex wetland ecosystem supporting a major inland fishery that provided jobs and a social framework for some 60,000 people in the immediate vicinity. The Aral was home to 20 fish species, of which some 60 percent were found nowhere else (Nikol'skii 1940; table 3.1). Today that ecosystem is decimated and its wetland areas and fish spawning grounds have diminished to a tiny fraction of their former extent. The fishery is devastated and the sea itself has been reduced by almost half its area (figure 3.3) and nearly two-thirds of its original volume. In the past 30 years water level has dropped some 15 m (from 53.4 m a.s.l. in

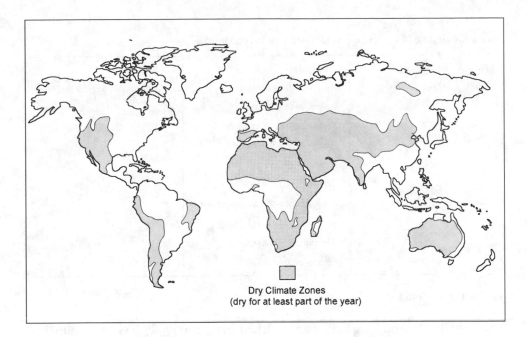

FIGURE 3.2

Global dry climate zones (dry for at least part of the year).

1960 to 38.6 m a.s.l. in 1993) and the average salinity of the waters has increased almost threefold (figure 3.3) and is even higher in shallow inshore regions. In a poignant memorial to the Aral Sea and its lost ecosystem, young Usbeki poet Mukhammed Salikh has written, "You cannot fill the Aral with tears," and the irony of his metaphor is profound. The Aral would be a lot less salty today if it could actually be replenished with tears!

It is all too evident that unless human usage patterns change drastically, there is little hope that anything of the sea will survive beyond the next 30 years or so. The Aral and its once vibrant ecosystem is being sacrificed as water is diverted from its two main tributaries (the Amu Dar'ya and Syr Dar'ya) to support massive cotton production in the region (Micklin 1988; Ellis 1990). Cotton is a particularly thirsty crop and the Aral is in a dry climatic zone (figure 3.2). From "shoot to suit" it takes as much as 2,700 tons of water to make a single ton of worsted material (Clarke 1993). The economic underpinnings of the decision to sacrifice all to grow cotton on the scale seen in Central Asia are hard to fathom, particularly given the ecological and social devastation that have ensued (Postel 1996).

In the present context, it is worth pointing out that the assault on the Aral began long before the devastating impacts of water diversion for irrigation and evaporative salinity. Beginning as early as 1927 and continuing through the early 1960s, the Aral, like so many other inland waters, has seen the sequential introduction of a long roster of exotic fish (table 3.2) and invertebrate species (Aladin

TABLE 3.1

Fishes Endemic to the Aral Sea and Its Affluent River Mouths

Salmonidae	Cyprinidae	Gasterosteidae
Salmo trutta aralensis	*Rutilus rutilus aralensis*	*Pungitius platygaster aralensis*
	Leuciscus idus oxianus	
	Scardinius erythrophthalmus	
	Aspius aspius iblioides	
	Barbus brachycephalus	
	Barbus capito conocephalus	
	Chalcalburnus chalcoides aralensis	
	Abramis brama bergi	
	Abramis sapa aralensis	
	Pelecus cultratus	

After Nikol'skii (1940).

and Potts 1992). And long before any marked change in water level or quality had been noted, the impact of these species introductions had caused considerable disruption of the native community. For example, early introductions of marine planktivorous fish species (e.g., *Clupea harengus membra, Alosa caspia*) had a profound impact on the composition and biomass of the zooplankton community (Aladin and Potts 1992; Aladin and Williams 1993) such that between 1959 and 1968 zooplankton biomass did not exceed 15 mg/m^3 (a reduction of nearly 90 percent from former levels). Similarly, introduced benthos feeders (e.g., *Acipenser stellatus*)[2] reduced native invertebrates biomass by almost two-thirds. Already by the early 1970s the native fishery had almost completely been replaced by exotics (figure 3.3). Yet despite all of the introductions of "fish food," fishery yields were down by almost two-thirds of their pre-1960s levels. By the mid-1970s, after salinity began to increase dramatically as rapidly dropping water levels radically reduced spawning areas, and as increased water diversion and weir construction in the affluent rivers impeded movements of anadromous fishes, the Aral fishery collapsed entirely. Today the fishery is gone and all that is left are three introduced marine species (an atherinid, a goby, and a flounder), and a single native, the Aral stickleback (*Pungitius platygaster aralensis*).

The social and economic dimensions of the ecological disaster of the Aral continue to plague the inhabitants of what were once the shores of that sea while affecting, in complex and often unexpected ways, the health and well-being of millions of people in the region. Whatever restoration may be possible, most of the fishes of the Aral are gone forever, as are all other components of this unique ecosystem. The lessons of the Aral are hard earned, yet it is clear that unless they are heeded this story is destined to be played out again and again as the competition for water intensifies.

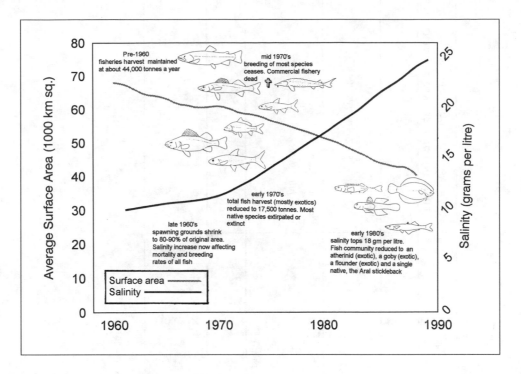

FIGURE 3.3

History of human impacts on the Aral Sea of Central Asia.

THE EARTH'S FRESHWATER RESOURCES: THE ILLUSION OF PLENTY

In a powerful introduction to the 1996 *State of the World,* Lester Brown makes the important observation that the late twentieth century is witness to a scale of human activity that for the first time has begun to impair the habitability of the planet and outstrip the capacity of the earth's natural systems to replenish themselves. In a reality that has yet to penetrate fully our collective consciousness, we are crossing resource thresholds never previously passed (Brown 1996). This is perhaps nowhere more evident than in the arena of freshwater usage and consumption (Clarke 1993; Gleick 1993; Postel 1996).

Despite the fragile limits of available freshwater, the persistence of this century's "water blindness" is probably a result of an illusion of plenty. After all, when viewed from outer space the planet is blue: the blue of the water that covers so much of its surface. Yet the great mass of that water is marine (figure 3.4),[3] and the seas and oceans of the planet, while driving world climate and powering the hydrologic cycle, are of little use for crop irrigation or most industrial and domestic use. It is to the remaining 2.5 percent of the earth's water that we must turn to satisfy human thirst. Yet of that small percentage, most is unavailable for direct human use or as habitat for aquatic freshwater life (figure 3.4). More than two-thirds of all freshwater is tied up as polar ice, another third is stored deep

TABLE 3.2

List of Fish Species Introduced Into the Aral Sea Between 1927 and 1965

INTENTIONAL INTRODUCTIONS	ACCIDENTAL INTRODUCTIONS	SECONDARY INTRODUCTIONS (FIRST INTRODUCED INTO AFFLUENT RIVERS)
Acipenseridae *Acipenser stellatus*	Syngnathidae *Syngnathus nigrolineatus*	Cyprinidae *Mylopharyngodon piceus*
Clupeidae *Alosa (Caspialosa) caspia* *Clupea harengus membras*	Atherinidae *Atherina* spp.	Channidae *Ophiocephalus argus warpachawski*
Hypophthalmidae *Hypophthalmus molitrix melanostomus*	Gobiidae *Pomatoschistus caucasisus* *Gobius (Apollonia)* *Gobius (Neogobius) fluviatilis pallasi*	Pleuronectidae *Platichthys flesus*
Cyprinidae *Ctenopharyngodon idella*		
Mugilidae *Mugil auratus* *Mugil saliens*		

Data from Nikol'skii (1940) and Aladin and Potts (1992); taxonomic nomenclature follows Blanc et al. (1971).

After Nikol'skii (1940).

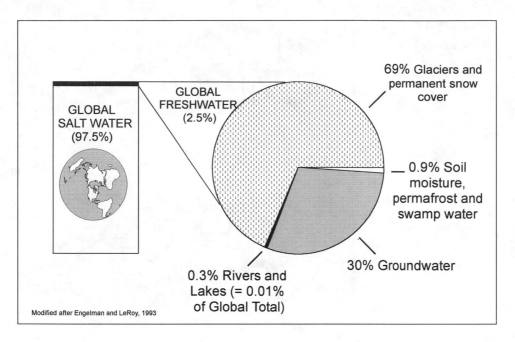

GLOBAL SALT WATER (97.5%)

GLOBAL FRESHWATER (2.5%)

69% Glaciers and permanent snow cover

0.9% Soil moisture, permafrost and swamp water

30% Groundwater

0.3% Rivers and Lakes (= 0.01% of Global Total)

Modified after Engelman and LeRoy, 1993

FIGURE 3.4

Distribution of the world's water.
Modified after Engelman and LeRoy (1993).

underground in aquifers (many of which hold "fossil water" with recharge rates measured in hundreds of years; globally groundwater is completely renewed only once every 1,400 years). The remainder is fixed as soil moisture and permafrost, and only about 93,000 km^3 (about 0.3 percent of all freshwater and less than 0.01 percent of total water) is available as freshwater riverine and lacustrine habitat.

Although 93,000 km^3 is certainly a lot of freshwater, only a small fraction of it is available for consumptive use. It is not human need (or greed) that determines the amount of water available for sustainable exploitation, but the global hydrologic cycle, which is fixed and unchanging (Shiklomanov 1993). According to most estimates, precipitation over land deposits about 113,000 km^3 of snow and rain, and about 72,000 km^3 of that evaporates back into the atmosphere. The remaining 41,000 km^3 is left to recharge aquifers, replenish lakes and rivers, and run off into the ocean (figure 3.5). This is the portion of the earth's freshwater that, theoretically at least, is sustainably usable and if this water were evenly distributed over the globe it would support a global population 10 times larger than today's. But of course it is not evenly distributed, either spatially or temporally, and therein lies one of the major problems with water as a harvestable resource. Clarke (1993) makes an important distinction between *total global runoff* (41,000 km^3), which cannot be considered stable or reliable,[4] and *reliable runoff*. Estimates of what constitutes this reliable (available) runoff vary, but a general consensus among water ex-

FIGURE 3.5

The hydrologic cycle over land.

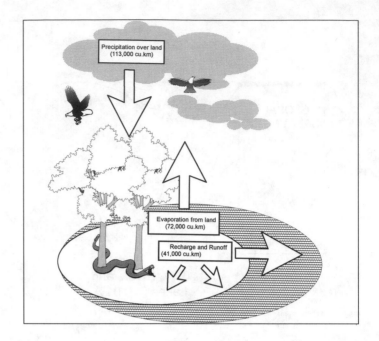

perts is that only approximately 9,000–14,000 km^3 is available for use on a sustainable basis (Clarke 1993; Engelman and LeRoy 1993; Gleick 1993).

As human populations grow, particularly in the arid and semiarid regions of the globe, where already some 600 million people live, competition for this precious resource is becoming increasingly intense (McCaffrey 1993; Postel 1996). The unsustainability of current usage patterns is evident in a consideration of the statistics presented in figure 3.6. In the 50 years since 1940 human population doubled while total water use almost quadrupled. Much of the accelerated water use is accounted for by agricultural extraction (figure 3.6), and increased irrigation helped power the drive for increased food production throughout the second half of this century. Consumption at such a rate cannot continue, however, and already the trend of increasing irrigation of crop land per capita has reversed. Since the late 1970s the area of land irrigated (per capita) has been in decline while the tally of ecological disaster and social disruption has increased. In a harrowing calculation of future water demand Postel (1996) has estimated that it will take an *additional* 780 billion m^3 of water (more than nine times the annual flow of the Nile River) simply to meet the grain requirements of the projected world population of 2025. But that water simply is not available, and we are already rapidly approaching the upper limit of the planet's renewable freshwater reserves (figure 3.6). Massive population growth, coupled with intensified habitat modification and disruption, is straining sustainability to its limits as freshwater resources are consumed and degraded (Gleick 1993; Brown et al. 1996). Although the prognosis for future human development is not good, it is considerably worse for much of the world's aquatic life.

Today, the finite nature of the earth's freshwater resources is revealed more and more as an intensified conflict between human consumptive usage and the

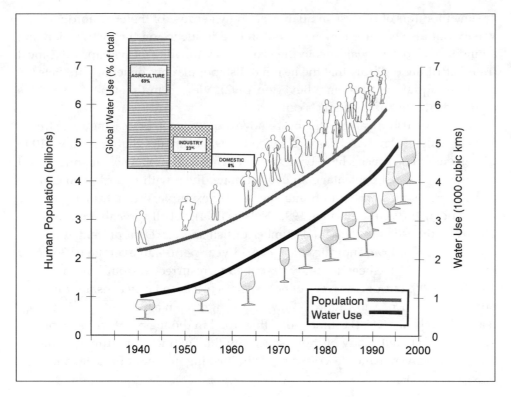

FIGURE 3.6

Changes in water consumption and global population between 1940 and 1990 (inset shows partitioning of global water use).

maintenance of aquatic health and biodiversity. In such a scenario, aquatic biodiversity is clearly losing out as it is sacrificed, often unwittingly, to human "need."

THE WEALTH OF AQUATIC BIODIVERSITY

Despite the fact that freshwater makes up less than one hundredth of a percent of the earth's water, the rivers, lakes, and coastal wetlands harbor truly exceptional concentrations of biodiversity. Recently it has been estimated that on a per unit basis the total biodiversity of freshwater is 65 times greater than that of the open oceans, and such a figure is not inconsistent with the higher net productivity and greater heterogeneity of freshwater habitats compared with open-water marine habitats. Yet because there have never been extensive biological inventories of most freshwater, our knowledge of the extent of aquatic biodiversity is woefully incomplete (Allen and Flecker 1993). Although all vertebrate groups include at least some aquatic members, in total they are few in number and fish dominate. Fishes are among the best-known aquatic organisms and often serve as faunal indicators of freshwater health and diversity. As Moyle and Leidy (1992) note, fish are key monitors of ecosystem quality, not only because of their extraordinary diversity but also because they have a profound influence on the distribution and abundance of other organisms in the waters they inhabit. Yet perhaps their most useful attribute is simply that the health of fish populations generally predicts the health of aquatic ecosystems (Beverton 1992). Viewed in this way, fish are the aquatic analogues of the miner's canary.

Although fish are among the best-known aquatic vertebrates, they are still far from completely known. Thus Greenwood (1992) estimates that only about 10 to 15 percent of fish species have been studied in any comprehensive manner, and I consider that small percentage an overestimate. Even with regard to such basic parameters as the number of living species, for example, estimates vary widely (e.g., Cohen 1970; Groombridge 1992; Nelson 1994), yet all agree that these tallies greatly underestimate the actual number of fish species. Data presented in Nelson (1976, 1984, 1994) indicate that in an 18-year period an average of 309 new species[5] a year have been formally described or resurrected from synonymy. Interestingly, the rate of discovery of new species has been fairly constant throughout this 18-year period (figure 3.7), with no sign of an asymptote having been reached. Nelson's (1994) estimate of a final total in the region of 28,500 is probably reasonable, but many of those species will be extinct in nature long before they are discovered and described by a diminishing number of fish taxonomists (Cotterill 1995).

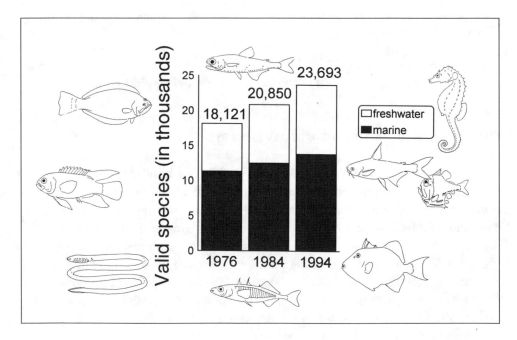

FIGURE 3.7

Number of fish species described between 1976 and 1994.

Whatever the total diversity may be, the current total of 23,693 species means that ray-finned fishes alone represent more than half of all vertebrate diversity. Because nearly half of all of these are restricted to or spend a significant portion of their life history in freshwater, the following statistic is even more impressive. To reflect the true dimension of freshwater biodiversity these totals may be restated as follows: globally some *25 percent of total vertebrate biodiversity is confined to less than one hundredth of a percent of the earth's water* . As freshwater species continue to be discovered at a higher rate than their marine counterparts (figure 3.7), even this startling statistic is probably an underestimate.

As counterintuitive as it may at first seem, freshwater ecosystems (particularly in the forested tropics) are among the biologically richest of habitats and, in terms of vertebrate biodiversity, certainly rival tropical rainforests.

ISLANDS IN A STREAM: VULNERABLE HABITATS

Highly diverse aquatic ecosystems are among the planet's most vulnerable and seem, once perturbated, to deteriorate at a faster rate than their terrestrial counterparts. For example, charting the recovery status of threatened and endangered species in the United States, Williams and Neves (1992) noted a marked difference between terrestrial and aquatic systems. Of 221 taxa for which approved recovery plans had been instituted, only 4 percent of aquatic species versus 20 percent of terrestrial species demonstrated an improved status. Figures for continuing decline show a similar trend, with 41 percent of the aquatic species still in decline (versus 30 percent in terrestrial habitats).

In seeking to understand the particular vulnerability of freshwater ecosystems, two main features emerge. The first is the essentially insular nature of these ecosystems and the second relates to properties of water itself. Groombridge (1992), in an overview of historical species extinction, recognizes a number of general patterns, the most striking of which is the preponderance of extinctions on islands. Over all taxa surveyed, 75 percent of extinctions since 1600 have occurred on islands, and of continental extinctions at least 66 percent represent aquatic species. Just like other insular ecosystems, freshwater habitats are extremely vulnerable to ecological disruption. The functional analogy between freshwater ecosystems and island habitats is particularly well illustrated by the data for molluscs (Groombridge 1992). A full 80 percent of recorded mollusc extinctions were terrestrial species isolated on islands, whereas of the 20 percent of continental extinctions, nearly all were freshwater taxa (figure 3.8). The most obvious reasons for the vulnerability of island species (restricted range and low population numbers) apply equally to freshwater species, which often have ranges limited to single watersheds, lakes, or river systems.

The problems for freshwater forms are exacerbated by the nature of the medium itself. Because water is the most universal solvent, we never see it in its pure liq-

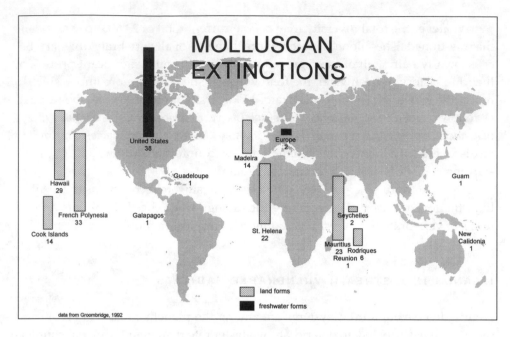

FIGURE 3.8

Record of molluscan extinction from 1600 to the present.
Data from Groombridge (1992).

uid form. The productivity of aquatic environments is driven largely by this capacity of water to act as a solvent, as is its role as a dilutant and sink for domestic, industrial, and agricultural wastes (Moyle and Leidy 1992). The effects of terrestrial inputs into aquatic systems are most acute where the concentration of humans is greatest and urbanization is increasing. The statistics are cause for concern: in 1975 some 38.5 percent of the world's population was concentrated in urban areas, and by 1990 this figure had risen to 42.7 percent. Surface and ground waters are intimately connected to the watersheds that surround them, and changes in land use (deforestation, conversion of grasslands and savannas, and loss of wetlands) have cascading negative effects on aquatic ecosystems. Concomitant with the loss of biotic communities is the loss of the ecological services these systems provide. As an example, Nash (1993) cites a marsh in Wisconsin that has received domestic waste and sewage since the early 1920s. Effluent waters, after passing through the marsh, have a decrease of 80 percent in biological oxygen demand, an 86 percent reduction in coliform bacterial load, and a 50 percent reduction in nitrate levels. As the real cost of replicating such services by artificial means (purification and water treatment plants, etc.) increases, so does the economic incentive for habitat preservation and restoration (see Dugan 1990; Wilcove and Bean 1994; Abramovitz 1996). In the meantime, however, the combined effects of large-scale land use changes along with increased industrial and domestic inputs are undermining all of these services while they continue to degrade water quality and habitat resilience worldwide.

From the perspective of aquatic biodiversity, perhaps even more important are the cumulative effects of human regulation of river flow and drainage (Covich 1993; Allen and Flecker 1993). It has been said that human beings are simply an invention of water as a device for transporting itself from one place to another (Farber 1994), and given the scale of effort that has been put into the resculpting of the world's waterways, one could almost be excused for subscribing to that view. The statistics are awesome: it has been estimated that *by the year 2000 more than 60 percent of the world's total stream flow will have been regulated,* primarily through damming, to provide power, flood control, irrigation, and domestic supplies to burgeoning human populations. The peak period of dam construction occurred between 1950 and 1980, and during those 30 years in North America alone about 6,000 large dams were constructed (Covich 1993). Today, of the nation's 5,200,000 km of rivers and streams, most have been so dramatically altered that only 2 percent are of a sufficiently pristine state to be considered worthy of federal protection status (Benke 1990). The sorry plight of the nation's wetlands is also well documented (Dahl 1990), and over a period of 200 years the lower 48 states have lost an estimated 53 percent of their original wetlands. This translates, on average, to a loss of about 60 acres (24 hectares) for every single hour that passed between the 1780 and 1980. The legacy of some 38,000 large dams around the world is felt as ecological, economic, and social problems surface, and the world's waters no longer flow (Covich 1993; Postel 1996; Abramovitz 1996). Despite all of these problems, water projects of pharaonic proportions are still being proposed throughout the developing world (Fearnside 1988, 1989; Ryder 1994; Heywood 1994; Abramovich 1995, 1996). Allen and Flecker (1993:36) aptly quote Fearnside's (1989) observation that many of these grandiose schemes are like "the pyramids of ancient Egypt," demanding "the effort of an entire society to complete but bring virtually no economic returns" while causing untold devastation to aquatic health and biodiversity.

The impact of humans on freshwater systems worldwide is unprecedented, and there can be little doubt that their effect on freshwater biodiversity worldwide has been profound. Clearly there remain many unknowns, but the rate of loss in some area is reasonably well documented and the tally is already daunting. Among North America's rich ichthyofauna, for example, some 364 taxa are currently considered in serious need of protection (Williams et al. 1989). This figure represents a 45 percent increase in endangerment over the previous decade, and depending on taxonomic convention it translates to some 27 to 36 percent of all native fishes being under threat. These figures for temperate North America are by no means unusual, and although accurate statistics are hard to come by, there is little question that the native fish fauna of Europe, for example, is in a considerably worse state (Stiassny 1996). Even in temperate regions once far from centers of human concentration, the rate of attribution is surprisingly high. For instance, the snow-fed waters of the Nepalese highlands harbor about 130 native species, of which 31 are rare (Shrestha 1990). Of these species, 42 percent are considered severely threatened and in danger of extinction.

Although there is no doubt that the extent of species loss in temperate waters is great and often underestimated (Lydeard and Mayden 1995), the situation in the tropics may be even worse, for it is in tropical regions (particularly in the forested tropics) that the greatest concentrations of fish diversity are found (figure 3.9). Considerably less has been documented in the freshwater tropics, but there is every indication that species loss resulting from the combined onslaught of habitat degradation and exotic species introduction is rapidly accelerating there also. A scattering of well-documented examples illustrate the dimension of loss already sustained, and although these examples represent only the tip of the iceberg, they indicate a bleak future for the world's tropical freshwater ichthyofaunas.

The once richly endemic ichthyofauna of peninsular Malaysia has undergone a profound attrition, apparently caused by habitat degradation resulting from deforestation throughout the region; in 1966 dry forest occupied some 68 percent of the peninsula but by 1990 that figure had dropped below 50 percent (Groombridge 1992). Despite a concerted 4-year collecting effort, only 45 percent of the fishes historically recorded from the peninsula could be found there (Whitten et al. 1987). A similar story has been documented on the island of Singapore, where more than 30 percent of the fish species recorded from the island in the mid-1930s are no longer found there. And in Mexico, for example, whereas only 36 fish species were considered at risk in the 1960s, within 10 years that figure had risen to 123 (Contreras-

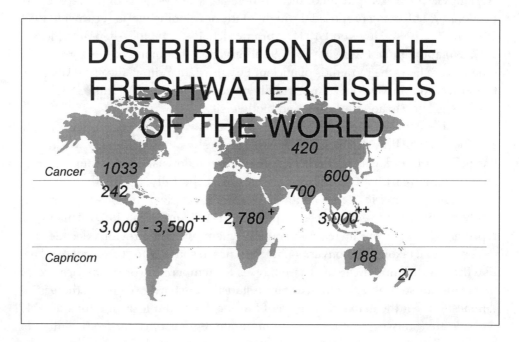

FIGURE 3.9

Distribution of the freshwater fishes of the world.

Plus signs indicate that figures are approximate and that many more species probably will be found in these regions.

Balderas and Lozano-Vilano 1994). As a final example, the freshwater ichthyofauna of the island of Madagascar typifies the precarious state of many freshwater systems worldwide. Over 80 percent of Madagascar's endemic ichthyofauna is restricted to freshwater and of these about 60 percent were once found in the rivers of the eastern coastal forests (Stiassny and Raminosoa 1994). It is precisely these remaining forests that are undergoing the most extensive degradation (Stiassny 1996:figure 5) and many of the species once recorded as abundant are no longer to be found at all.

Although the enormity of loss is manifest, there is still a bewildering amount we simply do not know. This lack of data, even of the most basic type, is a significant problem that must be met before the compilation of accurate inventories of threatened areas, endangered species lists, or programs for rational planning are possible. Elsewhere I have pointed out that for African freshwater fishes the publication of a series of checklists (CLOFFA I-IV, Daget et al. 1984, 1986, 1991) provides a valuable baseline for such studies (Stiassny 1996). However, there remain numerous lacunae, and in compiling taxonomic checklists and inventories for biodiversity assessment and conservation programs, it is imperative to keep in mind the nature of the data incorporated. For example, a review of CLOFFA reveals that a full 30 percent of species of the widespread genus *Chiloglanis* are known only from type series or a single type locality. Similarly, of the monotypic bagrid genera of Africa, about 80 percent are known only from their type localities, and of the 188 species of the family Mormyridae recorded in CLOFFA, a full quarter are also known only from types or type localities. Although it is possible that these figures accurately represent the actual distributional range of these species, it is much more likely that they simply reflect the fact that our knowledge of these taxa goes little further than the existence of a type series of specimens sequestered somewhere in a museum collection. Lack of reliable and comprehensive baseline data makes it difficult even to begin to estimate accurately the true dimension of the problems faced. This is particularly true when previous collecting efforts in a region have not been great and where subsequent collecting results in a dramatic increase in the number of fish species recorded. Thus a recent study by Teugels et al. (1992) found that fish diversity of the West African Cross River had been underestimated by as much as 73 percent. The collecting efforts of Roberts (1989) in Western Borneo (Barat Kalimantan) have essentially doubled the list of species recorded from the region, and Siebert (1994) estimates that his field studies of the freshwater fishes of Sumatra will probably result in a total of around 400 species (currently 272 species are recorded from the region). Such underestimation of taxic richness is widespread throughout tropical freshwater, and although our ignorance is undoubtedly greatest there, it is sobering to bear in mind that even in "well-known" temperate regions such as Europe, new fish taxa are still being discovered with some frequency. Bianco et al. (1987), for example, described a new genus of gobiid fish, *Economidichthys*, from the western Hellenic province of Greece. And more recently Economidis and Miller (1990) discovered yet another species of *Economidichthys, E. trichonis*, in a small freshwater lake within the range of *E. pygmaeus*. This new species, with females no larger than 27 mm, is probably Europe's smallest vertebrate.

All of these examples forcefully illustrate the pressing need to intensify efforts to survey freshwater biodiversity more fully before it is too late. Worldwide, Moyle and Leidy (1992) have estimated that some 20 percent of freshwater fish species are already extinct or are in serious decline. This percentage may even be an underestimate, with a figure of 30 to 35 percent perhaps being more accurate. Whatever the exact percentage, there can be no question that the degree of aquatic impoverishment is profound and increasing at an alarming rate (Miller et al. 1989; Contreras-Balderas and Lozano-Vilano 1994; Moyle 1995).

In an overview of threats to North American freshwater biodiversity, Wilcove and Bean (1994) remark that the word *crisis* may have been somewhat overused by environmentalists, but they go on to stress that it really is the only word that can be used to describe the current state of biological diversity in American freshwater ecosystems. This chapter can do little more than echo that sentiment while extending it to encompass a more global perspective of what can only be described as the crisis of freshwater.

ACKNOWLEDGMENTS

I am grateful to the many friends and colleagues who have shared their ideas and knowledge with me. Thanks to Vadim Birstein, who translated a Russian text and brought some additional references to my attention. I would also like to thank Lita Elvers, who has spent a great deal of time gathering references and helping in numerous ways with the preparation of this chapter. And last, but by no means least, thanks to Amy Simmons for her continued support.

NOTES

1. Much of what is characterized as food shortage may reflect underlying chronic water shortage: without sufficient water nothing will grow. And to grow an adequate diet for one person for 1 year requires about 300 metric tons of water (Clarke 1993). Not surprisingly, in the world's densely populated arid and semiarid areas food shortage is endemic and famine is a constant threat. It is worth noting also that waterborne disease is estimated to affect some 400 million people worldwide, and according to one estimate dirty water was the cause of some 27,000 deaths per day in 1986 (Clarke 1993). An excellent discussion of water quality and health is provided by Nash (1993).

2. Like most of the fish introduced into the Aral, the stellate sturgeon (*Acipenser stellatus*) was introduced in an attempt to increase fishery productivity. Between 1950 and 1960 many millions of larvae and fry were released into the Aral and tragically for the native ship sturgeon, *Acipenser nudiventris*, a parasitic gill trematode (*Nitzchia sturonis*) was introduced along with them. This parasite decimated the ship sturgeon's already stressed populations

and the precipitous decline of Aral *nudiventris* populations resulted in the closing of the fishery for some 25 years. Subsequently, the ship sturgeon has been largely extirpated from the lake by overexploitation in its breeding grounds (Zholdasova 1997). Birstein (1993) notes that the Aral ship sturgeon, extinct in the wild, is today known only from six individuals alive in captivity at the Moscow Aquarium.

3. There is some discrepancy in the literature regarding the actual figures for the total amount of global water, as well as for the balance of the hydrologic cycle (reviewed in Gleick 1993).

4. Of the total runoff (41,000 km^3), some 22,000–26,000 km^3 is lost in floods following torrential rains, another 5,000 km^3 or so falls on areas far from human settlement, and much evaporates before it can be efficiently harvested.

5. Throughout this chapter the term *fish* is used in a restricted sense to refer only to members of the Actinopterygii (i.e., the ray-finned fishes). Lampreys, hagfish, sharks, rays, coelacanths, and lungfish are not considered in the statistics presented herein.

REFERENCES

Abramovitz, J. N. 1995. Freshwater failures: The crises on five continents. *World Watch* 8: 26–35.

Abramovitz, J. N. 1996. Sustaining freshwater ecosystems. In L. R. Brown et al. *State of the world 1996 . A Worldwatch Institute report on progress toward a sustainable society*, 60–77. New York: WW Norton.

Aladin, N. V. and W. D. Potts. 1992. Changes in the Aral Sea ecosystems during the period 1960–1990. *Hydrobiology* 237: 67–79.

Aladin, N. V. and W. D. Williams. 1993. Recent changes in the biota of the Aral Sea, Central Asia. *Verhandlungen der Internationalen Vereinigung für Theoretische und Angewandte Limnologie* 25: 790–792.

Allen, J. D. and A. S. Flecker. 1993. Biodiversity conservation in running waters. *Bioscience* 43: 32–43.

Benke, A. C. 1990. A perspective on America's vanishing streams. *Journal of the North American Benthological Society* 9: 77–88.

Beverton, R. J. H. 1992. Fish resources; threats and protection. *Netherlands Journal of Zoology* 42: 139–175.

Bianco, P. G., A. M. Bullock, P. J. Miller, and F. R. Roubal. 1987. A unique teleost dermal organ in a new genus of fishes (Teleostei: Gobioidei). *Journal of Fish Biology* 31: 797–803.

Birstein, V. J. 1993. Sturgeons and Paddlefishes: Threatened fishes in need of conservation. *Conservation Biology* 7: 773–787.

Blanc, M., P. Banarescu, J.-L. Gaudet, and J.-C. Hureau. 1971. *European inland water fish. A multilingual catalogue*. London: FAO and Fishing News (Books).

Brown, L. R. 1996. The acceleration of history. In L. R. Brown et al. *State of the world 1996. A Worldwatch Institute report on progress toward a sustainable society*, 3–20. New York: WW Norton.

Brown, L. R. et al. 1996. *State of the world 1996. A Worldwatch Institute report on progress toward a sustainable society*. New York: WW Norton.

Clarke, R. 1993. *Water: The international crisis*. Cambridge, Mass.: MIT Press.

Cohen, D. M. 1970. How many recent fishes are there? *Proceedings of the California Academy of Science* 38(4): 341–345.

Contreras-Balderas, S. and M. L. Lozano-Vilano. 1994. Water, endangered fishes, and development perspectives in arid lands of Mexico. *Conservation Biology* 8: 379–387.

Cotterill, F. P. D. 1995. Systematics, biological knowledge and environmental conservation. *Biology and Conservation* 4: 183–205.

Covich, A. P. 1993. Water and ecosystems. In P. H. Gleick, ed. *Water in crisis: A guide to the world's fresh water resources* , 40–55. New York: Oxford University Press.

Daget, J., J. P. Gosse, and D. Thys van den Audenaerde. 1984. *Check-list of the freshwater fishes of Africa*. Vol. I. Tervuren/Paris: MRAC-ORSTOM.

Daget, J., J. P. Gosse, and D. Thys van den Audenaerde. 1986. *Check-list of the freshwater fishes of Africa*. Vol. II. Brussels/Tervuren/Paris: ISNB-MRAC-ORSTOM.

Daget, J., G. G. Teugels, and D. Thys van den Audenaerde. 1991. *Check-list of the freshwater fishes of Africa*. Vol. IV. Brussels/Tervuren/Paris: ISNB-ORSTOM-MRAC.

Dahl, T. E. 1990. *Wetlands losses in the United States 1780s to 1980s*. Washington, D.C.: U.S. Department of the Interior, Fish and Wildlife Service.

Dugan, P. J. 1990. *Wetland conservation*. Gland, Switzerland: IUCN.

Economidis, P. S. and P. J. Miller. 1990. Systematics of freshwater gobies from Greece (Teleostei: Gobioidei). *Journal of Zoology* 221: 125–170.

Ellis, W. S. 1990. A Soviet sea lies dying. *National Geographic*, February 1990: 73–92.

Engelman, R. and P. LeRoy. 1993. *Sustaining water. Population and the future of renewable water supplies*. Washington, D.C.: Population Action International.

Farber, T. 1994. *On water*. Hopewell, N.J.: Ecco Press.

Fearnside, P. M. 1988. China's Three Gorges Dam: "Fatal" project or step toward modernization? *World Development* 16: 615–630.

Fearnside, P. M. 1989. Brazil's Balbina Dam: Environment versus the legacy of the pharaohs in Amazonia. *Environmental Management* 13: 401–423.

Gleick, P. H. 1993. *Water in crisis: A guide to the world's fresh water resources*. New York: Oxford University Press.

Greenwood, P. H. 1992. Are the major fish faunas well-known? *Netherlands Journal of Zoology* 42: 131–138.

Groombridge, B., ed. 1992. *Global biodiversity: Status of the earth's living resources*. London: Chapman & Hall.

Hamer, M. 1990. The year the taps ran dry. *New Scientist* August 1990: 20–21.

Heywood, D. 1994. Reversal of fortune. Will Tonie Sap become a World Heritage Site? *Geographical* 66(11): 26–28.

Lydeard, C. and R. L. Mayden. 1995. A diverse and endangered aquatic ecosystem of the southeast United States. *Conservation Biology* 9: 800–805.

McCaffrey, S. C. 1993. Water, politics, and international law. In P. H. Gleick, ed., *Water in crisis: A guide to the world's fresh water resources*. 92–104. New York: Oxford University Press.

Micklin, P. P. 1988. Dessication of the Aral Sea: a water management disaster in the Soviet Union. *Science* 241: 1170–1176.

Miller, R. R., J. D. Williams, and J. E. Williams. 1989. Extinctions of North American fishes during the past century. *Fisheries* 14: 22–39.

Moyle, P. B. 1995. Conservation of native freshwater fishes in the Mediterranean-type climate of California, USA: A review. *Biological Conservation* 72: 271–279.

Moyle, P. B. and R. A. Leidy. 1992. Loss of biodiversity in aquatic ecosystems: Evidence from fish faunas. In P. L. Fielder and S. K. Jain, eds., *Conservation biology: The theory and practice of nature conservation, preservation and management* , 127–169. New York: Chapman & Hall.

Nash, L. 1993. Water quality and health. In P. H. Gleick, ed., *Water in crisis: A guide to the world's fresh water resources.* 23–39. New York: Oxford University Press.

Nelson, J. S. 1976. *Fishes of the world.* New York: Wiley.

Nelson, J. S. 1984 (2d ed.). *Fishes of the world.* New York: Wiley.

Nelson, J. S. 1994 (3d ed.). *Fishes of the world.* New York: Wiley.

Nikol'skii, G. V. 1940. *Fishes of the Aral Sea.* Moscow: Izdatelstvo.

Postel, S. 1992. Water and agriculture. In P. H. Gleick, ed., *Water in crisis: A guide to the world's fresh water resources,* 56–66. New York: Oxford University Press.

Postel, S. 1996. Forging a sustainable water strategy. In L. R. Brown et al. *State of the world 1996. A Worldwatch Institute report on progress toward a sustainable society,* 3–20. New York: W. W. Norton.

Roberts, T. R. 1989. The freshwater fishes of Western Borneo (Kalimantan Barat, Indonesia). *Memoirs of the California Academy of Science* 14: 1–210.

Ryder, D. 1994. Mekong. Overview of regional plans. *World Rivers Review* 9: 3–7.

Shiklomanov, I. A. 1993. World fresh water resources. In P. H. Gleick, ed., *Water in crisis: A guide to the world's fresh water resources* , 13–24. New York: Oxford University Press.

Shrestha, T. K. 1990. Rare fishes of the Himalayan waters of Nepal. *Journal of Fish Biology* 37: 213–216.

Siebert, D. J. 1994. Taxonomic databases in museums and zoos and their role in fish conservation. In *1994 Report of the Fish and aquatic invertebrate taxon advisory group,* 8–9. Upton by Chester: North of England Zoological Society.

Stansbery, D. H. 1973. Why preserve rivers? *The Explorer* 15: 14–16.

Stiassny, M. L. J. 1996. An overview of freshwater biodiversity: With some lessons from African fishes. *Fisheries* 21: 7–13.

Stiassny, M. L. J. and N. Raminosoa. 1994. The fishes of the inland waters of Madagascar. *Annales Musée Royale de l'Afrique Centrale, Zoologiques* 275: 133–149.

Teugels, G. G., G. M. Reid, and R. P. King. 1992. Fishes of the Cross River basin (Cameroon–Nigeria) taxonomy, zoogeography, ecology and conservation. *Annales Musée Royale de l'Afrique Centrale, Zoologiques* 266: 4–132.

Wetzel, R. G. 1992. Clean water: A fading resource. *Hydrobiology* 243/244: 21–30.

Whitten, A. J., K. D. Bishop, S. V. Nash, and L. Clayton. 1987. One or more extinctions from Sulawesi, Indonesia? *Conservation Biology* 1: 42–48.

Wilcove, D. S. and M. J. Bean. 1994. *The big kill: Declining biodiversity in America's lakes and rivers.* Washington, D.C.: Environmental Defense Fund.

Williams, J. E. et al. 1989. Fishes of North America endangered, threatened, or of special concern: 1989. *Fisheries* 14: 2–20.

Williams, J. E. and R. J. Neves. 1992. Introducing the elements of biological diversity in the aquatic environment. *Transactions of the 57th North American Wildlife & Natural Resources Conference,* 1992.

Zholdasova, I. 1997. Sturgeons and the Aral Sea ecological catastrophe. *Environmental Biology of Fishes* 48: 373–380.

CHAPTER 4

REQUIEM ÆTERNAM: THE LAST FIVE HUNDRED YEARS OF MAMMALIAN SPECIES EXTINCTIONS*

Ross D. E. MacPhee and Clare Flemming*

ABSTRACT

How many mammal species have become extinct in the past 500 years? Although an exact answer to this question is unlikely to be forthcoming, useful approximations can be achieved. However, it is first necessary to designate some reasonable criteria for defining extinctions at the species level. According to criteria adopted here, it may be said that the number lies between 60 and 88. The upper limit includes all instances of loss that meet certain basic evidential requirements. The lower limit excludes undescribed species whose validity is untested, as well as losses that may have occurred after 1950. The upper limit could be breached by ignoring some of our criteria, but only with concomitant losses in the empirical reliability of resulting estimates.

According to our database, approximately 2 percent of all modern-era mammalian species have been lost during the last 500 years. This is essentially identical to the estimate of 85 losses made in the most recent edition of the IUCN's (1996) *Red List*. Surprisingly, however, there is a 51 percent difference in the actual taxonomic content of the two lists. This high level of incompatibility is related to several factors, the most important of which is that the two datasets utilize different systematic and evidential criteria. It is recommended that systematists assume a much more active role in assembling future extinction lists, because many of the problems that must be solved in list construction are essentially taxonomic in nature.

*Originally appeared in R. D. E. MacPhee, ed., *Extinctions in Near Time: Causes, Contexts, and Consequences*. New York: Plenum Press (1999); reprinted here through invitation of the editors and the kind permission of the authors and Plenum Press.

INTRODUCTION

This contribution provides an analytical accounting of mammal species that have disappeared during the past 500 years: the "modern era" of this chapter. Our choice of date was dictated by several considerations, but two are paramount. A.D. 1500 marks more or less precisely the beginning of Europe's expansion across the rest of the world, a portentous event in human history by any definition. It is an equally momentous date for natural history because it marks the point at which empirical knowledge of the planet began to burgeon exponentially. These two factors, linked for both good and ill for the past half-millennium, have affected every aspect of life on earth, and are thus fitting subjects to commemorate in a record such as this.

A number of modern-era extinction lists (hereafter, "compilations") for Mammalia have been published in recent years (e.g., Goodwin and Goodwin 1973; Williams and Nowak 1986, 1993; World Conservation Monitoring Centre [WCMC] 1992; International Union for the Conservation of Nature and Natural Resources [IUCN] 1993, 1996; Cole et al. 1994). One may wonder why yet another recension is needed. Superficially, the empirical problem seems uncomplicated: A species either exists or it doesn't, so what is there to disagree about? Regrettably, things are not so simple. A target taxon that has not been collected or seen for an appreciable interval may indeed be extinct or it may simply not have been collected or seen. Unless someone has taken the trouble to search carefully and seriously for the target, mere lack of information on status should not be accepted blindly as evidence of extinction. Similarly, poor or outdated systematic studies are no more adequate for evaluating the status of a species than is incomplete inventorying or surveying. As we shall attempt to show, the quality of the evidence for evaluating whether a particular extinction has occurred varies widely from case to case, and even in the best circumstances a clear-cut decision is often not possible. Does any of this matter? We think so, because reliable databases are essential to any science. Extinction lists are likely to play an ever-increasing role in conservation policy debates, and it is important to ensure that such lists are not replete with erroneous data. As we shall proceed to show in the rest of this chapter, here we can offer only a preliminary database for studies of very recent mammalian extinctions.

If it were necessary to be absolutely certain that a loss had occurred before it could be counted, the list of modern-era extinctions would be very short indeed. This is the major reason that extinction rates are usually estimated by proxy measures, such as the one that relates annual loss of rain forest to some calculation of species richness (for an extended discussion, see May et al. 1995). The trouble is that proxy estimates are difficult to verify and can lead to rate estimates of enormous magnitude (see Brooks et al. 1997). For example, we regard as being beyond meaningful empirical verification the recent claim that "approximately 1800 populations per hour (16 million annually) are being destroyed in tropical forests alone" (Hughes et al. 1997).

Epistemologically, a more reasonable approach is to establish whether there is adequate justification for one's belief that any particular extinction has taken place. The efficient way to do this is to arrange knowledge about specific extinctions according to levels of apparent reliability. We have tried to achieve this goal by constructing a series of tables (tables 4.1–4.4), each defined by the degree to which available evidence relating to extinction appears to justify a specific conclusion.

An extinction (or, better, the evidence for an extinction) can be said to be resolved when it is found that it meets the following criteria: The name of the species thought to be extinct is valid systematically (e.g., the name is available and not preoccupied by another valid name); the species has been well characterized and, optimally, has been subjected to revisionary review; no individuals belonging to this species have been collected, inventoried, or observed reliably in the last 50 years; and there is evidence of a positive nature that places the loss date within the period spanning A.D. 1500–1950. Extinctions thus resolved are placed in table 4.1. Unresolved extinctions are those for which a meaningful prima facie case can be made. Because they currently fail to meet one or more of the stipulated criteria, however, their status cannot be resolved fully and must be held in abeyance until appropriate information is forthcoming. These potential instances of extinction are apportioned among tables 4.2–4.4 according to the nature of their deficiencies (additional descriptions of tables follows). We are aware that several taxa might have been placed, with about equal justification, in a table other than the one we selected. Our allocations reflect the merits of individual cases as we see them.

Obviously, the current organization of this extinction database is not immutable. Unresolved extinctions are potentially resolvable if evidence can be gathered that clarifies whatever point is in doubt. Similarly, a case that was thought to be fully adjudicated may be reopened if new evidence warrants it. This last point is important to note because we have taken special pains to identify taxa that should not be regarded as modern-era species-level extinctions according to our criteria, even though they have been so listed in other compilations (see table 4.7).

It is appropriate to end this introduction with a word about the causes of modern-era extinctions. This is a compendium of modern-era extinctions, not an explanation thereof (but see "Coda: Modern-Era Extinctions in Quaternary Perspective"). The choice of wording is exact: This work is an actuarial account, in which our concern is to establish what losses occurred, where, and when. How they occurred is a topic of undoubted importance, but in only a tiny fraction of cases is it reasonably clear why particular species disappeared (Caughley and Gunn 1996; Flemming and MacPhee 1997; MacPhee and Flemming in press). Systematic research on species that happen to be extinct is no more likely to uncover evidence regarding causes of loss than is ordinary armchair speculation. On the other hand, any investigation of causation that intends to stay grounded in reality requires reliable databases that assess the systematic facts as rigorously as possible.

The paper by Harrison and Stiassny (1999) should be consulted for a parallel treatment of extinctions among freshwater fishes, using criteria similar to the ones defined here. It is of interest to compare the nature of the two databases, which differ notably in some respects (e.g., absence of fossil material of fishes, very large number of apparently extinct taxa based on limited hypodigms, number of extinctions in last half-century).

ABBREVIATIONS AND CONVENTIONS

Abbreviations

C	central	NE	northeast
CITES	Convention on International Trade in Endangered Species of Wild Fauna and Flora	NW	northwest
		rcyrbp	radiocarbon years before present (i.e., before radiocarbon datum A.D. 1950)
E	eastern		
EED	effective extinction date (last reliable record of collection or observation)	SE	southeast
		W	west
		WCMC	World Conservation Monitoring Centre
I.	Island, Islands		
IUCN	International Union for the Conservation of Nature and Natural Resources of the World Conservation Union	WI	West Indies
		yrbp	years before present

Conventions Used in Tables 4.1–4.4

EFFECTIVE EXTINCTION DATE

*   Target taxon died out post-1500, but date of loss not otherwise constrained

Evidence Categories

1   Target taxon collected or seen in living state between A.D. 1500 and 1950

2   Target taxon associated with remains of exotics having a documented introduction date of A.D. 1500 (or later)

3   Target taxon dated radiometrically; calibrated 14 C date ± two sigma (standard errors) lying close to or within the period A.D. 1500–1950

HYPODIGM DESCRIPTIONS

H   Hypodigm consists of holotype only

H+   Hypodigm consists of two or three assigned specimens (including holotype)

‡   Hypodigm consists of four or more specimens

F   Hypodigm consists of fossil material only (no wild-caught specimens)
       Here, " fossil " is applied to bony remains of any age, such as mammal bones retrieved in modern-era deposits in caves

BODY SIZE CLASSES

BW	Body weight (in kg). Unless otherwise indicated, weight estimates are derived from data reported by Silva and Downing (1995) for the species stipulated in the tables or close relatives in the same body size range.
I	≤ 0.1 kg
II	0.1 to < 1.0 kg
III	1.0 to < 10 kg
IV	10 to < 100 kg
V	100 to < 1,000 kg
VI	≥ 1,000 kg

METHODS

One of the motivations for the present chapter comes from our own sense of frustration with previous lists of modern-era extinctions (Flemming and MacPhee 1995). In general, they have not been accompanied by the kinds of discussions that enable the interested but nonspecialist reader to interpret authors' justifications for including (or excluding) particular species. We consider this to be important. Although for reasons of space we present justifications in summary form in this chapter, the detailed assessments on which they are based are presented elsewhere (table 4.7; MacPhee and Flemming in press). We expect that our tables will be in a state of ongoing revision for the next several years until consensus is achieved on still-doubtful cases (see also "Notes on Major Taxa").

Table 4.1: Resolved Extinctions, A.D. 1500–1950

Listings in table 4.1 are limited to valid species of Recent mammals for which there is positive evidence of an EED within the period 1500–1950. Most of the species allocated to table 4.1 have been reviewed systematically in recent revisionary works, although inevitably professional opinion on the status of some species is not uniform. Criteria for identifying EEDs are explained more fully under "Organization of Tables."

Table 4.2: Unresolved Extinctions, A.D. 1500–1950: Missing Data

For a number of allegedly extinct taxa, the evidence for their systematic validity or EED (or both) is ambiguous, i.e., it neither confirms nor denies the possibility that they died out between 1500 and 1950. For example, the systematic validity of a taxon may be considered ambiguous when the species has never been reviewed adequately or when reasonable queries about its supposed distinctiveness have been raised in the professional literature. Until such queries are fol-

TABLE 4.1
Resolved Extinctions, A.D. 1500–1950 (N = 40)

VALID SPECIES NAME	COMMON NAME	RANGE	EED	CATEGORY	HYPODIGM	BW	NOTES
Hydrodamalis gigas	Steller's sea cow	Bering Sea	1768	1	‡	VI	
Prolagus sardus	Sardinian pika	Sardinia and Corsica	1777	(1)	F	II	
Hippotragus leucophaeus	Bluebuck	South Africa	1800	1	‡	V	
Pteropus subniger	Réunion flying fox	Mascarenes	Pre-1866	1	‡	II	
Pteropus pilosus	Large Palau flying fox	Carolines: Palau	1874	1	H+	II	
Conilurus albipes	White-footed rabbit rat	Australia	1875	1	‡	II	
Potorous platyops	Broad-faced potoroo	Australia	1875	1	‡	III	
Dusicyon australis	Falkland Islands dog	Falklands	1876	1	‡	IV	a
Oryzomys antillarum	Jamaican rice-rat	WI: Jamaica	1877	1	H+	I	b
Megalomys luciae	St. Lucia giant rice-rat	WI: St. Lucia	Pre-1881	1	H+	III	a
Uromys porculus	Little pig rat	Solomons: Guadalcanal	1887	1	H	II	c
Leimacomys buettneri	Groove-toothed forest mouse	Ghana	1890	1	H+	I	a
Lagorchestes leporides	Eastern hare-wallaby	Australia	1890	1	‡	III	
Pharotis imogene	Large-eared nyctophilus	Papua New Guinea	1890	1	‡	I	a
Oligoryzomys victus	St. Vincent pygmy rice-rat	WI: Saint Vincent	1892	1	H	I	a
Nyctimene sanctacrucis	Nendö tube-nosed fruit bat	Solomons: Santa Cruz	Pre-1892	1	H	I	d
Gazella rufina	Red gazelle	Algeria	Pre-1894	1	H+	IV	
Notomys amplus	Short-tailed hopping-mouse	Australia	Pre-1896	1	H+	I	
Notomys longicaudatus	Long-tailed hopping-mouse	Australia	1901	1	‡	I	
Megalomys desmarestii	Martinique giant rice-rat	WI: Martinique	1902	1	‡	III	a

(continued on next page)

TABLE 4.1 *(continued from previous page)*

Resolved Extinctions, A.D. 1500–1950 (N = 40)

VALID SPECIES NAME	COMMON NAME	RANGE	EED	CATEGORY	HYPODIGM	BW	NOTES
Rattus macleari	Maclear's rat	Christmas I. (Indian Ocean)	1903	1	‡	II	a
Rattus nativitatis	Bulldog rat	Christmas I. (Indian Ocean)	1903	1	‡	II	a
Peromyscus pembertoni	Pemberton's deer mouse	Mexico: San Pedro Nolasco I.	1931	1	‡	I	e
Caloprymnus campestris	Desert rat-kangaroo	Australia	1932	1	‡	II	a
Thylacinus cynocephalus	Thylacine	Tasmania	1936	1	‡	IV	f
Nesoryzomys darwini	Galapagos rice-rat	Galapagos: Santa Cruz	Pre-1940	1	‡	I	a
Nesophontes hypomicrus	Atalaye island-shrew	WI: Hispaniola	*	2, 3	F	I	a
Nesophontes micrus	Western Cuban island-shrew	WI: Cuba.	*	2, 3	F	I	a, g
Nesophontes paramicrus	St. Michel island-shrew	WI: Hispaniola	*	2, 3	F	I	a
Nesophontes zamicrus	Haitian island-shrew	WI: Hispaniola	*	(2)	F	I	g
Solenodon marcanoi	Marcano's solenodon	WI: Hispaniola	*	2	F	II	g
Megaladapis, cf. *M. edwardsi*	Tretretretre	Madagascar	*	(1), 3	F	IV	h
Palaeopropithecus, cf. *P. ingens*	Sloth lemur	Madagascar	*	3	F	IV	h
Xenothrix mcgregori	Jamaican monkey	WI: Jamaica	*	(1), 2	F	III	a, i
Megaoryzomys curioi	Curio's giant rat	Galapagos: Santa Cruz	*	2, 3	F	II	a, j
Brotomys voratus	Muhoy	WI: Hispaniola	*	(1), 2	F	I	a, k
Geocapromys columbianus	Cuban coney	WI: Cuba	*	2	F	III	a
Isolobodon montanus	Montane hutia	WI: Hispaniola	*	(1), 2	F	III	a
Isolobodon portoricensis	Allen's hutia	WI: Hispaniola and Puerto Rico	*	(1), 3	F	III	a
Plagiodontia ipnaeum	Johnson's hutia	WI: Hispaniola	*	2	F	III	l

(continued on next page)

TABLE 4.1 *(continued from previous page)*

Resolved Extinctions, A.D. 1500–1950 ($N = 40$)

a Weight estimate by authors.

b Morgan (1993) suggested that *O. antillarum* be revived as a species distinct from extant (and mainland) *O. couesi*, based on measurements and discrete traits.

c The related species *Ul. imperator* (table 4.4) is as poorly known as *Ul. porculus*. We treat these taxa differently because Groves and Flannery (1994) inferred that *imperator* may have been extant until the 1960s (even though no specimens of either taxon have been recovered since the collection of the type series).

d Possibly seen as late as 1907 in Nendö (Santa Cruz group), according to K. F. Koopman (personal communication).

e Extensive exploration and trapping activities by Lawlor (1983) in the 1970s failed to yield any additional specimens of *P. pembertoni*.

f There are frequent but unsubstantiated reports of thylacines having been seen in Tasmania and even on mainland Australia. Weight estimate (15–35 kg) is taken from Flannery et al. (1990).

g *Nesophontes zamicrus* not stated by Miller (1930) to have been found in association with *Rattus*, but included in a larger collection for which such an association was reported. Specimens of *N. micrus* from Cueva Martin (Sierra de Escambray, Cuba) have been dated to 590 ± 50 rcyrbp (corrected; MacPhee et al. 1999).

h Simons (1997) reported (?uncorrected) dates of 630 ± 50 rcyrbp for *Megaladapis* and 510 ± 80 rcyrbp for *Palaeopropithecus*, for samples from caves on the Manamby Plateau. Suggested species allocations are the present authors' and are based on known distributions of taxa concerned. Weight estimates are from Godfrey et al. (1997).

i Based on association with *Rattus* in caves in Jackson's Bay (MacPhee and Flemming 1997) and an apparent reference by Sloane (1707–1725); size estimate is by MacPhee and Fleagle (1991).

j Most recent ^{14}C date for this taxon is 210 ± 55 rcyrbp (Steadman et al. 1991).

k *Boromys* regarded as subgenus of *Brotomys*.

l Weight estimate based on Woods's 1984 estimate for another member of the genus (1984:438).

TABLE 4.2

Unresolved Extinctions, A.D. 1500–1950 (N = 20)

VALID SPECIES NAME	COMMON NAME	RANGE	EED	CATEGORY	HYPODIGM	BW	NOTES
Notomys macrotis	Big-eared hopping-mouse	Australia	1843	1	H	I	b1
Notomys mordax	Darling Downs hopping-mouse	Australia	1846	1	H	I	b2
Pteropus brunneus	Dusky flying fox	Percy I. (Coral Sea)	1874	1	H	II	b3, c
Kerivoula africana	Dobson's painted bat	Tanzania	Pre-1878	1	H	I	b4
Oryzomys nelsoni	Nelson's rice-rat	Maria Madre I. (Gulf of Calif.)	1897	1	‡	I	b5, c
Pseudomys gouldii	Gould's mouse	Australia	1930	1	‡	I	b2
Quemisia gravis	Quemi	WI: Hispaniola	*	(1)	F	III	a
Malpaisomys insularis	Volcano mouse	Canaries	*	(3)	F	II	a, d
Hippopotamus lemerlei	Lemerlé's Malagasy hippo	Madagascar	*	(1), 3	F	V	a, e
Hippopotamus madagascariensis	Common Malagasy hippo	Madagascar	*	(1)	F	V	a, e
Nesophontes longirostris	Long-nosed island-shrew	WI: Cuba	*	(2)	F	I	a, e
Nesophontes major	—	WI: Cuba	*	(2)	F	I	a
Nesophontes submicrus	—	WI: Cuba	*	(2)	F	I	a
Nesophontes superstes	—	WI: Cuba	*	2	F	I	a, b6
Megalomys audreyae	Barbuda giant rice-rat	WI: Barbuda	*	2, 3	F	II	a, f
Oryzomys hypenemus	Barbuda rice-rat	WI: Barbuda and Antigua	*	2, 3	F	II	f, b7
Brotomys offella	Cuban esculent spiny rat	WI: Cuba	*	2	F	I	a, g
Brotomys torrei	De la Torre's esculent spiny rat	WI: Cuba	*	2	F	I	a, g
Hexolobodon phenax	—	WI: Hispaniola	*	(2)	F	III	a
Rhizoplagiodontia lemkei	—	WI: Hispaniola	*	2	F	III	a, h

(continued on next page)

TABLE 4.2 *(continued from previous page)*

Unresolved Extinctions, A.D. 1500–1950 (N = 20)

a Weight estimate by authors.

b Species validity is challenged by: 1, Mahoney (1975); 2, Calaby and Lee (1989); 3, Koopman (1984); 4, Corbet and Hill (1991); 5, Hershkovitz (1971); 6, J. A. Ottenwalder (personal communication); 7, C. Flemming and D. A. McFarlane (unpublished data).

c Percy Island and Maria Madre Islands are Pacific islands.

d Cultural associations with remains of *M. insularis* indicate that this species became extinct "between 800 years and now" (Hutterer 1993). A ^{14}C date of ca. 1070 rcyrbp is available for the stratigraphic level that yielded this rodent (Hutterer et al. 1988).

e Although oral traditions indicate hippos may have survived in Madagascar until ca. A.D. 1500, the latest published date is 980 ± 200 rcyrbp (Dewar 1984).

f *Megalomys audreyae* and *Oryzomys hypenemus* may be identical, but we keep these taxa separated pending further analysis (Flemming and McFarlane unpublished data). Material assigned to *M. audreyae* from Darby Sink, Barbuda, has been AMS ^{14}C dated to 750 ± 50 rcyrbp (^{13}C/^{12}C corrected), close enough to warrant inclusion in this table.

g *Boromys* regarded as subgenus of *Brotomys*.

h According to Woods (1989a:67, table 3), *R. lemkei* occurs in association with *Rattus* and *Mus.*

lowed up in an appropriate systematic review, the status of a species cannot be considered as settled and is best regarded as unresolved. Similar reasoning applies to insecure dating. For example, we include some species of the Antillean insectivore genus *Nesophontes* (island-shrews) in table 4.1 because remains attributed to these species have been found in contexts that also yielded the black rat, *Rattus rattus*, introduced into the West Indies from Europe about A.D. 1500. By contrast, several other island-shrews have to be excluded from table 4.1 because no reliable associations with *Rattus* have been reported for them. Their EEDs are properly considered as unresolved, and will stay that way until evidence appropriate for a decision is available.

Table 4.3: Unresolved Extinctions, A.D. 1500–1950: Missing Names

The purpose of systematic research is to exhaustively organize the world's biota in a meaningful, hierarchical arrangement. The cornerstone of the systematic hierarchy is the species; until a biotic entity conforming to a species has a valid name, it does not have a secure identity or place in the hierarchy. We note this point because there are a number of Recent biotic entities that have not been named or diagnosed formally that nonetheless appear frequently on lists of extinct species. Unnamed, undiagnosed species are not truly open to systematic review, and for this reason their status must be considered unresolved. On the other hand, for most of these entities there is positive evidence for EEDs between 1500 and 1950, which means that they should not be ignored if they are, in fact, good species. The obvious solution is to list them separately, pending a final determination of their status and review by other systematists. Although nomina nuda these entities are best referred to as "taxa" because they have been assigned (with one exception) to valid genera, at least provisionally. It is to be hoped that most of these partially named taxa will have been formally described before publication of the next version of the extinction list.

Table 4.4: Unresolved Extinctions: Losses After A.D. 1950

How long after the last collection or sighting of a species should the taxon be declared extinct? The recommendation of the IUCN and CITES is 50 years, although this period is no more inherently meaningful than half or twice this term. However, database reliability clearly requires that some substantial "waiting period" be set before a species is posted formally to an extinction list. For example, the swamp deer *Cervus schomburgki*, believed to have become extinct more than 60 years ago, has apparently managed to survive in a small area in Laos: antlers referable to this species have turned up recently as hunting trophies (Schroering 1995). The Vietnamese warty hog, *Sus bucculentus* (often considered a subspecies of *Sus verrucosus*), also thought to be extinct, has recently been seen alive (Groves et al. 1997), as has the "extinct" woolly flying squirrel *Eupetaurus cinereus* of Pakistan (Zahler

TABLE 4.3

Unnamed Taxa Presumed Extinct Since A.D. 1500 ($N = 11$)

TAXON	COMMON NAME	RANGE	EED	CATEGORY	HYPODIGM	BW	NOTES
Oryzomys sp. 1	Barbados rice-rat	WI: Barbados	Pre-1890	(1)	F	I	b1, c
Notomys sp. 1	Great hopping-mouse	Australia	Pre-1900	2	F	I	b2, d
Nesophontes sp. 1	Grand Cayman island-shrew	WI: Grand Cayman	*	2	F	I	a, b3
Nesophontes sp. 2	Cayman Brac island-shrew	WI: Cayman Brac	*	2	F	I	a, b3
Megaoryzomys sp. 1	Isabela giant rice-rat	Galapagos: Isabela	*	3	F	II	a, b4
Nesoryzomys sp. 2	Isabela island rice-rat "A"	Galapagos: Isabela	*	2	F	I	a, b4
Nesoryzomys sp. 3	Isabela island rice-rat "B"	Galapagos: Isabela	*	2	F	I	a, b4
Oryzomyin sp. 1	Vespucci's rice-rat	Brazil: Fernando da Noronha I.	*	(1)	F	I	a, b5, e
Capromys sp. 1	Cayman hutia	WI: Cayman Is.	*	2, 3	F	III	a, b3
Geocapromys sp. 1	Grand Cayman coney	WI: Grand Cayman	*	2	F	III	a, b3
Geocapromys sp. 2	Cayman Brac coney	WI: Cayman Brac	*	(2)	F	III	a, b3

a Weight estimate by authors.

b Relevant information may be found in: 1, Ray (1964); 2, Flannery (1995b); 3, Morgan (1994); 4, Steadman et al. (1991); 5, Olson (1981).

c Physically known only from fossils, but Schomburgk may have seen animal in the 1840s (see Ray 1964). Because Fielden (1890) found no evidence of this rodent, Marsh (1985) raised the possibility that the rat died out between 1847 and 1890; for purposes of calculations, we have used the end of this time period (1890) as the EED. We consider Barbados to be a Caribbean rather than an Atlantic Ocean island.

d Flannery (1995b) suspected that this taxon became extinct in the last half of the 19th century; for purposes of calculations, we have used the end of this time period (1900) as the EED.

e Fernando da Noronha is an Atlantic Ocean island.

TABLE 4.4

Unresolved Extinctions After A.D. 1950 (N = 17)

VALID SPECIES NAME	COMMON NAME	RANGE	EED	CATEGORY	HYPODIGM	BW	NOTES
Monachus tropicalis	Caribbean monk seal	Caribbean Sea	1950s	1	‡	V	a
Geocapromys thoracatus	Little Swan Island coney	Little Swan I. (Caribbean Sea)	1950s	1	‡	II	b
Crateromys paulus	Small Ilin cloud rat	Philippines: Ilin I.	1953	1	H	II	
Onychogalea lunata	Crescent nailtail wallaby	Australia	1956	1	‡	III	c
Chaeropus ecaudatus	Pig-footed bandicoot	Australia	1960s	1	‡	II	a
Macrotis leucura	Lesser bilby	Australia	1960s	1	‡	II	
Perameles eremiana	Desert bandicoot	Australia	1960s	1	‡	II	
Lagorchestes asomatus	Central hare-wallaby	Australia	1960s	1	H	III	
Uromys imperator	Giant naked-tailed rat	Solomons: Guadalcanal	1960s	1	H+	II	d
Microtus bavaricus	Bavarian pine vole	Germany	1962	1	‡	I	
Nesoryzomys swarthi	San Salvador rice-rat	Galapagos: San Salvador	1965	1	‡	I	a, e
Mystacina robusta	Greater short-tailed bat	New Zealand: Big Cape I.	1965	1	‡	I	
Pteropus tokudae	Guam flying fox	Guam	1968	1	H+	II	f
Dobsonia chapmani	Dobson's fruit bat	Philippines: Negros I.	1970s	1	‡	II	g
Leporillus apicalis	Lesser stick-nest rat	Australia	1970	1	‡	II	
Macropus greyi	Toolache wallaby	Australia	1972	1	‡	IV	
Sylvilagus insonus	Omilteme cottontail	Mexico: Omilteme	1991	1	H+	III	

(continued on next page)

TABLE 4.4 *(continued from previous page)*

Unresolved Extinctions After A.D. 1950 ($N = 17$)

a Weight estimate by authors.

b Weight estimate reported by Morgan (1989).

c Flannery et al. (1990) cited 1956 as the "last unequivocal record" of the existence of *O. lunata*.

d Groves and Flannery (1994) reported "few or no sightings over last 40 years."

e *N. swarthi* is possibly part of *N. indefessus* according to Patton and Hafner (1983), although Musser and Carleton (1993) disagreed with this assessment.

f Possible sighting in 1984, but no further details (Flannery 1995c).

g L. Heaney (personal communication) was unable to secure further *D. chapmani* in Negros despite major collecting efforts in 1980s. Local informants said that a large bat, evidently this species, had not been seen in 20 years.

1996). Burbidge et al. (1988) have shown on the basis of interview evidence that a number of small marsupials thought to have disappeared more than 100 years ago may have survived until the 1960s or later. Although examples could be multiplied, the point here is that cases of *very* recent extinction should not be treated as resolved (in favor of complete loss) unless the evidence is overwhelming. We deal with this by separately tabulating the names of valid species whose extinction is thought to have occurred since 1950.

Other Tables

The remaining tables (tables 4.5–4.7) are concerned with additional analyses. Tables 4.5 and 4.6 present information on taxa not included (or interpreted differently) in the IUCN (1996) *Red List*. Disqualified taxa are separately noted in table 4.7, together with a brief statement of the reason for disqualification (e.g., species invalid, species still extant, species disappeared prior to our cutoff date). For additional information on disqualified taxa, see MacPhee and Flemming (in press).

ORGANIZATION OF TABLES

For each taxon considered, relevant information is presented in tabular format under eight headings: valid species name, common name, range, effective extinction date (EED), evidence category for EED, hypodigm size, body weight, and notes. Symbols used in the tables are noted under "Abbreviations and Conventions."

Valid Species Name

A valid name is defined as the correct name for a taxon according to the rules of zoological nomenclature, as codified in the most recent (1985) edition of the *International Code of Zoological Nomenclature*. Our basic authority file for valid species of Mammalia is Wilson and Reeder's (1993) edited compendium, with additions and corrections based on expert information from colleagues. All names in tables 4.1 and 4.4 are considered valid, while the validity of some names in table 4.2 has been questioned as indicated. The taxa noted in table 4.3 do not yet have formal binomina and have not been published properly. Table 4.7 lists names of taxa (whether valid or invalid) that cannot be shown to be modern-era, species-level extinctions.

Some compilers have listed extinctions at the subspecies as well as the species level (e.g., Goodwin and Goodwin 1973; Williams and Nowak 1986, 1993), but we avoid doing this here. Although any number of nominal mammalian subspecies may qualify as full species depending on the species concept used (see papers collected by Kimbel and Martin 1993 and Claridge et al. 1997), we prefer to see this determined in individual cases by original revisionary work.

TABLE 4.5

Comparison of Mismatches of Taxa Considered Extinct: Tables 4.1–4.4 vs. IUCN (1996) *Red List*

TAXON	IUCN (1996) CLASSIFICATION	THIS CHAPTER
Acerodon lucifer	Extinct	Disqualified
Pteropus loochooensis	Extinct	Disqualified
Phyllonycteris major	Extinct	Disqualified
Myotis milleri	Extinct	Disqualified
Myotis planiceps	Extinct	Disqualified
Nyctophilus howensis	Extinct	Disqualified
Procyon gloveralleni	Extinct	Disqualified
Equus quagga	Extinct	Disqualified
Sus bucculentus	Extinct	Disqualified
Cervus schomburgki	Extinct	Disqualified
Gazella arabica	Extinct	Disqualified
Canariomys tamarani	Extinct	Disqualified
Coryphomys buhleri	Extinct	Disqualified
Papagomys theodorverhoeveni	Extinct	Disqualified
Paulomys naso	Extinct	Disqualified
Spelaeomys florensis	Extinct	Disqualified
Nesoryzomys fernandinae	Extinct	Disqualified
Sphiggurus pallidus	Extinct	Disqualified
Brotomys contractus	Extinct	Disqualified
Heteropsomys antillensis	Extinct	Disqualified
Heteropsomys insulans	Extinct	Disqualified
Puertoricomys corozalus	Extinct	Disqualified
Plagiodontia velozi	Extinct	Disqualified
Plagiodontia araeum	Extinct	Disqualified
Amblyrhiza inundata	Extinct	Disqualified
Clidomys osborni	Extinct	Disqualified
Clidomys parvus	Extinct	Disqualified
Elasmodontomys obliquus	Extinct	Disqualified
Pharotis imogene	Critically endangered	Table 4.1
Leimacomys buettneri	Critically endangered	Table 4.1
Crateromys paulus	Critically endangered	Table 4.4
Isolobodon portoricensis	Critically endangered	Table 4.1
Sylvilagus insonus	Critically endangered	Table 4.4
Uromys imperator	Endangered	Table 4.4
Oligoryzomys victus	Endangered	Table 4.1
Kerivoula africana	Data deficient	Table 4.2

a See Table 4.7 and MacPhee and Flemming (in press). IUCN (1996) listed the following mammal species as Extinct in the Wild: *Mustela nigripes, Equus przewalskii, Gazella saudiya* (category not utilized in this chapter).

TABLE 4.6

Entities in Tables 4.1–4.3 Not Included in IUCN (1996) *Red List*

TABLE 4.1	TABLE 4.2	TABLE 4.3
Megaladapis, cf. *M. edwardsi*	*Nesophontes longirostris*	*Nesophontes* sp. 2[a]
Palaeopropithecus, cf. *P. ingens*	*Nesophontes major*	*Notomys* sp. 1
Xenothrix mcgregori	*Nesophontes submicrus*	*Oryzomys* sp. 1
Uromys porculus	*Nesophontes superstes*	Oryzomyin sp. 1
Oryzomys antillarum	*Hippopotamus madagascariensis*	*Megaoryzomys* sp. 1
Megaoryzomys curioi	*Hippopotamus lemerlei*	*Nesoryzomys* sp. 2
	Malpaisomys insularis	*Nesoryzomys* sp. 3
	Megalomys audreyae	*Geocapromys* sp. 2[a]
	Oryzomys hypenemus	

a Morgan (1994) identified two unnamed species each of *Nesophontes* and *Geocapromys* from the Cayman Islands; IUCN (1996) listed only one of each.

Species that are "extinct in the wild" raise a related concern. The IUCN (1996) applied this designation to species lacking free-living populations, and, in the mammal section of the current *Red List*, provided such a rating for Saudi gazelle (*Gazella saudiya*), Przewalski's horse (*Equus przewalskii*), and black-footed ferret (*Mustela nigripes*). Although "Extinct in the Wild" may be an important distinction for conservation purposes, it lies outside our system of evaluation. Consistency requires that if a recognized species is represented by one or more local populations (or even individuals), whether wild or captive, that species cannot be considered extinct.

Common Name

Because many extinct species are known only from accounts in specialist literature, it is not surprising that, for many of them, common names either do not exist or are not well fixed. Whether a species that was rarely or perhaps never seen by humans should have a "common" name is partly a matter of taste, although we argue that, given the widening interest in recently extinct species, there is nothing objectionable in manufacturing names that are somewhat more accessible to the nonspecialist than the typical neo-Latin binomen. For the sake of completeness we have included common names in this chapter if well fixed, but have generally avoided creating any new ones for taxa currently lacking such names.

Range

Although the range of a species is vital biological information, for reasons of space we supply only minimal geographical data in the tables. This permits the reader to glean basic biogeographical associations and compare losses by both major taxon and geographical area. "Range" is defined as apparent distribution in 1500 or later.

TABLE 4.7

Disqualified Taxa

NAME	COMMON NAME	RANGE[a]	REASON FOR REJECTION[b]
Antechinus apicalis	Dibbler	Australia	Extant as *Parantechinus apicalis*
Bettongia gaimardi	Gaimard's rat-kangaroo	Australia	Extant (Tasmania only)
Potorous gilbertii	Long-nosed potoroo	Australia	Extant as *P. tridactylus*
Acratocnus odontrigonus	Puerto Rican sloth	WI: Puerto Rico	Date of loss not established as within modern era[c]
Tolypeutes tricinctus	Northern three-banded armadillo	Brazil	Extant
Solenodon arredondoi	Arredondo's solenodon	WI: Cuba	Date of loss not established as within modern era
Solenodon cubanus	Almiquí	WI: Cuba	Extant
Nesophontes edithae	Edith's island-shrew	WI: Puerto Rico	Date of loss not established as within modern era
Cryptogale australis	Large-eared tenrec	Madagascar	Extant as *Geogale aurita*
Crocidura goliath	Goliath white-toothed shrew	C Africa	Extant
Soriculus corsicanus	Corsican shrew	Corsica and Sardinia	Date of loss not established as within modern era
Acerodon lucifer	Philippine fruit bat	Philippines: Panay I.	Extant as *A. jubatus*[d] (Luzon, Negros, and nearby islands)
Pteropus loochooensis	Marianas flying fox	Ryukyu Is.: Okinawa	Extant as *P. mariannus*[e] (Marianas, Carolines, and Ryukyu Is.)
Mormoops magna	—	WI: Cuba	Date of loss not established as within modern era
Pteronotus pristinus	—	WI: Cuba	Date of loss not established as within modern era
Reithronycteris aphylla	Jamaican pallid flower bat	WI: Jamaica	Extant as *Phyllonycteris aphylla*
Phyllonycteris obtusa	Poey's pallid flower bat	WI: Cuba and Hispaniola	Extant as *P. poeyi*
Phyllonycteris major	Puerto Rican flower bat	WI: Puerto Rico	Date of loss not established as within modern era

(continued on next page)

TABLE 4-7 (continued from previous page)

Disqualified Taxa

NAME	COMMON NAME	RANGE[a]	REASON FOR REJECTION[b]
Stenoderma rufum	Red fruit bat	WI: Puerto Rico and Virgin Is.	Extant
Phyllops vetus	—	WI: Cuba	Date of loss not established as within modern era
Natalus primus	Large funnel-eared bat	WI: Cuba, Hispaniola, and Jamaica	Extant as N. stramineus
Myotis milleri	Miller's myotis	Mexico: Baja California	Extant
Myotis planiceps	Flat-headed myotis	Mexico: Coahuila	Extant, probably extremely rare in nature
Nyctophilus howensis	Lord Howe Island bat	Australia: Lord Howe I.	Date of loss not established as within modern era
Pipistrellus sturdeei	Sturdee's Bonin pipistrelle	Japan: Bonin Is.	Extant[e]
Scotophilus borbonicus	Réunion vesper bat	Mascarenes: Réunion	Extant[f]
Cheirogaleus trichotis	Hairy-eared dwarf lemur	Madagascar	Extant as Allocebus trichotis
Ateles anthropomorphus	Cuban spider monkey	(WI: Cuba)	Extant as A. ?fusiceps (NW South America)
Antillothrix bernensis	Hispaniolan monkey	WI: Hispaniola	Date of loss not established as within modern era[g]
Mustela macrodon	Sea mink	Coastal NE North America	Extant as M. vison
Mustela nigripes	Black-footed ferret	USA: Midwest	Extant
Procyon gloveralleni	Barbados raccoon	(WI: Barbados)	Extant as P. lotor
Eschrichtius gibbosus	Atlantic gray whale	Atlantic Ocean	Extant as E. robustus[a] (Pacific Ocean)
Equus przewalskii	Przewalski's horse	China and Mongolia	Extant
Equus quagga	Quagga	South Africa	Extant as E. burchellii[h]
Sus bucculentus	Vietnamese warty pig	Viet Nam	Extant
Cervus schomburgki	Schomburgk's deer	Thailand	Extant (Laos)

(continued on next page)

TABLE 4.7 *(continued from previous page)*

Disqualified Taxa

NAME	COMMON NAME	RANGE[a]	REASON FOR REJECTION[b]
Rangifer dawsoni	Queen Charlotte Islands dwarf caribou	Queen Charlotte Is.	Extant as *R. tarandus*[b] (Holarctic)
Gazella arabica	Arabian gazelle	Saudi Arabia: Farasan Is.	Systematic status uncertain
Gazella bilkis	Queen of Sheba gazelle	Saudi Arabia	Systematic status uncertain[i]
Gazella saudiya	Saudi gazelle	Saudi Arabia	Extant
Bos primigenius	Aurochs (cattle)	Eurasia	Extant as *Bos taurus* (worldwide)
Bubalus mephistopheles	—	China	Extinction predates modern era
Epixerus ebii	African palm squirrel	W Africa	Extant
Geomys fontanelus	Sherman's pocket gopher	SE North America	Extant as *G. pinetis*
Microtus nesophilus	Gull Island vole	USA: New York	Extant as *M. pennsylvanicus* (E North America)
Tyrrhenicola henseli	Large Corsican field vole	Sardinia and Corsica	Date of loss not established as within modern era
Canariomys tamarani	—	Canaries: Grand Canary I.	Date of loss not established as within modern era
Coryphomys buhleri	Timor giant rat	Indonesia: Timor I.	Date of loss not established as within modern era
Papagomys theodorverhoeveni	Flores giant rat	Indonesia: Flores I.	Date of loss not established as within modern era
Paulamys naso	—	Indonesia: Flores I.	Extant
Pseudomys fieldii	Shark Bay mouse	Australia	Extant[j]
Pseudomys sp. 1	Basalt Plains mouse	Australia	Date of loss not established as within modern era[k]
Rhagamys orthodon	Hensel's field mouse	Sardinia and Corsica	Date of loss not established as within modern era
Solomys salamonis	Florida rat	Solomons: Florida Is.	Extant
Spelaeomys florensis	—	Indonesia: Flores I.	Date of loss not established as within modern era
Zyzomys pedunculatus	Central rock-rat	Australia	Extant[l]

(continued on next page)

TABLE 4.7 (continued from previous page)
Disqualified Taxa

NAME	COMMON NAME	RANGE[a]	REASON FOR REJECTION[b]
Megalomys curazensis	—	Netherlands Antilles: Curaçao I.	Date of loss not established as within modern era
Nesoryzomys fernandinae	Fernandina island rice-rat	Galapagos: Fernandina I.	Extant[m]
Nesoryzomys indefessus	Santa Cruz island rice-rat	Galapagos: Santa Cruz and Baltra	Extant[n]
Nesoryzomys sp. 1	Rabida island rice-rat	Galapagos: Rabida	Date of loss not established as within modern era[o]
Peromyscus nesodytes	Giant deer mouse	USA: Channel Is.	Date of loss not established as within modern era
Sphiggurus pallidus	West Indian prehensile-tailed porcupine	West Indies	Genus and species indet.[p]
Brotomys contractus	Small Hispaniolan esculent spiny rat	WI: Hispaniola	Date of loss not established as within modern era
Heteropsomys antillensis	Large Puerto Rican spiny rat	WI: Puerto Rico	Date of loss not established as within modern era[q]
Heteropsomys insulans	Hutia-like Puerto Rican spiny rat	WI: Puerto Rico and Vieques	Date of loss not established as within modern era[q]
Puertoricomys corozalus	Old Puerto Rican spiny rat	WI: Puerto Rico	Extinction predates modern era
Plagiodontia araeum	—	WI: Hispaniola	Date of loss not established as within modern era
Plagiodontia velozi	—	WI: Hispaniola	Extinct as P. ipnaeum[r]
Plagiodontia hylaeum	Dominican hutia	WI: Hispaniola	Extant as P. aedium
Alterodon major	—	WI: Jamaica	Extinct as Clidomys osborni
Clidomys osborni	Osborn's Jamaican plate-tooth	WI: Jamaica	Extinction predates modern era
Clidomys parvus	Small Jamaican plate-tooth	WI: Jamaica	Extinction predates modern era
Amblyrhiza inundata	Anguillan giant plate-tooth	WI: Anguilla and St. Martin	Extinction predates modern era
Elasmodontomys obliquus	Puerto Rican plate-tooth	WI: Puerto Rico	Date of loss not established as within modern era
Plesiorycteropus madagascariensis	Greater bibymalagasy	Madagascar	Date of loss not established as within modern era

(continued on next page)

TABLE 4.7 *(continued from previous page)*

Disqualified Taxa

a "Range" refers to distribution (known or alleged) of entity named in left-hand column; if distribution of current name combination is different, appropriate range is given in right-hand column (in parentheses).

b See "Methods." "Extant as . . ." means that the nominal taxon listed under "Name" is considered to be part of the valid and accepted species named in the right-hand column, "Reason for Rejection." This interpretation applies whether or not the nominal taxon is a named subspecies (e.g., *Rangifer tarandus dawsoni*) and whether or not it has been extirpated (e.g., *Eschrichtius robustus gibbosus*, the Atlantic gray whale).

c Williams and Nowak (1993) listed *A. odontrigonus* as having become extinct around 1500, mentioning that it "survived on Puerto Rico until around the time Columbus arrived." So far as we know, there are no published associations with *Rattus* or *Mus*, nor any radiocarbon dates suggesting survival into modern era. Woods (1989a:67, table 3) listed remains of "ground sloths (several species)" as having been found in association with *Rattus* and *Mus* in Trouing Jeremie #5, a Haitian cave locality. However, later in his text (p. 68) Woods stated that sloths were found in "deeper layers" in which introduced murids do not occur.

d Utzurrum (1992); L. Heaney (personal communication).

e Yoshiyuki (1989); Koopman (1993).

f Includes *S. leucogaster*; *S. borbonicus* has priority and is the valid name for this living species (see Cheke and Dahl 1981).

g Woods's (1989a) evaluation of remains of *Antillothrix* (= *Saimiri*) *bernensis* is ambiguous. The only published radiocarbon dates for *Antillothrix* are 3850 and 9550 rcyrbp (MacPhee et al. 1995), well beyond the cutoff for modern taxa.

h *E. quagga* has priority and is the valid name for this living species if this combination is accepted.

i May be part of *G. arabica*.

j Sometimes included in "*P. praeconis*", but *P. fieldii* has priority and is the valid name for this living species if this combination is accepted.

k Never recorded as a living animal, but Flannery (1995d) believed that it may have survived late into 19th century on the basis of fossil associations with other (still extant) endemic murids no longer found in the Basalt Plains mouse's known range.

(continued on next page)

TABLE 4·7 (continued from previous page)

Disqualified Taxa

l Extant according to T. Flannery (personal communication).

m Nesoryzomys fernandinae was originally described on the basis of owl-pellet material (Hutterer and Hirsch 1979); it has since been recorded as living (D. Steadman, personal communication).

n "N. indefessus" is only an extirpated subspecies, N. indefessus indefessus (Musser and Carleton 1993).

o Dated by Steadman et al. (1991) at 8540 ± 100 to 5700 ± 70 rcyrbp.

p Whether nominal taxon belongs in Sphiggurus or Coendu (and in which species thereof) is not settled (cf. Handley and Pine 1992); we see no reason to accept it as a separate, valid species (contra Woods 1993).

q Anthony (1916) reported that remains of H. insulans at the type locality were found in an intermediate sedimentary level, between older Acratocnus beds and younger Isolobodon beds. If Acratocnus died out well before modern era, this information does not constrain EED of H. insulans. No dating information is available for related species H. antillensis.

r See P. ipnaeum in Table 4.1.

Thus the range for *Thylacinus cynocephalus* is limited to Tasmania, even though its range earlier in the Quaternary included the Australian mainland as well as New Guinea (Rounsevell and Mooney 1995). The location of islands within their respective water bodies is as defined by the *Times Atlas* (1994:xiv).

Effective Extinction Date (EED)

The information included in this column is the date or interval when authenticated specimens of target taxa were last collected or observed scientifically, whether in nature or in captivity. Scientifically, the last observation of a species is an important datum—regardless of whether or not it straggled on for a few more years—because it provides a minimum date for final loss, here termed the effective extinction date. This is practical, as well as justifiable empirically, because the EED is the point at which opportunity to study the target taxon as a living entity terminates.

When does a species become extinct? The obvious answer—extinction occurs when the last member of a species dies—is unambiguous in its meaning but not in its application. Unless taxa are regularly censused in an exacting manner, to follow the demise of "last" survivors is impractical—and, of course, only applicable to species that are still extant. For older extinctions especially, estimates of the time of loss of rare taxa may be highly inaccurate (cf. Signor and Lipps 1982; McFarlane 1999; Alroy 1999). It may therefore strike the reader as odd that times of loss are often cited in calendar years in studies of modern-era extinctions. Yet the precision is not necessarily false, if it is properly understood that the date should correspond to the last authentic observation of living members of the taxon in question.

In the case of taxa known only from fossils, "last collected specimen" is not a meaningful concept and the EED is given as an interval. Because of dating uncertainties, this interval is usually indeterminate ("later than 1500") and signaled as such by an asterisk in the EED column.

For some taxa our EEDs will differ from "last observation" dates provided by other compilers, prompting the reader to wonder whether a lapsus has occurred. In this chapter, any date given in calendar years is the date of last collection of identified specimens, if this were possible to ascertain from museum records or reliable literature. Dates given to decade (e.g., "1960s") clearly imply that EEDs are less well constrained.

Other compilers have made recourse to the publication date of the species's name as equivalent to a last record, but we do this only when no other information is available. Thus the date of "last observation" of the St. Lucia giant rice-rat (*Megalomys luciae*) is sometimes given as 1901, which is the date of Forsyth Major's (1901) publication of the species name. However, the two specimens that comprise the entire hypodigm of this species are known to have been collected before 1881 (G. M.

Allen 1942). Its EED is therefore more precisely given as "pre-1881" rather than as "before 1901."

Evidence for EED

Acceptable evidence that a given extinction occurred in the modern era can be of three kinds, or categories. In the case of category 1, there must be evidence that the target species was actually collected or observed in the living state, at a specified time or times during the past 500 years. Category 2 evidence refers to associations of remains of target species with those of exotic species (known to have been introduced into the relevant area after 1500). Category 3 evidence consists of radiometric dating results that confirm the target taxon's survival into the modern era. An additional category, for fossils with attached hair or other soft tissues but no exotic associations, was rejected because it is not a reliable proxy for "extant within the last 500 years" (cf. permafrost mummies from Alaska and Siberia, which may be tens of thousands of years old; see Haynes and Eiselt 1999; MacPhee et al. 1999).

Although category 1 evidence is potentially the most detailed, it is basically limited to times and places in which literate observers were collecting specimens and keeping records. For this reason, evidence in categories 2 and 3 is often the only basis for adjudication; if clear-cut, such evidence should be accepted as dispositive.

Parentheses are used if there is some question about the quality or completeness of the evidence of a specific EED. For example, it is possible but not definite that Schomburgk (1848) saw a living example of the Barbados rice-rat (*Oryzomys* sp. 1). This taxon is otherwise known only from fossil evidence (Ray 1964). However, because of ambiguities in Schomburgk's observation, we rate this evidence as "(1)," implying that it is still unclear whether this species survived into the modern era (see table 4.3, footnote c).

Categories are defined additionally as follows.

CATEGORY 1

This category comprises collection-based or reliable observational evidence regarding the last appearance of the target taxon, with subsequent unfruitful efforts to relocate the target in nature.

"Collection-based evidence" consists of reliably identified, wild-caught specimens (vouchers). Usually, the vouchers of interest are ones that can be regarded as the "last"-collected specimens of the species in question. The amount of ancillary information accompanying vouchers is sometimes incomplete, especially for taxa last collected in the nineteenth century or earlier. However, as this part of the database is unlikely to improve, we have not excluded any taxon from being given a category 1 rating solely because last-collection records are imperfect.

"Reliable observational evidence" is the detection, by a qualified witness, of living specimens of the target taxon at stated times and places and the recording thereof. Only sightings by persons with intimate knowledge of the taxon in question should be accepted as reliable. Although we recognize that this stipulation may be too narrow, its real purpose is to control for the "thylacine effect," which we define as the tendency to accept the continued existence of a species according to the frequency with which unsubstantiated reports of sightings of it are made.

Optimally, before a determination of extinction is made, it should be settled beyond a reasonable doubt that vouchers have been identified securely and that observational evidence relates truly to the target taxon. This is particularly important in cases in which the taxon might be confused with another that is still extant.

Example: The extinction of the Falkland Islands dog, *Dusicyon australis*, was probably caused by overhunting in the nineteenth century. By the 1870s the Falkland Islands dog was very rare; the last confirmed killing took place in 1876 (Dollmann 1937). Because a comparatively large, island-dwelling vertebrate will not be overlooked easily, we regard the last reliable record for *D. australis* as a very good indicator of its final extinction date.

CATEGORY 2

Category 2 evidence requires a demonstrable association of remains assignable to the target taxon with those of the introduced species whose own date of introduction can be inferred reliably.

Category 1 evidence requires the existence of a collected specimen or observational materials concerning the EED of the target taxon. However, a relatively high proportion of mammal species included on modern-era extinction lists are known only as fossils and were never the subject of a recorded observation in the living state. Their inclusion on a list of resolved extinctions can be justified by proxy evidence, such as the climacteric provided by the introduction of *Rattus rattus* and *R. norvegicus* into the New World during the late fifteenth and mid-eighteenth centuries, respectively. (The *Rattus* proxy is not valid globally: some *Rattus* introductions into Mediterranean islands may have antedated 1500, and introduction of the kiore rat, *R. exulans*, onto South Pacific islands took place on a completely different schedule [Holdaway 1996, 1999]). The proxy provides a minimum estimate of the EED, although in many cases its relationship to the true extinction date may not be very close.

Ideally, it should be ascertainable from the excavator's report that the bones of the target were found in unimpeachable association with the introduced species's remains. This is an important consideration in cave sites with very low sedimentation rates, in which specimens of very different ages may occur at the same apparent level (MacPhee et al. 1999).

Example: Evidence for the extinction of *Solenodon marcanoi* meets category 2 requirements because bones referable to this species have been found in association with *Rattus rattus* (Morgan and Woods 1986), introduced into the New World at the start of the modern era.

CATEGORY 3

Evidence in this category derives from high-quality chronometric determinations, based optimally on bones or other tissues of the target taxon.

Within limits, radiocarbon determinations can be of assistance in determining EEDs (for excellent examples, see Steadman et al. 1991). Optimally, the determination should be made on bone or other tissue of the target itself, although this may not be feasible if the target is represented by few specimens. High-quality dates for bones are ones based on the collagen fraction, with $^{13}C/^{12}C$ ratios and standard errors within acceptable limits. Obviously, it is important to base one's conclusion on remains believed to be among the most recent available for the target taxon.

Because of the expense involved in radiocarbon dating, this method has not been used routinely for dating modern-era extinctions. Radiocarbon determinations offer some promise for better constraining EEDs that are based presently on category 2 evidence. For example, possible survival of *Nesophontes* into this century, as alleged by Miller (1930), G. M. Allen (1942), and several other workers on the basis of specimen "freshness," is not supported by available radiocarbon dates (cf. entry under *N. micrus*; MacPhee et al. 1999).

The EED provided by a radiocarbon date may be given as a range in calendar years equivalent to the calibrated mean ^{14}C date $\pm$ 2 standard errors (95 percent confidence level). Because the calibration curve is nonlinear (Stuiver and Reimer 1993), uncalibrated ^{14}C dates may correspond to more than one calendrical date, which has the effect of increasing the error. Given the narrow time frame of interest, it would be an excess of refinement to cite ^{14}C dates in the form of EEDs.

Example: Radiocarbon dating of the holotype (and only specimen) of the alleged endemic Cuban spider monkey *Ateles "anthropomorphus"* by MacPhee and Rivero de la Calle (1996) established that the specimen is less than three centuries old, not pre-Columbian as previously thought. However, because the species is not valid (specimen appears to be referable to the extant spider monkey, *Ateles fusciceps*), it therefore does not count as a modern-era extinction.

Hypodigm

The term *hypodigm* refers to the entire collected material of a species. Abundance of specimens in systematic collections is a measure of how well a particular taxon has been collected; it is not necessarily a measure of how common the species is or was in nature. This point affects the weighting of evidence in the construction of extinction lists in two ways. First, whether or not a taxon was originally rare in nature, it should be obvious that a comparative lack of specimens may complicate or limit revisionary work. For example, when a species in a poorly known but speciose group is represented by only one or two specimens, it may be exceedingly difficult to ascertain the species's validity if it differs little from close congeners. In such cases, the usual systematic practice is to give the target the benefit of the doubt and continue to recognize it. We follow this practice in most cases, except that extinct species whose validity has been questioned by appropriate authorities are placed on one of the unresolved lists. Secondly, for a species that may be truly rare in nature, it would seem important to know how efficient recent efforts to collect it have been, before pronouncing it extinct. Obviously, this point may be hard to evaluate in many cases. Here we rate specimen abundance in collections in order to invite reexamination of certain taxa through additional collecting or revisionary efforts, as appropriate (see "Abbreviations and Conventions"). Abundance in collections is not rated for fossil taxa.

Notations

Although detailed notes on each taxon cannot be presented here, in each table we separately footnote points of special importance for interpretation of evidence (see MacPhee and Flemming in press).

NOTES ON MAJOR TAXA

For many reasons (see Stiassny 1994) it is important to make a distinction between the loss of one or two species within a group that has many living representatives, and the complete obliteration of all members of a well-differentiated clade of substantial probable antiquity (cf. loss of the Guam flying fox, *Pteropus tokudae*, in a genus with almost five dozen extant species, versus loss of the entire insectivore family Nesophontidae). Some tragedies are, we believe, greater than others.

FIGURE 4.1

Mammalian species extinctions since A.D. 1500. Losses per order as a percentage of all losses (*N* = 88).

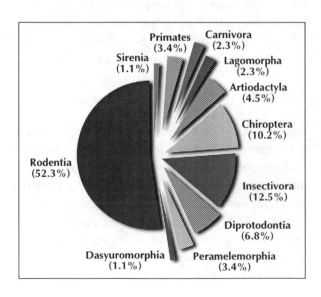

Distribution of modern-era losses by order is illustrated in figure 4.1. Of the 26 orders of Recent mammals recognized by Wilson and Reeder (1993), 11 have suffered species-level losses according to our criteria. The order most affected by modern-era extinctions is Rodentia (52.3 percent of all losses), followed distantly by Insectivora (12.5 percent), Chiroptera (10.2 percent), and three orders of Australian marsupials (accounting for another 11.3 percent).

Marsupialia

Of 10 marsupial extinctions listed in the tables (all of which are Australian), four are placed in table 4.1 and the rest in table 4.4. Even if some marsupials listed in table 4.4 still survive, this does not reduce the force of the argument that a number of marsupial species have been lost or brought to critically low numbers in very recent decades (Flannery 1995a). Note that a high proportion of the Australian (and Meganesian) total reflects extinctions among endemic placentals, particularly conilurine rodents (Smith and Quin 1996).

Insectivora

Although 10 species of Nesophontidae (Antillean island-shrews) are accepted as valid for present purposes, this family is being revised by H. Whidden (personal communication) and colleagues, and some major changes in the taxonomy of the group can be expected. (An eleventh species, Puerto Rican *N. edithae*, is valid and extinct; however, according to the available evidence it cannot be shown to have survived into the modern era and is therefore excluded; see table 4.7.) The quality of information respecting reported associations between island-shrews and exotics is quite variable, and this is reflected in table assignments.

Chiroptera

Many bats are island dwellers, and one might imagine that chiropteran losses on islands should have been substantial in the modern era, as that is the pattern for terrestrial mammals. However, it has proven difficult to verify how many species-level bat extinctions (as opposed to local extirpations and range reductions) have actually occurred, either on mainlands or on islands. Nine species meet our minimum criteria for inclusion in the tables. Of these, three are very recent losses (table 4.4), and at least one other has a validity problem (*Pteropus brunneus*, table 4.2). Although many bat species are undoubtedly threatened at present, documented losses of Chiroptera in the modern era are few.

Xenarthra

Xenarthra suffered major losses in the late Pleistocene (Martin 1984; Martin and

Steadman 1999), but bona fide Holocene extinctions within this order have been negligible except among the tardigrades (sloths). It appears likely that all Quaternary Antillean megalonychids—probably amounting to some dozen species—disappeared during the Holocene, but in fact, reliable dates are too few to place species EEDs more precisely. At least some megalonychids were still extant when humans first came to the Greater Antilles, 5,000–7,000 years ago, as a date of 6330 ± 50 rcyrbp (corrected) is available for *Megalocnus* sp. from the site of Cueva Beruvides in Matanzas Province, Cuba (MacPhee et al. 1999; see also Pino and Castellanos 1985; MacPhee 1997a). Despite the asseverations of Miller (1929a) and others, there are no conclusive observational records or exotic associations that suggest post-Columbian survival of any megalonychid species.

Primates

It has long been suspected that at least some of the so-called Malagasy "subfossil" taxa were still extant at the time of European discovery of Madagascar in 1502 (e.g., Grandidier 1905; Standing 1908; Godfrey 1986). This inference was based primarily on the evident freshness of bone from certain paleontological localities as well as oral traditions that seem to refer to hippos, elephantbirds, and giant lemurs (Flacourt 1661; Tattersall 1982). However, reliable evidence for EED assessment is still almost entirely lacking: Most published ^{14}C dates on identified remains are considerably earlier than 1500, which may indicate that current search images for very late sites are inadequate (D. A. Burney, personal communication, 1999). It is therefore of considerable interest that Simons (1997) has obtained dates close to 1500 on unspecified samples from cave sites containing skeletal material of the giant lemurs *Megaladapis* and *Palaeopropithecus*. (Attribution of this material to species in table 4.1 remains to be confirmed.) If these large-bodied lemurs managed to survive this late, then perhaps many other now-extinct Malagasy primates managed to do so as well (cf. Tattersall 1982).

Interestingly, a passage in Sloane's (1707–1725) natural history of Jamaica makes reference to monkeys living in the interior of that island. Recently, we recovered cranial and postcranial material of *Xenothrix mcgregori* in stratified contexts in southern Jamaica that also yielded *Rattus* (MacPhee 1997b; MacPhee and Flemming 1997), indicating that this monkey also survived later than 1500.

Carnivora

Carnivore losses in the modern era at the species level have been exceptionally light: a dog (*Dusicyon australis*) and a seal (*Monachus tropicalis*). Losses at the populational level, however, have been woefully serious in many groups, which is why threatened and extinct carnivore subspecies dominate compilations such as the IUCN *Red List*.

Cetacea

It is astounding that no whale species has disappeared in the past 500 years—this, despite the unchallengeable fact that many cetacean species have suffered incredible population depletions. At least one distinct population, the Atlantic gray whale, has died out in recent times (see table 4.7). Interestingly, this whale is very poorly known from contemporary records (Ellis 1991); it may have died out before the seventeenth century (Bryant 1996), well before the era of maximum overexploitation in the nineteenth and twentieth centuries.

Sirenia

The sole sirenian extinction of the modern era, *Hydrodamalis gigas*, occurred off the Commander Islands in the subarctic Bering Sea. The last accepted record of Steller's sea cow dates from 1768 (Forsten and Youngman 1982). Because this species would not have been confused with any other, we accept the apparent failure to find any individuals after 1768 as category 1 evidence. However, Nordenskiold (1881, vol. 2:276–278) recorded anecdotal information to the effect that Steller's sea cows were still in existence as late as 1780 or even 1854. This last date seems unlikely, given that whaling and seal hunting in the North Pacific were highly developed industries by the mid-nineteenth century (Ellis 1991).

Artiodactyla

Few artiodactyls have become extinct in the modern era, despite the fact that this group suffered copious losses earlier in the Quaternary (Martin 1984; Alroy 1999; Owen-Smith 1999). Our very brief roster includes only the red gazelle, *Gazella rufina*, about which virtually nothing is known (Groves 1985); the bluebuck, *Hippotragus leucophaeus*, usually mentioned on extinction lists; and the two species of Malagasy hippos, usually ignored.

Anecdotal information summarized by Godfrey (1986) strongly supports the notion that at least one kind of Malagasy hippo survived until the modern era. However, the youngest ^{14}C date, on material of *Hippopotamus lemerlei* from Itampolo, is 980 ± 200 rcyrbp (Mahé and Sourdat 1972; Dewar 1984). Nevertheless, the ubiquity of hippo remains in sites of apparently recent vintage and the widespread oral tradition of the hippolike lalomena (Flacourt 1661) justify our inclusion of this species and its undated relative *H. madagascariensis* (nec *Hexaprotodon*, contra Harris 1991) in table 4.2.

Rodentia

Rodentia is the mammalian order that has lost the most species in the modern era—

a total of 43 losses. The greatest loss by family amounts to 31 extinctions in Muridae, although no murid subfamily has lost more than a few percent of its Recent complement. According to our tables, 14.7 percent of all mammalian extinctions in the past 500 years have occurred within tribe Oryzomyini (13 species), but validity problems affecting several species in this group have not been satisfactorily resolved. According to Musser and Carleton (1993), almost two dozen nominal murid species are known from one or a few specimens. The status of these species is best described as utterly unknown. They are not included here because no reliable observer has claimed that they are actually extinct, although this inference has certainly been made in other cases where the evidence is just as poor.

To illustrate the slipperiness of anecdotal information, it is worth briefly noting the history of *Quemisia gravis*. Miller (1929b) believed that the distinctive fossil rodent to which he gave this name was the same as the *quemi* described by Oviedo y Valdez (1535), which therefore must have survived into post-Contact times. However, Woods (1989b) has argued that the term may refer instead to *Plagiodontia ipnaeum (= P. velozi)*, already known to have survived into the modern era (table 4.1). If Woods is correct, then *Quemisia gravis* lacks a definite EED.

Within Rodentia, island endemics have suffered proportionately more losses than continental species (36 insular extinctions, or 78.3 percent of all rodent losses in the modern era). The number of island losses in the modern era may be underestimated, if the poorly known efflorescences on various islands in the Lesser Antilles (Pregill et al. 1994) and Galapagos Islands (Steadman et al. 1991) prove to include additional very recent extinctions. On the other hand, fossil taxa in some groups (e.g., Capromyidae) seem to have been seriously oversplit (cf. Camacho et al. 1995).

Lagomorpha

The Sardinian pika, *Prolagus sardus*, is known apparently only from fossil material; the basis for the assumption that it was still alive in the last quarter of the eighteenth century is a single, rather vague observation by Cetti (1777; see also Vigne 1988).

The status of the Omilteme (or Omiltemi) cottontail *Sylvilagus insonus*, cited by MacPhee and Flemming (1997) as continental North America's only modern-era mammalian extinction, is difficult to evaluate. One "possible sighting" was reported in 1991 (Ceballos and Navarro 1991), but this species has escaped detection in several recent surveys of the mammals of Parque Omiltemi (Chapman and Ceballos 1990; Jiménez Almaraz et al. 1993). It is clearly a judgment call as to whether it should be recognized as extinct (as opposed to severely endangered, which is the current view of the Mexican government [Cervantes and Lorenzo 1997]). This example underlines the difficulty of rating putative extinctions that have occurred extremely recently.

DISCUSSION

How Many Mammalian Extinctions at the Species Level Have Occurred in the Modern Era?

Depending on criteria used, the question posed in this heading may have more than one answer. The least problematic group of losses consists of the 40 species listed in table 4.1. These taxa meet the basic criteria for recognition as resolved extinctions, meaning that nothing has been placed in evidence to contradict either their status as valid species or the inference that they were lost between 1500 and 1950. Nevertheless, in some cases the evidence for inclusion is not of especially high quality, and it is possible that future revision or renewed collecting may result in some deletions. The next group consists of the 31 unresolved extinctions listed in tables 4.2 and 4.3. Their credentials as examples of modern-era extinctions are often nearly as good as those of species in table 4.1, but they have been listed separately because of outstanding problems (date of loss, species validity, or nomenclature). The 17 unresolved losses in table 4.4 are also kept apart, because these species may—or may not—have died out since 1950.

If these results are combined to achieve a single estimate, then the number of species-level losses during the past 500 years for which there is at least some positive evidence comes to 88. This amount, by no means insignificant, could be pushed to 100 or more if one were to accept (as we do not) a wide variety of equivocal cases listed in some other compilations, and higher still if one were to make due allowance for lost species that never were, and may never be, detected empirically. However, for purposes of this discussion we shall stick with 88 as being the number of losses derived empirically from the accepted evidence.

SYSTEMATIC AND TEMPORAL CONSIDERATIONS

A large number of nominal species listed in the tables—18 taxa, or 20.5 percent—are based on extremely limited hypodigms ($\leq$ 3 specimens, H or H+). (The count would be even higher if fossil or "F" taxa, known from only one or a few specimens, were included.) Obviously, small hypodigms need not automatically raise suspicions concerning validity of species. But the fact that rarely collected, thinly documented taxa make up nearly one-fifth of suspected modern-era extinctions is worth worrying about, for roughly the same reason that the possibility of a type 2 error (incorrect acceptance of a false null hypothesis) in a statistical analysis is worth worrying about. We believe that the odds are against all such taxa being valid, but this is something to be worked out systematically by students of the groups in question.

We emphasize that it is dangerous to assume that an extinction has occurred merely because a species has not been recently seen or collected. Certainly, species not seen for some time could be extinct—or they could be rare in nature, or unusually cryptic, or behaviorally adept at escaping detection or trapping. For species that are collected rarely, an inference of extinction should be avoided until appropriate collecting methods to increase sample sizes have been tried repeatedly, without success. Thus, as noted above, we do not include some two dozen murids catalogued by Musser and Carleton (1993) as having very small hypodigms, because no reliable observer has claimed that they have become extinct. It would seem reasonable to keep this moratorium in place until appropriate surveys are made.

It is practically inevitable that poorly documented taxa will have highly suspect EEDs. For example, the groove-toothed forest mouse *Leimacomys buettneri*, collected only once in 1890, fulfills the EED criterion for entry in table 4.2, but the cloud rat *Crateromys paulus*, collected only once in 1953, has to be posted to table 4.4 because of the 50-year rule. Although table 4.1 is said to include only "resolved" extinctions, there is really nothing other than the spread of years that justifies the assumption that *L. buettneri* is more likely to be extinct than is *C. paulus*. We can only emphasize, once again, that the road to resolution is through collection and inventory.

GEOGRAPHICAL CONSIDERATIONS

It has long been recognized that the vast majority of vertebrate losses during the modern era have occurred on islands (e.g., Diamond 1984; Olson 1989). Our data reveal that extinctions among island mammals are even higher than previously reported. No fewer than 71.6 percent (63/88) of all suspected species-level extinctions among mammals in the past 500 years have occurred on islands, ac-

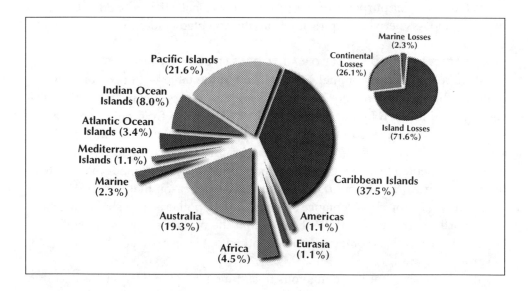

FIGURE 4.2

Mammalian species extinctions since A.D. 1500. Distribution of losses—continental, island, and marine species (*N* = 88).

cording to tables 4.1–4.4. If table 4.4 data (unconfirmed post-1950 losses) are omitted from the calculation, this figure rises to 78.9 percent, which compares closely with figures published for losses of birds (> 90 percent) and molluscans (80 percent) on islands (WCMC 1992; Steadman 1995).

It is often stated that the great majority of modern-era mammalian extinctions that have occurred on the planet took place in Australia, mostly within the past 150 years (Flannery 1995a; Smith and Quin 1996). However, we point out that it is the Caribbean region that has suffered the greatest concentration of losses (figures 4.2 and 4.3). According to the data in tables 4.1–4.4, 37.5 percent (33/88) of all modern-era extinctions occurred on West Indian islands. This is nearly equal to the combined total of losses on the Australian continent (19.3 percent, 17/88) plus the entire Pacific area (21.6 percent, 19/88).

The small number of losses on continents other than Australia is noteworthy (figures 4.2 and 4.3, top). In the Americas (excluding the West Indies), two of three documented modern-era extinctions occurred on nearshore islands. The single mainland loss (*Sylvilagus insonus*) for the Americas is not fully resolvable in our framework because of its recency (ca. 1991). Similarly, according to our data, Eurasia has lost only one continental species (a vole), while mainland Africa has witnessed only four species-level extinctions in the past 500 years (two antelopes, a mouse, and a bat). Despite an immense amount of habitat degradation in this cen-

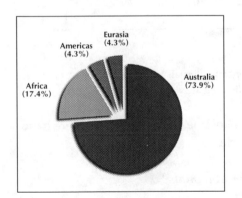

FIGURE 4.3

Top, Continental extinctions since A.D. 1500: percentage by area (*N* = 23). **Bottom,** Insular extinctions since A.D. 1500: percentage by area (*N* = 63).

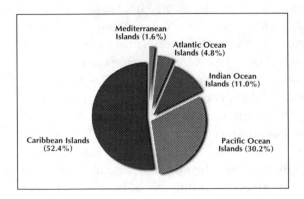

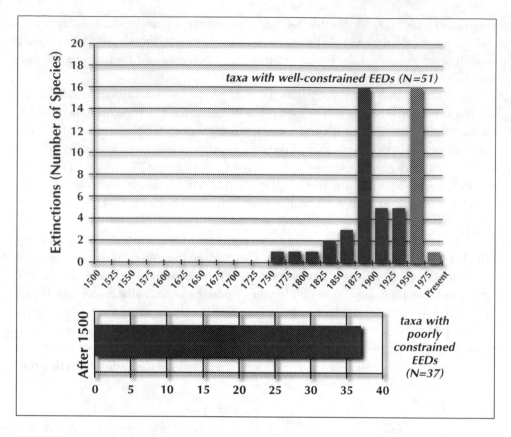

FIGURE 4.4

Mammalian species extinctions plotted at quarter-century intervals between A.D. 1500 and present (N = 51). Lower bar graph depicts known taxa that died out after A.D. 1500 (N = 37).

tury, at least at the empirical level continental biomes have suffered a mere handful of documented losses of mammalian species.

What Is the Modern-Era Extinction Rate for Mammals?

Figure 4.4 distributes the 88 losses in tables 4.1–4.4 along a time line divided into 25-year increments. Of this number, only 51 (58.0 percent) are dated sufficiently well to assign their EEDs to narrow intervals. (Quarter-century intervals were chosen for depiction, but the reader should be aware that some interval assignments are based on convention only.) Of these 51 species, only 2 (3.9 percent) apparently disappeared before the nineteenth century. Twenty-two species (43.1 percent) were lost in the nineteenth century (3 before 1850), whereas 27 species (52.9 percent) disappeared in the twentieth century (17 since 1950). It is noteworthy that the loss figure for the period 1850–1899 (19 species) is substantially higher than that for 1900–1949 (10 species). The cumulative curve for mammalian extinctions in the modern era (figure 4.5) thus has a distinctive appear-

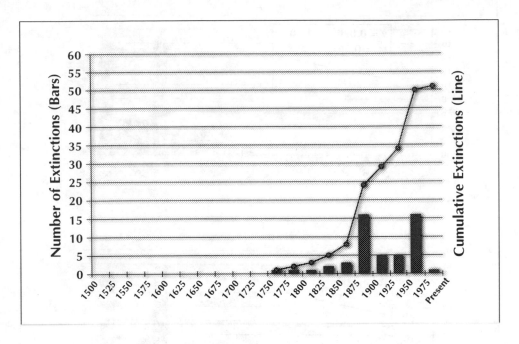

FIGURE 4.5

Mammalian species extinctions since A.D. 1500: losses with well-constrained EEDs (*N* = 51) by quarter-century (bars) and cumulatively (line).

ance, rising quickly through the nineteenth century and flattening notably during the past half-century.

The figures just quoted cannot be accepted at face value, because they do not include the large number of "fossil" taxa (*N* = 37) whose EEDs usually cannot be approximated even to the nearest century (separate bar in figure 4.4, "After 1500"). Large numbers of temporally unallocated losses may severely complicate the estimation of extinction rates from empirical data, as the following example illustrates. In figure 4.6A we portray 51 extinctions by quarter-century intervals from the beginning of the sixteenth century through the third quarter of the twentieth century. In examining the sharp slope of the regression line in the inset, it might be concluded that the extinction rate has shot upward exponentially in the past two centuries, from few losses before 1750 to dozens in the last several 25-year intervals. However, it is of course unlikely that there was only one mammalian extinction between ca. 1500 and 1775, for example. The graph merely reflects the fact that we cannot narrowly date more of the "fossil" extinctions, which is a different matter. To model the possibility that many imprecisely dated "fossil" species disappeared within this early period, we added 37 "fossil" taxa (figure 4.6B) randomly to the time line. Interestingly, these unbiased additions have a negligible effect on the regression expression (inset), which is driven by high values on the right side of the graph (although the line's intercept is shifted slightly, to the mid-sixteenth century). Nonrandom additions can have a quite

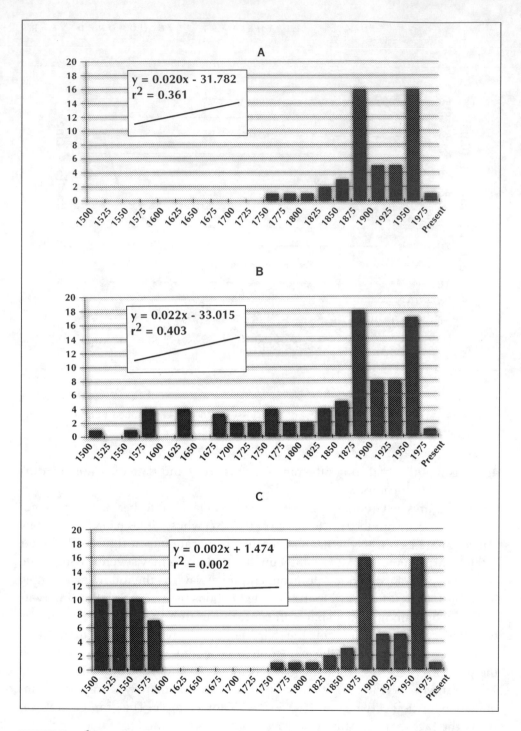

FIGURE 4.6

Mammalian species-level extinctions per century since A.D. 1500: **A,** modern-era extinctions with well-constrained EEDs ($N = 51$); **B,** $N = 88$ (as in A, with the addition of 37 fossil species placed at random in the interval 1500–1999); **C,** $N = 88$ (as in A, with the addition of 37 fossil species placed in the interval 1500–1599).

different effect, however: if we allot the 37 poorly dated extinctions in the tables to the period between 1500 and 1599, a distinctly bimodal graph results (figure 4.6C). Bimodality, combined with the much lower slope of the regression expression (inset), is inconsistent with the usual view that the modern-era extinction rate for mammals was very low until the mid-nineteenth century and then rose steeply thereafter. This scenario cannot be verified either, but that is not the point. The important thing is that extinction rate estimates and perceived patterns of loss are bound to be affected seriously by the quality of the data utilized. Given so many temporally unallocated losses (figure 4.4), the few facts in play can be made consistent with practically any extinction-rate scenario (Caughley and Gunn 1996).

How Does This Compilation Differ from Others?

We shall answer this question by comparing our results to the most recent major compilation, the IUCN (1996) *Red List of Threatened Animals*. This is the first *Red List* issued using the new, more elaborate set of criteria for ranking threat status (Mace 1993). Although the IUCN cutoff date is A.D. 1600 rather than 1500, the difference is unimportant inasmuch as there are no mammalian extinctions that can be dated explicitly to the sixteenth century according to our criteria.

The 88 losses counted in tables 4.1–4.4 amount to 1.9 percent of the 4,629 Recent species catalogued in Wilson & Reeder's (1993) *Mammal Species of the World* (which, however, excludes a number of the extinct species tabulated here). Our total differs trivially from the IUCN (1996) *Red List* estimate of 85 losses. At first glance, it might appear that such apparently close similarity in final totals indicates that basic data sets and methodologies must be highly compatible. Closer inspection reveals, however, that the similarity is almost accidental, because there is nearly a 2:1 difference in taxic content and status ratings in the two works.

This striking difference results from several factors (see tables 4.5 and 4.6). The IUCN (1996) *Red List* includes 28 instances of alleged extinction that we exclude for systematic or other reasons, whereas we include 23 species extinct in the modern era that the *Red List* does not mention at all. There are 8 additional species regarded by us as extinct that the IUCN classifies differently (as Critically Endangered, Endangered, or Data Deficient). Overall, the two lists classify only 57 species in the same way, that is, as instances of modern-era extinction. This yields a compatibility factor of only 49 percent (i.e., 57/116, number of status matches in the two works divided by matches + mismatches).

Two questions spring immediately to mind. First, if the study of modern-era extinctions is truly amenable to scientific inquiry, how can assessments of the status of individual taxa differ as substantially as they do between our lists and the *Red List*? Secondly, where do we go from here?

It should be evident from remarks peppered throughout this chapter that we do not, in fact, have much of a "science" of modern-era extinctions of mammals at this time. It is therefore no wonder that differing criteria for assessing extinction status exist, and that this produces incompatible results. The biggest single source of disagreement is that the IUCN (1996) *Red List* and compilations published previously accept many instances of extinction that we reject because they do not meet minimal objective criteria (see table 4.7). A possible reason for this discrepancy is that the burden of assembling databases like the *Red List* falls disproportionately to researchers whose interest and training in systematics may be limited. Scientists who are not systematists themselves, or are not adequately advised by systematists, face substantial hurdles in dealing with and interpreting basic taxonomic information. For example, it seems to us that the category "Extinct" was applied rather uncritically in the compilation of current and past *Red Lists*. Possibly, compilers relied too much on offhand, unsubstantiated guesses in the literature about the extinction status of specific taxa (see discussion by Harrison and Stiassny 1999). It is also possible that it was difficult to get qualified systematists to scrutinize such claims.

The second question concerns the future of lists like the ones offered here. Dramatic differences among compilations indicate that systematic and evidentiary criteria are being applied unevenly. For future compilations, we recommend better use of the revisionary literature than is ordinarily done, because this literature is the only arbiter of what is valid and what is not. We think it should also be mandatory for compilers to search paleontological sources for relevant "fossil" taxa, although we recognize that this point has pertinency only for certain groups (e.g., vertebrates, molluscans). In this way it should be possible to reach a consensus among conservation biologists, systematists, and paleontologists working to the same end.

In this vein it is worth remarking that much better use of radiometric dating should be made for sorting out doubtful cases. Because of the paucity of published dates, category 3 information played only a very small role in the construction of our lists (figure 4.7). In the future, a substantial effort should be undertaken to resolve the status of the large number of extinct species with poorly constrained EEDs (currently 37 taxa).

We would like to end this section by answering an objection raised by some of our colleagues, that extinction lists like the ones presented here do an actual disservice to the cause of conservation biology. Such lists, it is claimed, cannot be comprehensive because they capture only a fraction of all species extinctions that have occurred in the modern era. In the wrong hands, the little knowledge that we do provide can become a truly dangerous thing. Although we acknowledge the force of the sentiments that prompt this view, they are ultimately beside the point. We cannot control for varying tastes in the taxonomic recognition of species, and we cannot control for species extinctions that have left no accessible record. Nor can we control for the use others may make of extinction lists. What

FIGURE 4.7

Mammalian species extinctions since A.D. 1500: availability of evidence of extinction, by category.

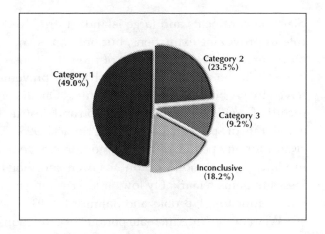

we can do is present and discuss the evidence, and identify where the existing database is probably adequate and where it is defective. Only in this way will extinction studies attain a scientific footing (see recommendations outlined by Harrison and Stiassny 1999).

CODA: MODERN-ERA EXTINCTIONS IN QUATERNARY PERSPECTIVE

Compared with losses that occurred earlier in the Quaternary (Martin and Steadman 1999), there are many things that modern mammalian extinctions are not.

Not Just Another Case of Overkill

The familiar mantra concerning human-induced extinctions intones that it is only logical to assume that currently recognized causes of species endangerment will ineluctably lead to extinction—unless, of course, remedial measures are undertaken. As is well known, habitat loss is usually rated as the leading cause of endangerment, followed by overexploitation and introduction of exotic species (cf. WCMC 1992; Hanski et al. 1995; Baillie 1996; Czech and Krausman 1997). "Other causes," such as incidental taking and disease, are often not rated or rated as having effects on isolated taxa only (cf. WCMC 1992). It is also accepted that many, if not most, endangered taxa are threatened by more than one cause (Czech and Krausman 1997). Such frameworks for threat analyses are usually self-limited to anthropogenic influences and therefore do not include "natural" factors like climate change (except "global warming," which is viewed as having an anthropogenic origin).

Interestingly, when Late Quaternary extinctions are under discussion, the relative ranking of causes is essentially inverted. Overkill is often identified as the chief, if not the only, cause of these extinctions, especially those that took place on continents and large islands. Introduced species are allowed a strong role in provoking extinctions, but only on islands. Habitat destruction is much more rarely invoked as a cause of prehistoric losses, but lack of evidence pointing to it in the earlier Quaternary has not prevented the assertion that humans have always been destroyers of nature: "On all continents, there is an extensive record of prehistoric human alteration and destruction of habitat coinciding with a high rate of species extinctions" (Primack 1995:59). In actuality, there is no such record for any period before the modern era. Today, even in continental settings in which habitat modification has been enormous, the number of documented losses remains remarkably low, at least among mammals and birds (e.g., Brazilian Atlantic forest; Brooks and Balmford 1996).

What of the "intermediate period" of extinction, the modern era that we focus on here that lies between the first major European voyages of discovery and the advent of the postindustrial biodiversity crisis? No more than a handful of recent mammalian extinctions can be plausibly attributed to overexploitation in any of its guises (here including overhunting and persecution). The best supported cases of overexploitation culminating in extinction are *Hydrodamalis gigas* (hunted for meat), *Dusicyon australis* (for fur), and *Thylacinus cynocephalus* (persecuted as a pest). *Monachus tropicalis* was hunted intensively for its fat, which was rendered into lamp oil (cf. G. M. Allen 1942). Demand for seal oil collapsed in the mid-nineteenth century with the appearance of better alternatives, but by this time the Caribbean monk seal was already so extremely rare that J. A. Allen (1887) described it as being nearly "mythical." Whether its population collapse was related solely to earlier overhunting is, however, completely unexplored. Likewise, the wallaby *Macropus greyi* may have been hunted incidentally for its fur, but there is no good evidence to indicate that fur-trapping (or any other specific pressure) caused its extinction (Smith 1995). Why rarely seen *Hippotragus leucophaeus* disappeared is similarly obscure. Overhunting may have had some effect if, as often claimed, the bluebuck's range was limited to a single district in Cape Colony (Swellendam). But there is no evidence on this point.

Some small mammals—especially ones living on islands—may also have been brought to the point of extinction through overexploitation. However, as with large mammals, the evidence is limited and tenuous. *Pteropus subniger* was evidently hunted for its meat, but so were (and are) many other pteropodid species (Heaney and Heideman 1987). Yet among bats, such exploitation has led at most to local extirpations, not complete species extinctions (e.g., bat losses on 'Eua, Tonga [Koopman and Steadman 1995]). In reference to extinctions in the West Indies, it is reasonable to assume that many of the endemic rodents were large enough to make the hunting of them worthwhile, although the archeological evidence for direct impacts of this sort is negligible. The capromyid *Isolobodon portoricensis* —frequently encountered in archeological sites—may even have been brought into semidomestication on some islands (Miller 1929b; Flemming and MacPhee in prep.). Yet, if a domesticant, why did it become extinct (cf. Alcover et al. 1999) in reference to the Balaeric goat *Myotragus*)?

Not a Climate Change Event

"Climate change," a grab-bag category, has come to mean any kind of environmental change that some author or another believes could be offered seriously as a cause of extinction. Often, it has an essentially non-specific, negative meaning: It is whatever human-induced change is not (e.g., habitat destruction, overkill). To find it, one looks wherever people are not, which is not an easy thing.

It is of special interest that none of the serious climatic disruptions of recent centuries is known to have induced a documented mammal or bird extinction. Thus, the Little Ice Age, which resulted in average temperatures well below normal for the better part of four centuries, has not been implicated in any extinctions, despite clear effects on species distributions (Luckman 1986; Pielou 1991). There is ample documentation that the El Niño/Southern Oscillation causes severe short-term depressions in sizes of Pacific seabird populations, but this phenomenon does not seem to have induced any recent species losses (Martin and Steadman 1999). Extreme pulses, in the form of record highs/lows in temperature, snowfall, precipitation, and a multitude of other phenomena have been recorded for all parts of the planet, but in no case have these effects been implicated in a correlated vertebrate extinction. How much more severe must such occurrences be, and on what time scale, to cause outright extinction?

Not a Continental Event

As noted above, nearly three quarters of all mammal extinctions in the past 500 years occurred on islands (figures 4.2 and 4.3). Of the remainder, most occurred on just one continent, Australia. In the Americas 11,000 rcyrbp, by contrast, virtually all losses were mainland. No extinctions are known to have taken place at this time on the Greater and Lesser Antilles or the Galapagos, the New World's principal groups of oceanic islands.

Not a Megafaunal Event

As may be seen by examining the body weight (BW) column in tables 4.1–4.4, the vast majority of species that have disappeared since 1500 were rather small (figure 4.8). Only 12.5 percent (11/88) of modern-era extinctions involved "large" species (defined here as body size ≥ 10 kg). Although empirical weights based on substantial samples are unavailable for many of the species concerned, it is unlikely that *Dusicyon australis*, *Gazella rufina*, *Thylacinus cynocephalus*, or *Palaeopropithecus*, cf. *P. ingens* met Martin's Limit for megafauna (i.e., body mass ≥ 44 kg). If these species

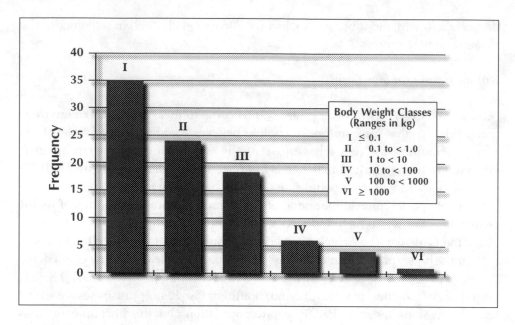

FIGURE 4.8

Mammalian species extinctions since A.D. 1500: body weights, by class (*N* = 88).

are excluded, only 8 percent (7/88) of modern-era extinctions can be considered to lie within the "megafaunal range" as defined traditionally (in rough rank-order by size: *Hydrodamalis gigas*, *Hippopotamus lemerlei* and *H. madagascariensis*, *Monachus tropicalis*, *Hippotragus leucophaeus*, *Macropus greyi*, and *Megaladapis*, cf. *M. edwardsi*). By contrast, during the end-Pleistocene extinctions in North America, almost three-quarters (70.1 percent) of all losses occurred among mammal species having body sizes this large (Graham and Lundelius 1984).

Not the Last Word

In end-Pleistocene extinctions in North America, artiodactyl and perissodactyl ungulates were highly affected (28 losses, for 41.2 percent of 68 species-level extinctions; data of Martin 1984; Graham and Lundelius 1984). By contrast, to the degree that they can be considered to have been fully detected by paleontological methods, losses among rodents (14.7 percent) and bats (5.7 percent) contributed relatively little to total wastage.

   This situation has been effectively reversed in the past 500 years. Only 4.5 percent (4/88) of all documented extinctions in this period have occurred among ungulates (exclusively within Artiodactyla), whereas losses among bats and rodents contribute, respectively, 10.2 percent and fully 54.5 percent of the total.

   It is important to recognize that our numbers would not change much even if certain taxa (e.g., several unlisted "giant" lemurs and ungulates rejected here because of insufficient information) were counted in the modern-era tally. Proboscidea and Perissodactyla, which have (or recently had) many members in the megafaunal size range, lost no species in the modern era despite suffering substantial wastage during the late Pleistocene (Martin and Steadman 1999; Alroy 1999). The low proportion of megafaunal extinctions in post-Pleistocene times may be partly artifactual, because, compared with earlier epochs, we live in a world largely bereft of big mammals (Owen-Smith 1988, 1999). Yet the fact that so few megafaunal species have become extinct in recent times, despite egregious and persistent levels of high endangerment, remains a puzzle of the first order and demands a meaningful explanation (see discussion by MacPhee and Marx 1997).

Additional information on the topics covered in this chapter may be found at these websites: www.creo.org and www.amnh.org/science/biodiversity/extinction/.

ACKNOWLEDGMENTS

We thank Tim Flannery, Sara Gilson, Colin Groves, Larry Heaney, the late Karl Koopman, Don McFarlane, Veronica Mahanger, Guy Musser, Ronald Nowak, David Steadman, and Don Wilson for reading, criticizing, and improving the manuscript version of this chapter. We are especially grateful to Steven Kometani for the creativity he demonstrated in drafting the figures.

We were surprised and gratified by the request of our colleagues, Joel Cracraft and Francesca Grifo, that this paper be reprinted here so that it could reach another audience. We extend our thanks to the editors of Plenum Press and Columbia University Press for making this possible.

REFERENCES

Alcover, J. A., B. Seguí, and P. Bover. 1999. Extinctions and local disappearances of vertebrates in the western Mediterranean islands. In R. D. E. MacPhee, ed., *Extinctions in near time: Causes, contexts, and consequences.* New York: Kluwer/Plenum.

Allen, G. M. 1942. *Extinct and vanishing mammals of the Western Hemisphere with the marine species of all the oceans.* New York: Cooper Square.

Allen, J. A. 1887. The West Indian seal (*Monachus tropicalis*). *Bulletin of the American Museum of Natural History* 2(1).

Alroy, J. 1999. Putting North America's end-Pleistocene megafaunal extinction in context: Large-scale analyses of spatial patterns, extinction rates, and size distributions. In R. D. E. MacPhee, ed., *Extinctions in near time: Causes, contexts, and consequences.* New York: Kluwer/Plenum.

Anthony, H. E. 1916. Preliminary report on fossil materials from Porto Rico, with descriptions of a new genus of ground sloth and two new genera of hystricomorph rodents. *Annals of the New York Academy of Sciences* 27: 193–203.

Baillie, J. 1996. Analysis. In J. Baillie and B. Groombridge, eds., *1996 IUCN red list of threatened animals*, 24–43. Gland, Switzerland: IUCN.

Brooks, T. and A. Balmford. 1996. Atlantic forest extinctions. *Science* 380: 115.

Brooks, T. M., S. L. Pimm, and N. J. Collar. 1997. Deforestation predicts the number of threatened birds in insular Southeast Asia. *Conservation Biology* 11(2): 382–394.

Bryant, P. J. 1996. Dating remains of gray whales from the eastern North Atlantic. *Journal of Mammalogy* 76(3): 857–861.

Burbidge, A. A., K. A. Johnson, P. J. Fuller, and R. Southgate. 1988. Aboriginal knowledge of the mammals of the central deserts of Australia. *Australian Wildlife Research* 15: 9–39.

Burney, D. A. 1999. Rates, patterns, and processes of landscape transformation and extinction in Madagascar. In R. D. E. MacPhee, ed., *Extinctions in near time: Causes, contexts, and consequences.* New York: Kluwer / Plenum.

Calaby, J. H. and A. K. Lee. 1989. The rare and endangered rodents of the Australasian region. In W. Z. Lidicker, ed., *Rodents: A world survey of species of special concern.* IUCN Occasional Paper no. 4: 53–57.

Camacho, P. A., R. Borroto, and I. Ramos Garcia. 1995. Los capromidos de Cuba: Estado actual y perspectivas de las investigaciones sobre su sistematica. *Marmosiana* 1: 43–56.

Caughley, G. and A. Gunn. 1996. *Conservation biology in theory and practice.* Cambridge, Mass.: Blackwell Science.

Ceballos, G. and Navarro, D. 1991. Diversity and conservation of Mexican mammals. In M. A. Mares and D. J. Schmidly, eds., *Latin American mammalogy: History, biodiversity and conservation*, 167–198. Norman: University of Oklahoma Press.

Cervantes, F. A. and C. Lorenzo. 1997. *Sylvilagus insonus. Mammalian Species* 568, 4 pp. American Society of Mammalogists.

Cetti, F. 1777. *Appendice alla storia naturale dei quadrupedi di Sardegna.* Sassari.

Chapman, J. A. and G. Ceballos. 1990. The cottontails. In J. A. Chapman and J. E. C. Flux, eds., *Rabbits, hares and pikas: Status survey and conservation action plan*, 95–110. Gland, Switzerland: IUCN.

Cheke, A. S. and J. F. Dahl. 1981. The status of bats on western Indian Ocean islands, with special reference to *Pteropus. Mammalia* 45: 205–238.

Claridge, M. F., H. A. Dawah, and M. R. Wilson. 1997. *Species: The units of biodiversity.* London: Chapman & Hall.

Cole, F. R., D. M. Reeder, and D. E. Wilson. 1994. A synopsis of distribution patterns and the conservation of mammal species. *Journal of Mammalogy* 75: 266–276.

Corbet, G. B. and H. E. Hill. 1991 (3d ed.). *A world list of mammalian species.* London: British Museum and Oxford University Press.

Czech, B. and P. R. Krausman. 1997. Distribution and causation of species endangerment in the United States. *Science* 277: 1116.

Dewar, R. E. 1984. Extinctions in Madagascar: The loss of the subfossil fauna. In P. S. Martin and R. G. Klein, eds., *Quaternary extinctions: A prehistoric revolution*, 574–593. Tucson: University of Arizona Press.

Diamond, J. M. 1984. Historic extinction: A Rosetta Stone for understanding prehistoric extinctions. In P. S. Martin and R. G. Klein, eds., *Quaternary extinctions: A prehistoric revolution*, 354–403. Tucson: University of Arizona Press.

Dollmann, J. G. 1937. Mammals which have recently become extinct and those on the verge of extinction. *J. Soc. Preserv. Fauna Empire* 30: 67–74.

Ellis, R. 1991. *Men and whales.* New York: Knopf.

Fielden, H. W. 1890. Notes on the terrestrial mammals of Barbados. *Zoologist* 3(14): 52–55.

Flacourt, E. 1661. *Histoire de la Grande Isle Madagascar.* Paris: Pierre L'Amy.

Flannery, T. F. 1995a. *The future eaters.* New York: George Braziller.

Flannery, T. F. 1995b (2d ed.). Great hopping-mouse. In R. Strahan, ed., *Mammals of Australia,* 582. Washington, D.C.: Smithsonian Institution Press.

Flannery, T. F. 1995c. *Mammals of the south-west Pacific and Moluccan islands.* Ithaca, N.Y.: Comstock/Cornell University Press.

Flannery, T. F. 1995d (2d ed.). Basalt plains mouse. In R. Strahan, ed., *Mammals of Australia,* 619. Washington, D.C.: Smithsonian Institution Press.

Flannery, T. F., P. Kendall, and K. Wynn-Moylan. 1990. *Australia's vanishing mammals, endangered and extinct native species.* New South Wales: RD Press.

Flemming, C. and R. D. E. MacPhee. 1995. The quick, the dead, and the uncertain: How many recently extinct mammals are there? Abstract with program, 75th Annual Meeting of the American Society of Mammalogists.

Flemming, C. and R. D. E. MacPhee. 1997. Then and now: 20,000 years of mammalian extinctions in the New World. *Journal of Vertebrate Paleontology* 17(3 suppl.): 46A.

Flemming, C. and R. D. E. MacPhee, in prep. Re-determination of holotype of *Isolobodon portoricensis* (Cariomorpha, Capromyidae), with notes on Recent mammalian extinctions in Puerto Rico.

Forsten, A. and P. Youngman. 1982. *Hydrodamalis gigas. Mammalian Species* 165, 3 pp. American Society of Mammalogists.

Forsyth Major, C. I. 1901. The musk-rat of Saint Lucia (Antilles). *Annals and Magazine of Natural History* 7(7): 204–206.

Godfrey, L. R. 1986. The tale of the tsy-aomby-aomby. *The Sciences* (NY) (Jan.–Feb): 49–51.

Godfrey, L. R., W. L. Jungers, K. E. Reed, E. L. Simons, and P. S. Chatrath. 1997. Subfossil lemurs: Inferences about past and present primate communities in Madagascar. In S. Goodman and B. Patterson, eds., *Natural change and human impact in Madagascar,* 218–256. Washington, D.C.: Smithsonian Institution Press.

Goodwin, H. A. and J. M. Goodwin. 1973. List of mammals which have become extinct or are possibly extinct since 1600. IUCN Occasional Paper no. 8.

Graham, R. W. and E. L. Lundelius. 1984. Coevolutionary disequilibrium and Pleistocene extinctions. In P. S. Martin and R. G. Klein, eds., *Quaternary extinctions: A prehistoric revolution,* 223–249. Tucson: University of Arizona Press.

Grandidier, A. 1905. Recherches sur les lémuriens disparus et en particulier sur ceux qui vivaient á Madagascar. *Nouvelles Archives du Museum d'Histoire Naturelle* 4(7): 1–140.

Groves, C. P. 1985. An introduction to the gazelles. *Chinkara* 1: 4–16.

Groves, C. P. and T. F. Flannery. 1994. A revision of the genus *Uromys* Peters, 1867 (Muridae: Mammalia) with descriptions of two new species. *Records, Australian Museum* 46: 145–169.

Groves, C. P., G. B. Schaller, G. Amato, and K. Khounboline. 1997. Rediscovery of the wild pig *Sus bucculentus. Nature* 386: 335.

Handley, C. O. and R. H. Pine. 1992. A new species of *Coendou* Lacépède, from Brazil. *Mammalia* 56: 237–244.

Hanski, I., J. Clobert, and W. Reid. 1995. Ecology of extinction. In V. H. Heywood ed., *Global biodiversity assessment.* London: Cambridge University Press.

Harris, J. M. 1991. Family Hippopotamidae. In J. M. Harris, ed., *Koobi Fora research project,* vol. 3., 31–85. London: Clarendon Press.

Harrison, I. J. and M. L. J. Stiassny. 1999. The quiet crisis: A preliminary listing of freshwater fishes of the world that are extinct or "missing in action." In R. D. E. MacPhee, ed., *Extinctions in near time: Causes, context, and consequences.* New York: Kluwer/Plenum.

Haynes, G. and B. S. Eiselt. 1999. The power of Pleistocene hunter–gatherers: Forward and backward searching for evidence about mammoth extinction. In R. D. E. MacPhee, ed., *Extinctions in near time: Context, causes, and consequences*. New York: Kluwer / Plenum.

Heaney, L. R. and P. D. Heideman. 1987. Philippine fruit bats: Endangered and extinct. *Bats* 5(1): 3–5.

Hershkovitz, P. 1971. A new rice rat of the *Oryzomys palustris* group (Cricetinae, Muridae) from northwestern Colombia, with remarks on distribution. *Journal of Mammalogy* 52: 700–709.

Holdaway, R. N. 1996. The arrival of rats in New Zealand. *Nature* 384: 225–226.

Holdaway, R. N. 1999. Differential vulnerability in the New Zealand vertebrate fauna. In R. D. E. MacPhee, ed., *Extinctions in near time: Context, causes, and consequences*. New York: Kluwer / Plenum.

Hughes, J. B., G. C. Daily, and P. R. Erlich. 1997. Population diversity: Its extent and extinction. *Science* 278: 689–692.

Hutterer, R. 1993 (2d ed.). Order Insectivora. In D. E. Wilson and D. M. Reeder, eds., *Mammal species of the world: A taxonomic and geographic reference*, 69–130. Washington, D.C.: Smithsonian Institution Press.

Hutterer, R. and U. Hirsch. 1979. Ein neuer *Nesoryzomys* von der Insel Fernandina, Galápagos (Mammalia, Rodentia). *Bonner Zoologische Beitraege* 3: 276–283.

Hutterer, R., N. López-Martínez, and J. Michaux. 1988. A new rodent from Quaternary deposits of the Canary Islands and its relationships with Neogene and Recent murids of Europe and Africa. *Palaeovertebrata* 18: 241–262.

International Trust for Zoological Nomenclature. 1985 (3d ed.). *International Code of Zoological Nomenclature*. Berkeley: University of California Press.

IUCN. 1993. *1994 IUCN red list of threatened animals* (B. Groombridge, ed.). Gland, Switzerland: IUCN.

IUCN. 1996. *1996 IUCN red list of threatened animals* (J. Baillie and B. Groombridge, eds.). Gland, Switzerland: IUCN.

Jiménez Almaraz, T., J. Juarez Gomez, and L. Leon Paniagua. 1993. Mamíferos. In I. Luna Vega and J. Llorente Bousquets, eds., *Historía natural del parque ecológico estatal Omiltemi, Chilpancingo Guerrero, México*, 503–549. Mexico City: Universidad Nacional Autonóma de México.

Kimbel, W. H. and L. B. Martin, eds. 1993. *Species, species concepts, and primate evolution*. New York: Plenum.

Koopman, K. F. 1984. Taxonomic and distributional notes on tropical Australian bats. *Amer. Mus. Novitates* 2778.

Koopman, K. F. 1993 (2d ed.). Order Chiroptera. In D. E. Wilson and D. M. Reeder, eds., *Mammal species of the world: A taxonomic and geographic reference*, 137–232. Washington, D.C.: Smithsonian Institution Press.

Koopman, K. F. and D. W. Steadman. 1995. Extinction and biogeography of bats on 'Eua, Kingdom of Tonga. *Amer. Mus. Novitates* 3125.

Lawlor, T. 1983. The mammals. In T. J. Case and M. L. Cody, eds., *Island biogeography in the Sea of Cortez*, 265–289. Berkeley: University of California Press.

Luckman, B. H. 1986. Reconstruction of Little Ice Age events in the Canadian Rocky Mountains. *Géog. Phys. Quat.* 40: 17–28.

Mace, G. M. 1993. The status of proposals to redefine the IUCN threatened species categories. In B. Groombridge, ed., *The 1994 IUCN red list of threatened animals*, xlviii–lv. Gland, Switzerland: IUCN.

MacPhee, R. D. E. 1997a. Digging Cuba: The lessons of the bones. *Natural History* 106(11): 50–54.

MacPhee, R. D. E. 1997b. Vertebrate paleontology of Jamaican caves. In A. Fincham, ed., *Jamaica underground: The caves, sinkholes, and underground rivers of the island.* (2d ed.), 47–56. Kingston, Jamaica: The Press University of the West Indies.

MacPhee, R. D. E. and J. G. Fleagle. 1991. Postcranial remains of *Xenothrix mcgregori* (Primates, Xenotrichidae) and other late Quaternary mammals from Long Mile Cave, Jamaica. *Bulletin of the American Museum of Natural History* 206: 287–321.

MacPhee, R. D. E. and C. Flemming. 1997. Brown-eyed, milk-giving, and extinct: Losing mammals since A.D. 1500. *Natural History* 106(3): 84–88.

MacPhee, R. D. E. and C. Flemming. In press. Crossing the irremeable stream: An annotated catalogue of mammalian species extinctions since A.D. 1500.

MacPhee, R. D. E. and P. A. Marx. 1997. The 40,000-year plague: Humans, hyperdisease, and first-contact extinctions. In S. Goodman and B. Patterson, eds., *Natural change and human impact in Madagascar*, 169–217. Washington, D.C.: Smithsonian Institution Press.

MacPhee, R. D. E. and M. Rivero de la Calle. 1996. Accelerator mass spectrometry ^{14}C age determination for the alleged "Cuban spider monkey," *Ateles* (= *Montaneia*) *anthropomorphus. Journal of Human Evolution* 30: 89–94.

MacPhee, R. D. E., I. Horovitz, O. Arredondo, and O. Jiménez Vasquez. 1995. A new genus for the extinct Hispaniolan monkey *Saimiri bernensis* Rímoli, 1977, with notes on its systematic position. *Amer. Mus. Novitates* 3134.

MacPhee, R. D. E., C. Flemming, and D. P. Lund. 1999. "Last occurrence" of the Antillean insectivoran *Nesophontes*: New Radiometric dates and their interpretation. *Amer. Mus. Novitates* 3264.

Mahé, J. and M. Sourdat. 1972. Sur l'extinction des vertébrés subfossiles et l'aridification du climat dans le Sud-Ouest de Madagascar. *Bulletin Société Géologique de France* 14: 295–309.

Mahoney, J. A. 1975. *Notomys macrotis* Thomas, 1921, a poorly known Australian hopping mouse (Rodentia: Muridae). *Journal of the Australian Mammalogy Society* 1: 367–374.

Marsh, R. E. 1985. More about the Bajan mouse. *Journal of the Barbados Museum and Historical Society* 37(3): 310.

Martin, P. S. 1984. Prehistoric overkill: The global model. In P. S. Martin and R. G. Klein, eds., *Quaternary extinctions: A prehistoric revolution*, 354–403. Tucson: University of Arizona Press.

Martin, P. S. and D. W. Steadman. 1999. Prehistoric extinctions on islands and continents. In R. D. E. MacPhee, ed., *Extinctions in near time: Causes, contexts, and consequences.* New York: Kluwer/Plenum.

May, R. M., J. H. Lawton, and N. E. Stork. 1995. Assessing extinction rates. In J. H. Lawton and R. M. May, eds., *Extinction rates*, 1–24. London: Oxford University Press.

McFarlane, D. A. 1999. A comparison of methods for the probabilistic determination of vertebrate extinction chronologies. In R. D. E. MacPhee, ed., *Extinctions in near time: Causes, contexts, and consequences.* New York: Kluwer/Plenum.

Miller, G. S. 1929a. A second collection of mammals from caves near St. Michel, Haiti. *Smithsonian Miscellaneous Collections* 81(9).

Miller, G. S. 1929b. Mammals eaten by Indians, owls, and Spaniards in the Coast Region of the Dominican Republic. *Smithsonian Miscellaneous Collections* 82(5).

Miller, G. S. 1930. Three small collections of mammals from Hispaniola. *Smithsonian Miscellaneous Collections* 82(15).

Morgan, G. S. 1989. *Geocapromys thoracatus. Mammalian Species* 341. *American Society of Mammalogists.*

Morgan, G. S. 1993. Quaternary land vertebrates of Jamaica. In R. M. Wright and E. Robinson, eds., *Biostratigraphy of Jamaica. Memoir, Geological Society of America* 182: 417–442.

Morgan, G. S. 1994. Late Quaternary fossil vertebrates from the Cayman Islands. In M. A. Brunt and J. E. Davies, eds., *The Cayman Islands: Natural history and biogeography,* 465–508. Boston: Kluwer.

Morgan, G. S. and C. A. Woods. 1986. Extinction and the zoogeography of West Indian land mammals. *Biological Journal of the Linnaean Society* 28: 167–203.

Musser, G. G. and M. D. Carleton. 1993 (2d ed.). Family Muridae. In D. E. Wilson and D. M. Reeder, eds., *Mammal species of the world: A taxonomic and geographic reference,* 501–756. Washington, D.C.: Smithsonian Institution Press.

Nordenskiold, A. E. 1881. *The voyage of the Vega round Asia and Europe,* 2 vols. (trans. by A. Leslie). London: Macmillan.

Olson, S. L. 1981. Natural history of vertebrates on the Brazilian islands of the mid South Atlantic. *Nat. Geogr. Soc. Res. Rep.* 13: 481–489.

Olson, S. L. 1989. Extinction on islands: Man as a catastrophe. In D. Western and M. Pearl, eds., *Conservation for the twenty-first century,* 50–53. New Haven: Oxford University Press.

de Oviedo y Valdez, G. F. 1535. *Natural history of the West Indies* (trans. and ed. by S. A. Stroudemire, 1959). University of North Carolina Studies in Romance Language Literature 32: 1–140.

Owen-Smith, R. N. 1988. *Megaherbivores: The influence of very large body size on ecology.* London: Cambridge University Press.

Owen-Smith, R. N. 1999. The interaction of humans, megaherbivores and habitats in the late Pleistocene extinction event. In R. D. E. MacPhee, ed., *Extinctions in near time: Causes, contexts, and consequences.* New York: Kluwer/Plenum.

Patton, J. L. and M. S. Hafner. 1983. Biosystematics of the native rodents of the Galapagos archipelago, Ecuador. In R. I. Bowman, M. Berson, and A. E. Leviton, eds., *Patterns of evolution in Galapagos organisms,* 539–568. San Francisco: AAAS.

Pielou, E. C. 1991. *After the Ice Age: Return of life to glaciated North America.* Chicago: University of Chicago Press.

Pino, M. and N. Castellanos. 1985. Acerca de la asociación de perezosos cubanos extinguidos con evidencias culturales de aborígenes cubanos. *Reporte de Investigación del Instituto Ciencias Sociales (La Habana)* 4.

Pregill, G. K., D. W. Steadman, and D. R. Watters. 1994. Late Quaternary vertebrate faunas of the Lesser Antilles: Historical components of Caribbean biogeography. *Bulletin of the Carnegie Museum of Natural History* 30: 1–51.

Primack, R. B. 1995. *A primer of conservation biology.* Sunderland, Mass.: Sinauer Associates.

Ray, C. E. 1964. A small assemblage of vertebrate fossils from Spring Bay, Barbados. *Journal of the Barbados Museum and Historical Society* 31(1): 11–22.

Rounsevell, D. E. and N. Mooney. 1995 (2d ed.). Thylacine. In R. Strahan, ed., *Mammals of Australia,* 164–165. Washington, D.C.: Smithsonian Institution Press.

Schomburgk, R. H. 1848. *The history of Barbados.* London: Longmans.

Schroering, G. B. 1995. Swamp deer resurfaces. *Wildlife Conservation* (Dec.): 22.

Signor, P. W. and J. H. Lipps. 1982. Sampling bias, gradual extinction patterns and catastrophes in the fossil record. *Special Paper, Geological Society of America* 190: 291–296.

Silva, M. and J. A. Downing. 1995. *CRC handbook of mammalian body masses*. Boca Raton, Fla.: CRC Press.

Simons, E. L. 1997. Lemurs: Old and new. In S. M. Goodman and B. D. Patterson, eds., *Natural change and human impact in Madagascar*, 142–168. Washington, D.C.: Smithsonian Institution Press.

Sloane, H. 1707–1725. *A voyage to the islands Madera, Barbados, Nieves, S. Christophers and Jamaica*, 2 vols. London: British Museum.

Smith, A. P. and D. G. Quin. 1996. Patterns and causes of extinction and decline in Australian conilurine rodents. *Biology and Conservation* 77: 243–267.

Smith, M. J. 1995 (2d ed.). Toolache wallaby. In R. Strahan, ed., *Mammals of Australia*, 339–340. Washington, D.C.: Smithsonian Institution Press.

Standing, H. 1908. On recently discovered subfossil primates from Madagascar. *Transactions of the Zoological Society of London* 18: 69–162.

Steadman, D. W. 1995. Prehistoric extinctions of Pacific island birds: Biodiversity meets zooarchaeology. *Science* 267: 1123–1131.

Steadman, D. W., T. W. Stafford, D. J. Donahue, and A. J. T. Jull. 1991. Chronology of Holocene vertebrate extinctions in the Galápagos Islands. *Quaternary Research* 36: 126–133.

Stiassny, M. L. J. 1994. Systematics and conservation. In G. K. Meffe and C. R. Carroll, eds., *Principles of conservation biology*, 64–66. Sunderland, Mass.: Sinauer Associates.

Stuiver, M., and Reimer, P. J. 1993. Extended ^{14}C data base and revised CALIB 3.0 ^{14}C age calibration. *Radiocarbon* 35: 215–230.

Tattersall, I. 1982. *The primates of Madagascar*. New York: Columbia University Press.

Times Atlas of the World. 1994 (9th ed.). Toronto: Random House of Canada.

Utzurrum, R. C. B. 1992. Conservation status of Philippine fruit bats (Pteropodidae). *Silliman Journal* 36: 27–45.

Vigne, J.-D. 1988. *Les Mammifères post-glaciaires de Corse: Etude archéozoologique*. Paris: Centre National de la Recherche Scientifique.

WCMC. 1992. *Global biodiversity: Status of the Earth's living resources*. London: Chapman & Hall.

Williams, J. D. and R. M. Nowak. 1986. Vanishing species in our own backyard: Extinct fish and wildlife of the United States and Canada. In L. Kaufman and K. Mallory, eds., *The last extinction*, 107–140. Boston: Massachusetts Institute of Technology Press.

Williams, J. D. and R. M. Nowak. 1993 (2d ed.). Vanishing species in our own backyard: Extinct fish and wildlife of the United States and Canada. In L. Kaufman and K. Mallory, eds., *The last extinction*, 115–148. Boston: Massachusetts Institute of Technology Press.

Wilson, D. E. and D. M. Reeder, eds. 1993 (2d ed.). *Mammal species of the world: A taxonomic and geographic reference*. Washington, D.C.: Smithsonian Institution Press.

Woods, C. A. 1984. Hystricognath rodents. In S. Anderson and J. K. Jones, eds., *Orders and families of recent mammals of the world*, 389–446. New York: Wiley.

Woods, C. A. 1989a. A new capromyid rodent from Haiti: The origin, evolution, and extinction of West Indian rodents and their bearing on the origin of New World hystricognaths. In C. C. Black and M. R. Dawson, eds., *Papers on fossil rodents in honor of Albert Elmer Wood*, 59–89. Los Angeles: Natural History Museum of Los Angeles County.

Woods, C. A. 1989b. Endemic rodents of the West Indies: The end of a splendid isolation. In W. Z. Lidicker, ed., *Rodents: A world survey of species of special concern*. IUCN Occasional Paper no. 4: 11–19.

Woods, C. A. 1993 (2d ed.). Suborder Hystricognathi. In D. E. Wilson and D. M. Reeder, eds., *Mammal species of the world: A taxonomic and geographic reference*, 771–806. Washington, D.C.: Smithsonian Institution Press.

Yoshiyuki, M. 1989. *A systematic study of the Japanese Chiroptera*. Tokyo: National Science Museum.

Zahler, P. 1996. Rediscovery of the woolly flying squirrel (*Eupetaurus cinereus*). *Journal of Mammalogy* 77: 54–57.

PERSPECTIVE

SOIL BIODIVERSITY: LIFE IN SOIL

Diana H. Wall

Soils, like air and water, are natural resources that integrate terrestrial ecosystems in the biosphere and, like water and air, have been continually degraded by humans. Franklin D. Roosevelt once stated, "The nation that destroys its soil destroys itself." In the generations since this declaration, without knowing how resilient or stable soil ecosystems are, we have caused soil pollution, groundwater contamination, and erosion, resulting in a loss of carbon and nutrients from the soil. Today, scientists realize that soil is not a buffer, that soils and their teeming biota unite terrestrial and aquatic ecosystems.

Soils are a critical and dynamic center for the majority of processes in both natural and managed ecosystems. Soil fertility, nutrient turnover, nutrient uptake by plants, formation of soil organic matter, nitrogen fixation, methane production, CO_2 production, soil development, and production of organic acids that weather rocks are all processes involving soil biota that are necessary for the sustainability of the planet. The major global storage reservoir for carbon in the form of organic matter is in soils (estimates of about $1,500 \times 10^{15}$ g C are stored in soils). For comparison purposes, this translates to the top meter of soil having twice the organic matter found above ground. The living microbes, fungi, invertebrates, and vertebrates in soils are responsible for changing carbon and nitrogen, through several complex steps of decomposition to forms available for plant growth, while contributing to the rate of production and consumption of CO_2, methane, and nitrogen. The annual flux of CO_2 that returns to the atmosphere as a result of decomposition and other soil processes amounts to approximately 68×10^{15} g C/yr (Schlesinger 1991).

Soil biota remain among the vast number of unknown life forms on our planet. For most taxa, our best estimates of the extent of diversity are provisional, yet estimates indicate that the soil biota are as diverse as the above-ground taxa. Almost every phylum that exists above-ground can be found represented in soils. It is therefore difficult to say which organisms or groups of organisms are most important in relation to critical ecological processes and soil attributes, such as stability and resilience, or to determine the impact of species loss. Experiments in microcosms have generally been limited to five or fewer species, providing a valuable but limited estimate of the role of species interactions in soils. Field experiments have historically studied processes by assigning organisms to large functional groups based on size, biomass, and trophic group.

As an example of the vast diversity of organisms below-ground, it was estimated, using molecular methods, that 100 g of a beech forest soil from western Norway contained more than 10,000 bacterial genomic species (more than 70 percent homology in their DNA). Yet because many of these organisms cannot be cultured with present techniques, their ecology remains largely unrevealed. Even with groups that have been well studied, we still have only an elementary knowledge of their geographic diversity. For example, estimates indicate that only 30 to 35 percent of oribatid mites in North America have been described; if they are added to estimates of undescribed oribatids from tropical zones, there may be as many as 100,000 undescribed species within this single group (Heywood 1995). One fact emerges from these examples: our knowledge of the baseline biodiversity of soil biota is extremely limited. Without accurate knowledge of soil biodiversity, our understanding of species distributions, composition, and community structure, the interactions of these species, and how these taxa constrain or control ecosystem processes (i.e., carbon cycling, global change, nutrient cycling) will remain rudimentary. Questions of endemism, introductions of species, and species extinction are unexplored for many kingdoms of the soil biota.

Traditionally, there have been many obstacles to the study of key soil species and their ecological roles in ecosystems. Soil is an opaque medium, making the in situ identification of most organisms impractical. The taxa of soil food webs change with the physicochemical environment, the quality of organic matter, climate, and geography, resulting in few comparisons of the ecological roles of soil taxa in different ecosystems. It is not surprising, then, that the greatest barriers to research in soil biology are methodological. Sampling and identification methods are taxon specific and many techniques are in their developmental infancy. No single extraction or collection method will quantitatively extract or collect all soil organisms, or even one phylum, presenting problems for identification and enumeration. Molecular techniques are now widely used by soil biologists and they promise faster advances in the identification of species.

There are other constraints to the study of biota in soils. The number of phyla in soils (microbes to earthworms) makes interactions and ecological roles in soil much more complex to study than their larger counterparts above-ground. The sizes of the organisms in soil are microscopic to macroscopic, and a taxon's morphology can change throughout the life cycle. Scale, both temporal and spatial (soil particles to landscape), varies with the taxa, further making analyses difficult. Perhaps more than in above-ground research, the lack of emphasis and funding has created an extinction of the few systematists who can identify the organisms involved in critical roles in soil. However, this has contributed to a collaboration between ecologists who have considerable expertise in determining genus and some species-level taxa, and systematists who maintain collections and use molecular techniques for the study of relationships. Nevertheless, ecological roles attributed to the soil biota are most often based on trophic groups or groups of species with similar morphology, not to species. Consequently, the study of soil biota has limited ability to establish the cause-and-effect relationships between the loss of species and the impact on terrestrial and global ecosystem processes.

The present taxonomic knowledge of soil biota has been recently summarized (Groombridge 1992; Freckman 1994; Hawksworth and Ritchie 1993; O'Donnell et al. 1994; Systematics Agenda 2000 1994; Heywood 1995). The status of identification and ecological roles of the soil biota, by size, could be assessed as follows (Freckman 1994):

• *Larger soil fauna (invertebrates).* These can be collected quantitatively and qualitatively from soils, many may be identified to species, and their ecological roles are known in general. These roles include direct processors of organic matter (e.g., snails, earthworms, enchytraeid worms, wood lice, millipedes, silverfish, bristletails, termites), predaceous regulators (e.g., spiders, centipedes, true bugs, carabid beetles, ants), secondary consumers (e.g., springtails, mites, other beetles), and creators of soil structure (earthworms, millipedes, termites, and many members of other categories). Taxa that cannot currently be reliably identified include enchytraeid worms, many mites, larval beetles and flies, parasitic wasps, and bark lice. Knowledge of these soil organisms varies dramatically with locale. Only a few locations have well-described invertebrates.

• *Micro- and mesofauna (invertebrates).* Assays vary in the ability to quantitatively and qualitatively extract these organisms from the soil. Knowledge of the ecological functions of this group is generally lacking. Many are predators, consumers of bacteria and fungi, and are involved in regulating the rate of decomposition. Springtails (Collembola) and other insects appear to have a reasonable base of taxonomic specialists, although in some taxa only one or two such individuals may exist. However, reliable identifications may be impossible or difficult to obtain for many groups of protozoa, rotifers, tardigrades, nematodes, and mites. Few molecular methods are available for these diverse taxa and their ecological roles are based primarily on trophic group estimates of their ecological function in ecosystem processes.

• *Microbes.* Advances over the past 10 years have been substantial and additional methods are available for the assessment of bacterial and fungal biodiversity. However, each method suffers from technical or interpretative limitations, and no single method provides an unequivocal estimate of bacterial or fungal diversity. Species that can be identified by culture techniques or visual techniques are not necessarily important in situ. Some new research uses indirect methods to identify, in situ, which microbial groups are linked to important processes and then, ex situ, uses new methods to determine the species diversity of these groups. Traditional morphological methods combined with molecular identifications provide important tools for the assessment of fungal diversity. As with bacteria, assays of fungal chemical diversity and new technology can contribute to identification.

Soil biodiversity generally is greater than above-ground diversity, even after human disturbance. Species composition within soil food webs may change because of this disturbance, making the impact of species loss more difficult to determine. However, that anthropogenic activities can decrease soil biodiversity is well documented, particularly in the fauna of agroecosystems, where the addition of fertilizers increases plant productivity but masks the importance of soil biota in providing nutrients to plants. Complementary evidence from experiments has shown that increases in biodiversity can enhance plant growth, nutrient mineralization, and resistance to stress (Clarholm 1989; Couteaux et al. 1991; Elliott et al. 1979; Lavelle et al. 1992). A few experiments have indicated that a loss of soil biodiversity can diminish the functioning of ecosystem processes (Verhoef and Brussaard 1990).

Nevertheless, even the limited knowledge available on the importance of soil biodiversity has immense economic significance. Agricultural, forestry, and environmental research has long been directed toward maintaining soil fertility and enhancement of beneficial soil organisms (earthworms, mycorrhizae, rhizobia, entomophagous nematodes, rhizosphere bacteria, and bioremediation using bacterial species) while decreasing soil plant pathogens and the introduction of exotic organisms. Global modifications such as land use and climate change have major effects on soils, ecosystems, our economic base, and, ultimately, our future. These changes alter the balance of the carbon fluxes, affecting the spatial and temporal distribution of ecosystem resources, vegetation diversity and landscape vegetation patterns, and the soil biota involved in processes such as decomposition and the release of greenhouse gases. Future research studies of soil biodiversity and the ecological role of taxa at different scales of resolution (genetic to landscape) and in different biomes would provide a model for comparison with other systems and would transfer knowledge about soils into studies of conservation and resource use in both managed and unmanaged systems.

REFERENCES

Clarholm, M. 1989. Effects of plant–bacterial–amoebal interactions on plant uptake of nitrogen under field conditions. *Biology and Fertility of Soils* 8: 373–378.
Couteaux, M. M., M. Mousseau, M.-L. Celerier, and P. Bottner. 1991. Increased atmospheric CO_2 and litter quality: Decomposition of sweet chestnut leaf litter with animal food webs of different complexities. *Oikos* 61: 54–64.

Elliott, E. T., D. C. Coleman, and C. V. Cole. 1979. The influence of amoebae on the uptake of nitrogen by plants in gnotobiotic soil. In J. L. Harley and R. Scott Russell, eds., *The soil-root interface*, 221–229. London: Academic Press.

Freckman, D. W., ed. 1994. Life in the soil. Soil biodiversity: Its importance to ecosystem processes. Report of a workshop held at the Natural History Museum, London.

Groombridge, B., ed. 1992. *Global biodiversity: Status of the earth's living resources*. London: Chapman & Hall.

Hawksworth, D. L. and J. M. Ritchie. 1993. *Biodiversity and biosystematic priorities: Microorganisms and invertebrates*. Wallingford, U.K.: CAB International.

Heywood, V. H., ed. 1995. *Global biodiversity assessment*. Cambridge, U.K.: Cambridge University Press.

Lavelle, P., E. Blanchart, A. Martin, A. V. Spain, and S. Martin. 1992. Impact of soil fauna on the properties of soils in the humid tropics. In R. Lal and P.A. Sanchez, eds., *Myths and science of soils of the tropics*. SSSA Special Publication no. 29. Madison, Wisc.: Soil Science Society of America.

O'Donnell, A. G., M. Goodfellow, and D. L. Hawksworth. 1994. Theoretical and practical aspects of the quantification of biodiversity among microorganisms. *Philosophical Transactions of the Royal Society of London* 345B: 65–73.

Schlesinger, W. H. 1991. *Soil warming experiments in global change research*. Report of a workshop held in Woods Hole, Massachusetts, September 27–28, 1991. National Science Foundation Ecosystem Studies Program.

Systematics Agenda 2000. 1994. *Systematics Agenda 2000: Charting the biosphere*. New York: American Museum of Natural History.

Verhoef, H. A. and L. Brussaard. 1990. Decomposition and nitrogen mineralization in natural and agroecosystems: The contribution of soil animals. *Biogeochemistry* 11: 175–211.

PERSPECTIVE

PETRUCHIO'S PARADOX: THE OYSTER OR THE PEARL?

G. Carleton Ray

Petruchio's impassioned speech to Katharina in *Taming of the Shrew* (Act IV, Scene iii) encapsulates a conservation dilemma: "What, is the jay more precious than the lark, / Because his feathers are more beautiful? / Or is the adder better than the eel, / Because his painted skin contents the eye?"

Is conservation in the eye of the beholder? Is the song of the lark more to be valued than the silence of a rose? Should the richness of the painted coral reef take precedence over the reefs formed by the succulent and possibly endangered oyster? What portion of the landscape can be defined as rich? Do little things run the world (Wilson 1987) or is it the big things that really matter (Terborgh 1988)?

EARTH OR SEA?

Does our evolution as giant land animals cloud our view? Human terrestrial evolution has caused this blue, water-planet to be called Earth, a slip of mind over matter that can lead to egregious bias. For example, we have been told that the 7 percent of land that is tropical forests houses more species than any other place. With images of forests being torn and burnt before our eyes, these places have rightly gained high conservation priority. But behind this choice is an ironic process of rejection. The cover of *Bioscience* (October 1994) displays striking images of postagricultural changes of the terrestrial earth at human hands, over the title, "Where in the *World* Are the Conservation Crises" (emphasis mine). The accompanying article, by Stanford University's Center for Conservation Biology, purports to be a "global analysis of the distribution of biodiversity and the expansion of the human enterprise." Yet their globe is one without water, fresh or salt.

Which systems are really most diverse, and by what measure? Although most *described* species are terrestrial—no surprise there—the greatest phyletic, thus also genetic, diversity is marine (Ray 1985, 1988; May 1988; Grassle et al 1991). The terrestrial realm is occupied by only 11 phyla (only one, perhaps, endemic, the status of the Onychophora being arguable). This compares with about 28 marine (13 endemic), 17 symbiotic (4 endemic), and 14 freshwater (none endemic). Put in more tangible terms, a single clump of oysters in the Chesapeake Bay may host, attached to its shells, more animal phyla than all of the land, worldwide (Ray 1996).

More specifically, Gleick (1993) offers that of all the world's water, 97.6 percent is salty and only 0.0093 percent fresh, translating to a teaspoon in a bathtub. Yet 12 percent of all animal species live in this small, freshwater piece of the globe, which contains a proportion of biodiversity to space that is exceeded nowhere else. If that were not enough, consider that freshwater extinction rates are probably higher than for any other Earth system and that depletion of freshwater resources continues to be catastrophic. Gleick reminds us that even in "healthy" North America, 30 percent of fish species are threatened, endangered, or of special concern, and of the 108 species of mussels in the Ohio River basin alone, 39 percent are either extinct or endangered.

What does the logic of naming this planet Earth seemingly omit? Merely water, where life originated, where it is richest, and where it may be most endangered. It appears that our problem is perception. We earthly creatures can witness the depletion and destruction of the land before our eyes. The seas present an umbral surface into which we peer only with difficulty, and usually with ignorance. As I have previously said (Ray 1988:45), "The last fallen mahogany would lie perceptibly on the landscape, and the last black rhino would be obvious in its loneliness, but a marine species may disappear beneath the waves unobserved and the sea would seem to roll on the same as always." Furthermore, on the land we have long observed the massive depletion of Earth's biological capital, notably of soils, forests, and large mammals, but only recently have world-wide increases of "dead," anoxic zones and toxic phytoplankton blooms become obvious in the sea. We still can only guess what the effects might be of the "strip-mining" of marine fishes, which may be at least the equal of deforestation in its effect on biodiversity. In sum, what is gone we may not know, and what seems healthy may not be so.

I do not bring these matters to attention to exacerbate a terrestrial–marine competition or to question whether the harpy eagle or the whale is more to be valued, or even to suggest that we rename our planet Sea. The whole world is too rich and interdependent for that. But we must admit that we have no consistent way to establish one place or species over another as highest in priority, other than bias and familiarity. Even the data for extinction are highly uncertain (Smith et al 1993). It seems that unless we devise priorities on a sound scientific basis, Petruchio will remain with us.

SEEKING THE MAGIC BULLET

New discoveries of species and the facts and fears of extinction have recently brought the diversity of life on our small planet into critical, worldwide focus. We have been brought to astonishment by the discovery of seemingly unending life forms, including those within the ocean's depths. Because loss of species is so tangibly tragic and permanent, conservation of species, particularly the charismatic and endangered, has become an end in itself. From this perspective, it follows that we should identify hot spots for protection of maximal species diversity.

Hot-spot criteria appear in many conservation publications (e.g., species richness, high endemism, and critical habitats for reproduction, nurseries, feeding, or other biological and ecological processes). An outstanding example of the application of the species-richness criterion is provided by the ICBP (1992:6), whose contention is that birds have "dispersed to, and diversified in, all regions of the world" and "occur in virtually all habitat types and attitudinal zones." Therefore, birds are assumed to be surrogates for biodiversity in general. There are two serious problems with this approach. First, the general assumption behind hot spots is ecological stasis, which is to say that environmental change is not sufficiently considered. All places will inevitably change location and character, probably in a shorter time than we may think. Second, the ICBP's own analysis shows clearly that birds are not surrogates for other vertebrates, such as reptiles and amphibians, Further, Prendergast et al. (1993) found that in Britain, where data are probably as good as can be found anywhere, only 12% of the bird and butterfly hot spots coincide. Interestingly, in their analysis, "cold spots" do not coincide either.

Does it not seem obvious that if hot spots were to be identified for *all* taxa—ants, termites, nematodes, plants, and the rest—the entire earth would be included, probably more than once over? Put in reverse, there are no walruses in the

Amazon, no parrots in Greenland, no snow "fleas" (springtails of the order Collembola) on the desert, no mighty oaks in the sea's abyss, and no flashlight fishes in the lofty Alps. That is, through more than four billion years of biological evolution, from prokaryotes to monkeys and whales, every portion of Earth has become critical habitat for some form of life. Clearly, a different conservation paradigm is required.

THE OYSTER OR THE PEARL?

Most people, including scientists, are drawn inexorably to warm and user-friendly places, where species are presumably most varied and beautiful. Thus the tropics have become almost synonymous with richness, productivity, the good life, and diversity. But some of us are just as drawn to the cool clarity, color, and diversity of the higher latitudes. As Seton (1911:244) said for arctic prairies, "I never before saw such a realm of exquisite flowers so exquisitely displayed, and the effect at every turn throughout the land was colour, colour, colour. . . . What Nature can do only in October, elsewhere, she does here all season through, as though when she set out to paint the world she began on the Barrens with a full palette and when she reached the tropics had nothing left but green."

Or, consider the lowly, brainless, songless oyster, about which Jonathan Swift remarked early in the eighteenth century, "He was a bold man that first eat an oyster." More optimistically, Richard Sheridan declared, once eaten, "An oyster may be crossed in love." Even Shakespeare was ambivalent about oysters: "Rich honesty dwells like a miser, sir, in a poor house, as your pearl in your foul oyster" (*As You Like It*) or "Why, the world's mine oyster, / Which I with sword will open" (*Hamlet*). In more modern times, oysters have made fortunes, as have oyster pearls. Of course, it must be noted that temperate oysters produce the flesh and a tropical oyster, blessed with a nacre-producing organ, produces the pearl. But both molluscs have been valued principally as products, with little thought given to the environments that support them.

Oysters filter and clarify water. They also build reefs, which are as important structural components of estuaries as coral reefs are to tropical waters. Through their activities, oysters can change the very nature of the estuaries in which they live. Thus they have the potential to influence the metapopulation dynamics of fishes and invertebrates, both of estuaries and throughout the near-shore coastal zone. We hypothesize that in the continuum of watershed and estuarine dynamics, in which this creature plays such an important ecological and economic role, the oyster is key (Ray et al. 1997). It is the metaphoric, ecological pearl. Yet following two centuries of overharvest, greed, and mismanagement (or "dismanagement") the oysters of many estuaries have become insignificant, both economically and ecologically. We now know that the oyster-rich Chesapeake Bay of John Smith was a very different place than now exists (McCormick-Ray 1998).

The Chesapeake Bay has become, over much of its expanse, nearly oysterless, a murky, erosion-prone body (a cause–effect relationship is suspected, but not proven). Surely, the bay still produces much and still bears semblances of richness and solitude. It is still valued for commerce and remains a playground for boaters and for swimmers, except during the summertime of stinging nettles. Seen only from a modern perspective, there is much value left, but the historical retrospect is dismal. This signifies that a principal culprit for the demise of marine living resources and biodiversity over the centuries is not merely greed, politics, and indecision, but also science itself. Fishery science has been devised to serve commerce and remains largely driven by output-side yield models. That is, the input-side, the ecosystem itself, still receives too little attention. To purloin Moore's title, "science as a way of knowing" is overwhelmed:

> We have reached the point in history, therefore, when biological knowledge is the *sine qua non* for a viable human future. Such knowledge will be especially necessary for the leaders of society—in government, industry, business, and education—but the tough decisions will have to be supported by an informed electorate. A critical subset of society will have to understand the nature of life, the interactions of living creatures with their environment, and the strengths and limitations of the data and procedures of science itself. The acquisition of biological knowledge, for so long a luxury except for those concerned with agriculture and the health sciences, has now become a necessity for all. (Moore 1993:4)

The reality is that knowing life and its processes is indeed the sine qua non of sustaining life, including biological diversity. Therefore, we urgently need to reexamine our biases and apply the rapidly growing knowledge of both species natural history and ecosystem dynamics, and to look beyond the narrowly bounded traditions of conservation toward entire regional ecosystems to achieve our conservation goals.

ABOUT FACE!

I contend that hot-spot conservation tactics reveal two fatal flaws. The first is the way we tend to think. An insight into Petruchio's questions may lie in Alfred North Whitehead's notion of the "fallacy of misplaced concreteness [that] flourishes because the disciplinary organization of knowledge requires a high level of abstraction" (Daly and Cobb 1989:25). In this case, the abstraction is the species. The second flaw is the way conservation often operates. We can hardly know how fast extinction is progressing, especially because we do not know how many species exist. Despite this limitation, we must accept the ubiquity of critical habitat. The obvious way out of this problem is to look beyond species to ecosystem management, defined by Franklin (1997) as "managing ecosystems so as to assure their sustainability."

Ecosystem sustainability is key, as history tells us. Only a millennium or two ago, humans were inconsequentially fragmented over the land and the oceans were barely touched or known. Now, humans dominate the planet, ecosystems are fragmented, and the "Marine Revolution" is under way (Ray 1970; Parsons and Seki 1995). Functionally impaired ecosystems are now less able to sustain the products, species, or ecological functions that constitute biodiversity or that contribute to human well-being. This recognition inevitably alters our management and conservation imperatives. Set-aside preservation is still essential for specific, narrow purposes, but much more comprehensive management and even hands-on restoration are now higher priorities. The conservationist's scope must expand, upscale and hierarchically, from species to land- and seascapes and to entire regions and biogeographic provinces. Furthermore, humans must be treated as part and parcel of the land and seascape.

Surely we dare not let go of condors, whales, or the Appalachian Blue Ridge, where fall colors and timber rattlesnakes can exist side by side. There is simply no excuse to eliminate a species knowingly, either by exploitationist development, social causes (jobs), conservation bias, or ignorance. The endangered species paradigm must remain, but with acknowledgment of what it can do and what it cannot. Nor can we ignore the need to take a broader view. Species and protected areas alike may hang on within a sea of pollution, people, and misguided environmental engineering. But the simple truth is that pockets of species and habitats are not viable in the long term unless major changes in social and conservation perspectives occur.

Biodiversity is a very old portion of our intellectual history. We have inherited an extraordinary lineage from at least Aristotle and Pliny the Elder to the works of Linnaeus (taxonomy), Cuvier (paleontology), Humboldt (biogeography), Darwin (evolution), Mendel (genetics), Elton (ecology), and others. Despite this long lineage, scientists have only recently conjectured that sustainability will ultimately depend on knowledge of how nature works, which means how biodiversity is distributed, how it originates, how it is maintained, and, most importantly, what the relationship is between species and community diversity and ecological function.

Thus, while keeping a keen eye on species, we must also view biodiversity as more ecological than biological, more evolutionary than descriptive, and more dynamic than static. As Bartholomew (1986) reminded us, "Indifference to a phenomenon's natural context can result in a paralyzing mismatch between the problem and the questions put to it." I interpret this to mean that it is essential to recognize that ecosystems are what we must manage, not species one at a time, but together as functioning assemblages, interacting with their abiotic world.

Franklin (1993) is emphatic on this score: "Efforts to preserve biodiversity must focus increasingly at the ecosystem level because of the immense number of species, the majority of which are currently unknown. . . . The ecosystem approach is the only way to conserve organisms and processes in poorly known or unknown habitats and ecological subsystems." In other words, the sustainability of systems is the goal, encompassing all scales and based on concepts of adaptive management. Not everyone agrees. Tracy and Brussard (1994) assert that the identification of umbrella species as "coarse filters" is the challenge, and besides, that it is "terrifically difficult to define an ecosystem." Franklin (1994) responds, I think correctly, that umbrellas do not really work and that we must maintain as much as we can of biodiversity and "not simply those relatively few species that we choose to recognize."

Just on the face of the matter, it seems not logical, and certainly not eco-logical, to assume that a small-scale attribute (species) can cover a large-scale one (ecosystem). For example, which oceanic species might we choose: a whale or a diatom? What may each cover? In fact, species (great or small) do not necessarily travel from habitat to habitat via corridors, as the chosen species or umbrella design often emphasizes, but have many ways to "percolate" through the land- or seascape (Gardner et al. 1992). As for ecosystems, land- and seascape pattern is the surrogate for biodiversity. Ecological pattern is describable in time and space, and biodiversity may be measured and its significance assessed within these dimensions. Strong evidence is emerging that if the focus in on patterned systems and the biodiversity within, then an insight into processes may be achieved, whereas the reverse is not true. For example, Tilman (1996) found that biodiversity may stabilize community and ecosystem processes, but perhaps not population processes; that is, populations may exhibit great variability, even when processes are stable. This finding, if substantiated, is a portion of the essence for sustainable conservation and management.

In conclusion, biological diversity is everywhere. Even city folk seem to crave it. I can think of no better expression of this than Dasmann's (1968:179) classic, *A Different Kind of Country:*

> After spending an hour or more examining the wonders of this new city center I felt depressed—by the absence of people and of life except for trees, shrubs, and flowers growing in greater or lesser concrete pots. I moved to what was left of old Philadelphia, into the narrow streets, the dirty old converted town houses, the jumble of shops and theaters and the mixtures of older tall buildings. Here, and not in the new malls, was where the people were—crowding the sidewalks, moving into theaters or pubs, traveling to or from church, window-shopping. I ate lunch in the back of an old delicatessen and worried about the future.

Returning to Petruchio, perhaps if we can change our paradigm from "things" to higher-order patterns and process, from stasis to dynamics, and from an economic view of worth to a socioecological expression of value, then perhaps we can save biological diversity and, ultimately, ourselves. Biodiversity is not only a set-aside issue, but part of a larger task: "stewardship of *all* of the species on *all* of the landscape with every activity we undertake" (Franklin 1993). The answer to Petruchio's paradox lies somewhere between the heart and head, and in the nexus of our caring for life on Earth–Sea and the understanding of ecological systems, wherein the beholder becomes the student and the policymaker the perpetrator.

ACKNOWLEDGMENTS

I wish to thank the American Museum of Natural History, and Michael Novacek and Joel Cracraft of that institution, for the opportunity to participate in the conference on "The Living Planet in Crisis." I also wish to thank Sara Shallenberger Brown, the Munson Foundation, the Henry Foundation, and Edward M. Miller for their support of the Global Biodiversity Fund of the University of Virginia. Herbert F. Bormann, Dennis H. Grossman, and M. G. McCormick-Ray made useful comments.

REFERENCES

Bartholomew, G. A. 1986. The role of natural history in contemporary biology. *Bioscience* 36: 324–329.

Daly, H. E. and J. B. Cobb, Jr. 1989. *For the common good.*Boston: Beacon.

Dasmann, R. F. 1968. *A different kind of country.* New York: Macmillan.

Franklin, J. F. 1993. Preserving biodiversity: Species, ecosystems, or landscapes? *Ecological Applications* 3(2): 202–205.

Franklin, J. F. 1994. Response. *Ecological Applications* 4(2): 208–209.

Franklin, J. F. 1997. Ecosystem management: An overview. In M. A. Boyce, ed., *Ecosystem management*, 21–53. New Haven, Conn.: Yale University Press.

Gardner, R. H., M. G. Turner, V. H. Dale, and R. V. O'Neill. 1992. A percolation model of ecological flows. In A. J. Hansen and F. di Castri, eds., *Landscape boundaries*, 259–269. Berlin: Springer-Verlag.

Gleick, P. H. 1993. *Water in crisis: A guide to the world's fresh water resources.* New York: Oxford University Press.

Grassle, J. F., P. Lasserre, A. D. McIntyre, and G. C. Ray 1991. Marine biodiversity and ecosystem function. *Biology International* Special Issue 23: 1–16. Paris: International Union of Biological Sciences.

International Council for Bird Preservation. 1992. *Putting biodiversity on the map; Priority areas for global conservation, summary* . Cambridge, U.K.: ICBP.

May, R. M. 1988. How many species are there on Earth? *Science* 241: 1441–1449.

McCormick-Ray, G. 1998. Oyster reefs in 1878 seascape pattern—Winslow revisited. *Estuaries* 21 (4B):784–800.

Moore, J. A. 1993. *Science as a way of knowing.*Cambridge, Mass.: Harvard University Press.

Parsons, T. R. and H. Seki. 1995. A historical perspective of biological studies in the ocean. *Aquatic Living Resources* 8: 113–122.

Prendergast, J. R., R. M. Quinn, J. H. Lawton, B. C. Eversham, and D. W. Gibbons. 1993. Rare species, the coincidence of diversity hotspots and conservation strategies. *Nature* 365: 335–337.

Ray, G. C. 1970. Ecology, law, and the "Marine Revolution." *Biological Conservation* 3(1): 7–17.

Ray, G. C. 1985. Man and the sea: The ecological challenge. *American Zoology* 25: 451–468.

Ray, G. C. 1988. Ecological diversity in coastal zones and oceans. In E. O. Wilson and F. Peter, eds., *BioDiversity* ,36–50. Washington, D.C.: National Academy Press.

Ray, G. C. 1996. Conservation of coastal-marine biodiversity. In F. diCastri and T. Younès, eds., *Biodiversity, Science and Development*. Wallingford, U.K.: CAB International, fig. 22.3, p. 231.

Ray, G. C., B. P. Hayden, M. G. McCormick-Ray, and T. M. Smith. 1997. Landseascape diversity of the U.S. East Coast coastal zone with particular reference to estuaries. In R. F. G. Ormond, J. D. Gage, and M. V. Augel, eds., *Marine biodiversity: Patterns and processes*. Cambridge, U.K.: Cambridge University Press.

Seton, E. T. 1911. *Arctic prairies.*New York: Harper & Row.

Smith, F. D. M., R. M. May, R. Pellew, T. H. Johnson, and K. S. Walter. 1993. Estimating extinction rates. *Nature* 364: 494–496.

Terborgh, J. 1988. The big things that run the world: a sequel to E. O. Wilson. *Conservation Biology* 2(4): 402–403.

Tilman, D. 1996. Biodiversity: Population versus ecosystem stability. *Ecology* 77 (2): 350–363.

Tracy, C. R. and P. F. Brussard. 1994. Preserving biodiversity: Species in landscapes. *Ecological Applications* 4(2): 205–207.

Wilson, E. O. 1987. The little things that run the world (The importance and conservation of invertebrates). *Conservation Biology* 1(4): 344–346.

PART II

CONSEQUENCES OF BIODIVERSITY LOSS: SCIENCE AND SOCIETY

REGIONAL AND GLOBAL PATTERNS OF BIODIVERSITY LOSS AND CONSERVATION CAPACITY: PREDICTING FUTURE TRENDS AND IDENTIFYING NEEDS

Joel Cracraft

What countries are most at risk of losing their biodiversity? What is the capacity of these countries to meet these threats? In which countries and regions will societal impacts be most severe as biodiversity is progressively lost? What criteria are most relevant for assessing threat and capacity?

These and similar questions are of great concern to those seeking to promote the conservation and sustainable use of biodiversity. A link between the long-term health of biological diversity and societal well-being is widely accepted—it is, after all, the core of the rationale for the Convention on Biological Diversity—but this relationship is complex and generally depends on the economic and political context of the countries in question. Seen from the perspective of a single country, the relationship may even seem paradoxical at times. On the one hand, the wealth of virtually all advanced societies has been built on and remains supported by the use of biological resources, hence the widespread belief that the exploitation of natural resources will lead to prosperity. On the other hand, many countries are using their resources at an ever-increasing rate but are realizing only a little, or moderate, reciprocal increase in wealth in return. Whereas the use of biological diversity has supported the prosperity of some countries, for others current patterns of resource exploitation are easily seen to constitute a threat to the long-term well-being of their citizens.

There are many reasons why it is important to peer into the future and attempt to assess the status of the natural resources of individual nations and regions. For one thing, there is mounting evidence that the loss of biological diversity will have severe consequences for the prosperity of societies everywhere. It is critical to identify the nations and regions that are likely to be most affected by the loss of biodiversity and to assess its potential consequences. Reciprocally, the current well-being of a society and our expectations for its future prospects tells us much about the outlook for biological diversity itself. Indeed, the growing political and economic reality is that we cannot save biodiversity without saving humanity at the same time: the two are inextricably linked.

Numerous government, intergovernment, and nongovernment organizations and institutions are increasingly concerned with documenting the loss of biodiversity, identifying threats to it, and predicting the consequences of this loss on societal structure and function. These organizations want quantitative assessments of the threats to biodiversity and measures of the capacity of countries to respond to those threats. These assessments are needed for various reasons, from predicting which regions and countries are most likely to face increased challenges and thus pose threats to other countries in the same region, as well as which might threaten the economic and political security of all nations. Moreover, because it is widely recognized that the need for conservation action and developmental assistance far outweighs the resources available, such assessments are also relevant for choosing among recipient nations when allocating these scarce resources. In a real sense, setting priorities is a statement about the future of the recipient nation (the expectation that problems will be addressed and solved) and the cost-effectiveness of the investment being made.

Much more dialogue is needed about assessing global trends in biodiversity loss and conservation priorities, as well as the dynamics of the linkages between biodiversity loss and the well-being of societies. This study is an effort to bridge what we think we know about patterns of biological diversity, the threats to its survival, and the social and economic responses to those threats. The core of this paper is a quantification of these relationships for 77 countries in Tropical America, sub-Saharan Africa, and South Asia, most of which have high species diversity.

In recent years several studies have attempted to assess diversity and environmental threat and to propose priorities for conservation action. Dinerstein and Wikramanayake (1993) devised a conservation potential/threat index (CPTI) for 23 Indo-Pacific countries. Each country was ranked according to the percentage of remaining forest cover, the amount expected to persist over the next decade using current deforestation rates, and percentage of the country already protected. The "index" referred to the placement of the country in a graph space in which the percentage of the country protected was plotted against the percentage of the unprotected forest that was expected to remain in 10 years. This perspective of threat is predicated on the presence of protected areas as well as the amount of area remaining unprotected. However, it is debatable whether this is the best measure, or index, of threat. Whether areas are protected does not ensure biodiversity against threat, and factors that are causally related to biodiversity loss must still be identified. Moreover, having or not having protected areas may reflect the capacity of a country to respond to threat rather than the degree of threat per se.

A second, more comprehensive analysis is that of Sisk et al. (1994), who attempted to identify the extinction threat to 89 countries. Sisk et al. created two indices, one measuring the amount of biodiversity in a given country, the other the threat to that diversity. Their index for species richness used quantitative information for relative mammalian and butterfly species diversity (based on comparison to the known global numbers for these taxa), whereas their endemism index measured the proportion of the global biota for these two groups endemic to each country. Threat to biodiversity was evaluated in two ways: by a forest loss index that reflected the annual percentage lost and by a population pressure index that measured the relative rate of increase in population density.

A more complicated index to biodiversity priority-setting has been proposed by Moran et al. (1996). Their cost-effective priority investment index (CEPII) takes the product of a success ratio (percentage of forest currently protected) and a threat ratio (unprotected forest in 10 years using current deforestation rates) and then multiples that value by a species diversity index (the summation of higher plant, mammal, bird, reptile, and amphibian diversity normalized to square kilometers), with the result being divided by the amount of economic aid to biodiversity, also normalized to square kilometers. Within the context of the CEPII, diversity is seen as the benefit that accrues with a certain economic investment (the latter estimated from the amount of United States aid to biodiversity projects in each country).

The focal unit of each of these studies was the individual country. In contrast, another recent approach to priority-setting (Dinerstein et al. 1995) assessed diversity, threat, and prioritization at the ecoregional level within South America. Ecoregions, which often cut across political boundaries, were ranked according to their degree of biological distinctiveness using species richness, number of endemics, beta diversity, rarity of habitat types, and the presence of rare or endangered species as measures. Conservation status for each region was evaluated using data about loss of original habitat, degree of fragmentation, conversion rate, and the degree of protection.

ASSESSING BIOLOGICAL DIVERSITY, THREAT, AND CAPACITY

Developing a methodology to evaluate and quantify the threat to biodiversity, as well as a country's capacity to respond to that threat, is particularly complex because many indicators of human development can be seen as measuring threat under one social context but as measuring capacity to conserve or respond to diversity loss under another. One example, as noted earlier, might be the amount of natural resources consumed. It would generally be supposed that high use implies high threat, but high use could also reflect a highly developed economy that has resources to address threats to biodiversity. Such situations can be found in countries that import natural resources to substitute for in-country resources that have already been depleted or significantly reduced. Whereas high resource use, in this example, suggests that the exporting country may have threats to its biodiversity, this will not necessarily be the case for the importing country. Given these complex relationships, it is important to choose measures of threat and capacity that are largely independent of one another.

Indices

Three indices are used to measure the relative position of a country within a region and globally among all countries (table 5.1): an estimate of its species diversity

TABLE 5.1

Components of the Species Diversity Index, Biodiversity Threat Index, and Capacity Response Index

Each index, for each country within each of the three regions as well as globally among regions, is formed by the summation of the rank orders of the following components:

SPECIES DIVERSITY INDEX (SDI)	BIODIVERSITY THREAT INDEX (BTI)	CAPACITY RESPONSE INDEX (CRI)
Rank order of higher plant species diversity (WRI 1994)	Rank order of population pressure index (annual increase in density) (Sisk et al. 1994)	Rank order of the UNDP Human Development Index (HDI: life expectancy, adult literacy, GDP / capita) (UNDP 1991)
Rank order of butterfly species diversity (Sisk et al. 1994)	Rank order of the land area that has been subjected to medium and high disturbance (WRI 1994)	Rank order of total fertility rate 1990–1995 (WRI 1994)
Rank order of land bird species diversity (Slud 1976)	Rank order of percentage of cropland change between 1979 and 1991 (WRI 1994)	Rank order of percentage of labor force in agriculture 1990–1992 (UNDP 1994)
Rank order of mammalian species diversity (Sisk et al. 1994)	Rank order of annual percentage of forest lost (Sisk et al. 1994)	

(species diversity index; SDI), the threat to that diversity (biodiversity threat index; BTI), and the capacity of the country to respond to threats to its diversity (capacity response index; CRI). As a graphical convention, each index is divided into thirds (figures 5.1–5.8) to facilitate a description of low, medium, and high relative rankings for the different countries.

To construct an index, the relative ranks of a country were determined for a restricted set of variables. The ranks for each variable were then summed to construct an overall score that constituted the index in question. There are alternative ways to create indices, and each has its own advantages and disadvantages. In this study no variable is taken to have more weight than another, so it is assumed that each variable forming an index is equally accurate at estimating diversity, threat, or capacity, as the case may be. In fact, when patterns of variation are examined multivariately, no single variable within an index was found to be overwhelmingly more important than any other (data not shown).

The most problematic aspects of such an exercise are the accuracy and comparability of the data themselves from one country to the next; the choice and interpretability of any variable with respect to one's concept of what diversity, threat, and capacity mean; and the assumption that diversity, threat, and capacity can be quantified by a single numerical value (its index score). Whereas choice of a particular variable to estimate diversity, threat, or capacity may seem eminently reasonable, another variable may be just as good even though it tracks that estimate differently. For example, relative population pressure might be considered an important estimator of threat to biodiversity, but so may the rate of growth in land area devoted to agriculture. Moreover, those two variables might not be closely correlated with one another. At the same time, a given variable—such as population pressure—may not be an equally effective predictor of threat across a spectrum of countries because its impact on biodiversity depends on a number of other economic and developmental characteristics specific to each country.

Concerns over the accuracy of raw variables will arise. Many of these values, especially those for species diversity, will always be in a state of flux as scientific knowledge increases. Nearly all variables for threat and capacity are derived from well-known institutional analyses undertaken regularly by the World Resources Institute (WRI), United Nations Environment Programme (UNEP), United Nations Development Programme (UNDP), and the World Bank. Many of the data are reported to those institutions by individual countries, so there are bound to be questions about accuracy and comparability. But such compilations are all we have to work with at a global and regional scale. The expectation here is that by using a number of variables to form an index and by using relative ranks for those variables to derive a summary index value, the *relative* position of a particular country in a graph space will not be distorted.

Broadly defined indices such as those proposed here have an advantage in that they should be applicable across a wide range of countries. For example, estimating threat solely in terms of deforestation rate may underestimate threat within countries having habitats that are not heavily represented by forests. An index based on a number of variables also has an advantage over one constructed from a single variable because the former will average out idiosyncrasies in the value of any given variable that has been assigned to an individual country. A similar advantage is gained by using ranks for each variable rather than absolute measures.

It should be stressed that all three indices are constructed from the relative ranks within a region as well as globally (depending on the comparison being made), so the value for an index depends on whether one is comparing that country regionally or globally, with the index score increasing or decreasing as the set of countries changes. Because these measures of diversity, threat, or capacity are rank orders, measures of variables classified as high capacity within one region may be low capacity within another.

Some previous assessments have created indices by normalizing the variables measuring diversity, threat, or capacity to the size of the country. The approach adopted here is to have these indices reflect a direct comparison among countries with respect to diversity, threat, or capacity, thus absolute, not normalized, measures are employed in establishing ranks.

SPECIES DIVERSITY INDEX (SDI)

The SDI quantifies the relative diversity of a country by summing the relative ranks of four selected groups of organisms: higher plants, butterflies, land birds, and mammals (table 5.1). A high score on the SDI implies greater relative diversity. In general, diversities of birds, mammals, and butterflies are all correlated with one another, but the relationships among these variables and plant diversity tend to be much weaker. This may result from the poor estimates that are available for plant species diversity in many countries. Data on plant species diversity (WRI 1994:table 20.4) were lacking for a few countries. In these cases, species number was estimated using figures from nearby countries having similar ecological conditions and correcting for country size.

BIODIVERSITY THREAT INDEX (BTI)

The BTI uses the rank orders of the annual increase in human population density, the percentage of the country's land area subjected to medium to high disturbance, the percentage of cropland change between 1979 and 1991, and the annual percentage of forest loss (table 5.1). Although these measures are not entirely independent of one another, the correlations among them are weak across the set of countries. It is therefore assumed that they measure threat somewhat differently and can be summed. This index is a broader measure of threat than is presented in most previous studies, which were often restricted to population growth rate or a measure of deforestation. By themselves, population growth rate and deforestation are incomplete measures because they imply different consequences for biodiversity in countries that have different absolute populations or amount of forest, or have different developmental capacities. Thus two additional variables are incorporated into the BTI. The amount of land that has been subjected to medium to high disturbance and the amount of land converted to agriculture are generally good indicators of the degree to which a country's wildlands have already been degraded; taken in concert with population growth rate and annual deforestation, they provide a more refined measurement of threat into the future. A high score on the BTI implies a greater threat to biodiversity.

CAPACITY RESPONSE INDEX (CRI)

The CRI estimates the capacity of a country to respond to environmental degradation and the loss of biodiversity. It is based on the supposition, often articulated in the literature (IUCN/WWF/UNEP 1980; McNeely et al. 1990; WRI/IUCN/UNEP 1992) and codified in the Convention on Biological Diversity, that the conservation of biodiversity and increased human development go hand in hand.

The CRI is therefore designed to measure a country's overall developmental capacity. The index is formed from the rankings of three variables (table 5.1), and a high score on the index implies a greater capacity to respond to threat and undertake conservation actions. The Human Development Index (HDI) of UNDP (1991) is a general measure of a country's capacity in terms of health, education, and economic well-being, and numerous aspects of developmental capacity are encapsulated in the HDI. To this is added a measure of fertility rate in order to more directly factor in the effects of short-term changes in population growth. Finally, the index also includes the percentage of the labor force in agriculture as a fairly direct measure of what proportion of the population is *not* in industry or services. Virtually all developed economies devote 5 percent or less of their labor force to agriculture; most of the countries discussed in this study have much more of their labor force in agriculture, which usually correlates negatively with economic well-being and developmental capacity. Other variables that also could be used to measure capacity, such as direct measures of economic activity, were also examined, but for most of these comparable data were not available for all countries.

The major imponderable when measuring threat and capacity is the political and economic stability of the country in question. Many countries considered in this study have greater threat to biodiversity than may be implied by their score on the BTI because of ongoing political instability or the significant possibility of future instability. Likewise, political instability is likely to have a negative effect on the developmental and economic well-being of a country and therefore influence its ability to respond to environmental threats. Although identifying particular countries as being susceptible to political instability is qualitatively possible, a quantitative measure that is comparative across the countries considered in this analysis is not widely available. Hence, although critical for assessing the future threat to biodiversity, political instability was excluded from the BTI and CRI.

This analysis also does not include the amount of aid devoted to conservation, as other studies have done (e.g., Dinerstein and Wikramanayake 1993; Moran et al. 1996). Often the reasons why some countries get aid and others do not have less to do with the capacity of the country to respond to the threats to its biodiversity than with the political alliances that are formed among governments. It might be assumed that the total amount of economic aid is a more important indicator of whether a country is going to be able to meet its conservation objectives, but comparable data are often not available. Much economic aid comes in the form of military and security assistance, which some might see as a threat to biodiversity rather than as a basis for responding to threat. The CRI measures overall development capabilities, which have long been related to the amount of resources that countries devote to conservation activities.

Regions and Countries

The analysis is restricted primarily to countries that contain forested habitats of various kinds, and most of the countries are located in tropical regions. In a few cases countries that extend into temperate zones have been included. A central concern was to include countries that are characterized by high diversity and those in which the consequences of biodiversity loss might be considered more acute from a global perspective. The study is restricted to the level of countries—although comparisons within and among regions are the focus of the discussion—because it is at that level that decisions affecting biodiversity loss and conservation action are most politically relevant.

The countries included in this study were those of Tropical America, sub-Saharan Africa, and South Asia. The database for Tropical America includes all 20 countries south of the United States except French Guiana, for which it was not possible to obtain comparative data for all variables. The database for sub-Sahara Africa includes 40 countries. Finally, 17 countries of South Asia were incorporated into the analysis. In total, data were collected for 77 countries.

Data

Quantitative data were collected for 46 variables describing patterns of biological diversity, demography, and the economic and human development of each country in each region. After preliminary analysis, these variables were used to construct the three indices. Table 5.1 identifies the components of the indices and the sources of the data. Appendices 5.1–5.3 contain the values and rank orders of all 11 variables within the three indices within the three primary regions. Appendix 5.4 gives the three scores for each index for each country when compared globally across all regions.

RESULTS

Figures 5.1–5.6 plot the positions of countries for the three indices within each region, and figures 5.7 and 5.8 do this for the global comparisons. Within each, the graph space is divided into thirds (high, medium, low) for ease of discussion.

Regional Patterns

TROPICAL AMERICA

The data identify seven countries in the top third of the SDI for Tropical America (figure 5.1). All are South American except for Mexico. Of these, three (Mexico, Venezuela, and Ecuador) rank high on the BTI and four fall in the middle third of that index. Given the criteria used to measure threat, none of these countries could be placed in the low-threat category. In fact, Colombia and Brazil had BTI scores that nearly placed them in the high-threat category. The highest threat scores for Tropical America are for countries that did not rank as being of highest diversity at the regional level: Costa Rica (but see the global-level discussion later in this chapter), El Salvador, and Guatemala. With the exceptions of Belize and Panama, all the countries in the low- and medium-threat categories are South American. From these data, therefore, the biodiversity of Central America is seen to be under greater threat than is that of most of South America.

Figure 5.2 compares the BTI of the 20 Tropical American countries with their scores on the CRI. Among the seven countries showing the greatest diversity, all except Bolivia exhibit medium to high capacity. Indeed, Bolivia has the second lowest CRI score in all Tropical America; only Guatemala scored lower. Moreover, Peru's score on the CRI just barely falls into the middle third, as does Mexico into the upper third. Depending on how these data are interpreted, Bolivia, Peru, and

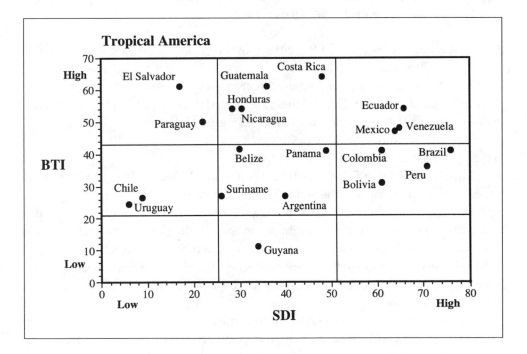

FIGURE 5.1

Position of 20 Tropical American countries by their scores on the SDI and BTI (see appendix 5.1)

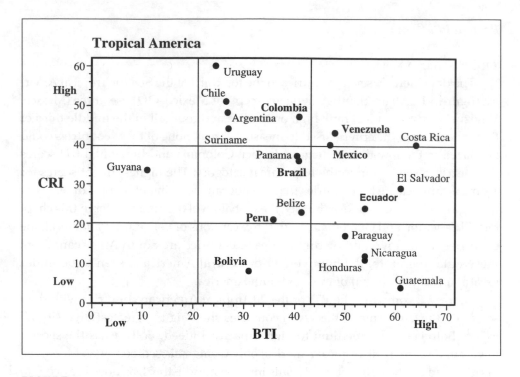

FIGURE 5.2

Position of 20 Tropical American countries by the CRI and the BTI (appendix 5.1). Countries in bold scored in the highest third of the SDI (figure 5.1).

Ecuador would be considered to be the most critically threatened high-diversity countries in Tropical America.

Four countries having the highest threat ranking are also in the lower third in terms of capacity ranking. Three of these are located in Central America (Guatemala, Honduras, and Nicaragua) and the fourth (Paraguay) is South American.

SUB-SAHARA AFRICA

Among the 40 sub-Saharan African countries, 13 were identified as having scores in the upper third of the SDI (figure 5.3). With respect to these high-diversity countries, three (Nigeria, Kenya, and Uganda) score in the upper third of the BTI and thus are under high threat, eight are in the middle third of the BTI, and two (Central African Republic and Zambia) are considered to be subject to low threat. Ghana is very nearly in the high-threat category, and South Africa and Congo (former Zaire), currently identified as being under medium threat, are nearly in the low-threat category.

Figure 5.4 compares the 40 African countries with respect to their ranking on both the BTI and CRI Five of the 13 high-diversity countries (figure 5.3) are identified as having high capacity, six with medium capacity, and two as having low capacity. Cameroon and Nigeria are just barely classified as being high capacity, and

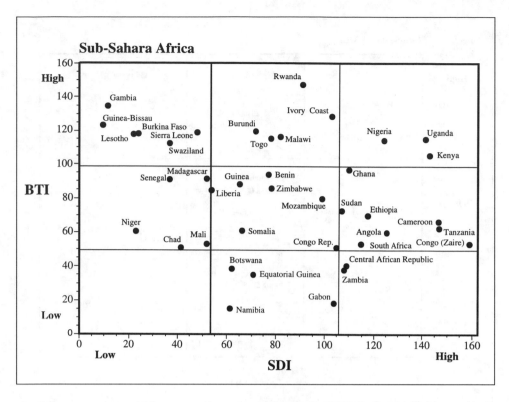

FIGURE 5.3

Position of 40 sub-Sahara African countries by their scores on the SDI and BTI (see appendix 5.2).

at least four countries (Angola, Somalia, Mozambique, and Gambia) are at the low end of the medium-capacity category.

Five countries are categorized as having their biodiversity under the highest threat but as having the lowest capacity to respond: Uganda, Burkina Faso, Burundi, Malawi, and Rwanda. Gambia also nearly falls into this category. Four countries, in contrast, are at the opposite end of the rankings, those with the lowest threat and highest capacity: Botswana, Namibia, Gabon, and Zambia, with South Africa and Congo Republic nearly in this category also.

SOUTH ASIA

Six high-diversity countries are recognized within the South Asian region (figure 5.5). Two other countries—Vietnam and Papua New Guinea—also have substantial diversity but fell just short of the upper one-third boundary for high-diversity ranking.

Of these six countries, the diversity of India and Thailand was categorized as being under high threat, whereas the diversity of the remaining four countries—China, Myanmar, Malaysia, and Indonesia—was judged as being subject to medium threat. Significantly, none of the six was subject to low threat. Six additional

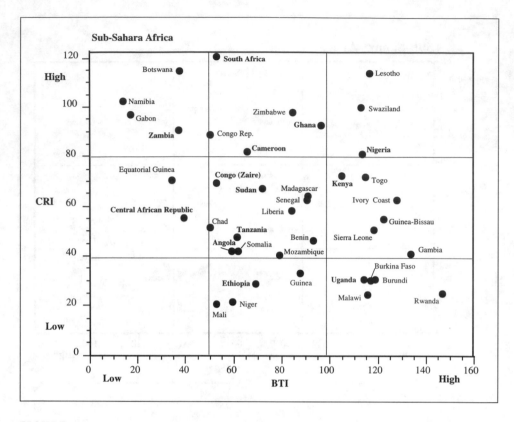

FIGURE 5.4

Position of 40 sub-Sahara African countries by their scores on the SDI and BTI (see appendix 5.2).

countries join Thailand and India in the highest threat category, three of these (Nepal, Philippines, and Vietnam) having moderate levels of species diversity.

With respect to the variables examined in this study, all six of the highest-diversity countries fortunately have medium to high capacity (figure 5.6). The Philippines, classified here as having medium-diversity ranking, also has high capacity. Four countries—Afghanistan, Laos, Bhutan, and Nepal—have the lowest capacity in the region, although Cambodia and Papua New Guinea are also close to having low-capacity ranking. The most serious situation in the region is with the medium-diversity nation Nepal, which was the only country scoring low on the capacity index and high on the threat index.

Global Patterns

Figure 5.7 shows the rank of each country in terms of its global rank score for the SDI and the BTI (see appendix 5.4). When viewed against other nations outside their region, some countries shift categories compared to regional perspectives (figures 5.1–5.6), but most retain their same relative positions.

Two particularly strong patterns emerge. First, of the 21 countries ranked in the

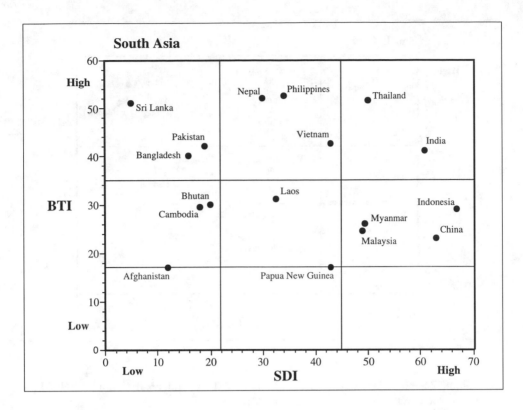

FIGURE 5.5

Position of 17 South Asian countries by their scores on the SDI and BTI (see appendix 5.3).

high-diversity category, 9 are located in Tropical America, whereas both sub-Saharan Africa and South Asia each have six. The second pattern is that of the 25 countries ranked highest on the BTI, 14 are located in Africa. Of all the countries surveyed, the diversity of Rwanda is under highest threat. Importantly, if diversity were normalized to country size, Rwanda would be the country with the highest diversity.

Five countries with high diversity ranks also have high ranks on the BTI: Thailand from South Asia, Costa Rica and Ecuador from Tropical America, and Uganda and Kenya from sub-Saharan Africa.

Figure 5.8 plots all 77 countries by their global ranking for the CRI and BTI. Several patterns can be identified. First, 13 of the 21 high-diversity countries have high CRI ranks and 6 have medium CRI ranks. Only 2 (Tanzania and Uganda) are placed in the low-capacity category. With the sole exception of South Africa, all 13 high-diversity countries with high capacity are located in either Tropical America or South Asia. This draws attention to a second major pattern: of the 26 low-capacity countries, 22 are from sub-Saharan Africa, and of the 25 medium-capacity countries, 15 are from sub-Saharan Africa. Botswana, South Africa, and Lesotho are the only sub-Saharan countries to have high CRI rank. Globally, then, countries in sub-Saharan Africa have much less capacity to respond to biodiversity loss than their counterparts in South Asia and Tropical America. Most of the African coun-

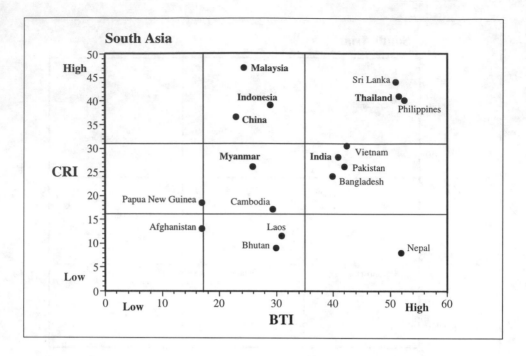

FIGURE 5.6

Position of 17 South Asian countries by the CRI and BTI (appendix 5.3). Countries in bold are those that scored in the highest third of the SDI (figure 5.5).

tries ranked as having high capacity within that continent (figure 5.4) are ranked as having medium capacity when compared globally.

DISCUSSION

Biodiversity Under Threat: Patterns and Predictions

Across the spectrum of countries considered in this analysis, 25 out of 77 (32 percent) rank in the upper third of the BTI. Nine of these 25 high-threat countries (36 percent) have low capacity. Nine more are ranked as having medium capacity. This means that of the countries whose biodiversity is most under threat, only 7 (28 percent) have high capacity; that is, they rank in the upper third of the CRI.

As previously noted, most countries under the highest threat (14 of the 25) are from sub-Sahara Africa (figures 5.7 and 5.8). Two of them, Uganda and Kenya, rank as high-diversity countries, although Nigeria is very nearly in that category also. In Africa, 8 countries count as being in both the high-threat and low-capacity categories and 13 as being in the medium-threat and low-capacity categories (figure 5.9). One conclusion seems obvious from this analysis: the regions of greatest conservation and societal concern are in West Africa (including the Sahel) and East

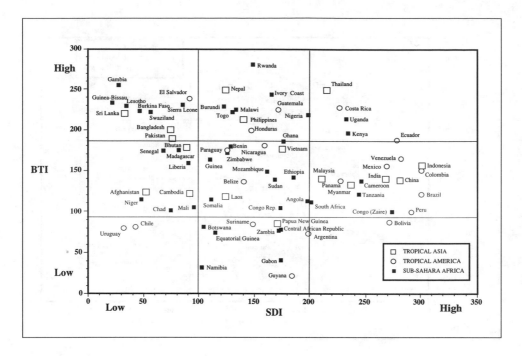

FIGURE 5.7
Position of all 77 countries by their scores on the SDI and BTI (see appendix 5.4).

Africa/Madagascar. Southern Africa and Central Africa, by the criteria adopted here, are under less threat (although it is apparent from figure 5.8 that several countries in these regions are close to being categorized as low capacity).

Not unexpectedly, different approaches to evaluating threat arrive at different results. Sisk et al. (1994:table 1) identified six African countries as areas of critical concern, using population density and growth rate as their measures of threat (1994:figure 1): Angola, Ghana, Ivory Coast, Kenya, Nigeria, and Uganda. All six countries were also recognized in this study, but the more comprehensive measure of threat adopted here, considering countries that have both medium and high species diversity, expands the list of countries recognized as having severe threat to their biodiversity (figure 5.9). That this approach provides a more realistic estimate of threat for sub-Sahara African countries is supported by the fact that none of the high-threat countries identified in figure 5.8, nor the countries shown on figure 5.9 with the exception of Liberia, have more than 20 percent of their moist forest remaining (Sayer et al. 1992). Indeed, most of their levels of moist forest remaining are 10 percent or less.

Compared to sub-Sahara Africa, the countries of South Asia rank much more favorably on the BTI and CRI (figure 5.8). The most severely threatened country in South Asia, if one takes into account its lack of capacity rather than its level of diversity, is Nepal, followed by Pakistan and Bangladesh (figure 5.10). Sisk et al. (1994:table 2) did not recognize any of these three countries as being areas of criti-

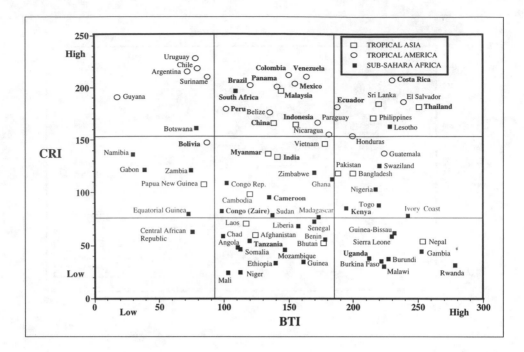

FIGURE 5.8

Position of all 77 countries by the CRI and BTI (appendix 5.4). Countries in bold are those that scored in the highest third of the SDI (figure 5.7).

cal concern, primarily because they house much less diversity than other nations in the region. Their list instead included China, India, Philippines, Sri Lanka, and Thailand (all of these except India are ranked as high-capacity countries in this analysis). Based on somewhat different criteria (having low percentages of habitat protected as well as the rate of habitat loss) Dinerstein and Wikramanayake (1993:figure 2) identified Vietnam, China, Philippines, and Bangladesh as being the most threatened countries in the region.

Five South Asian high-diversity countries rank as having medium threat (figure 5.7): Malaysia, Myanmar, India, China, and Indonesia. All of those, except Myanmar and India, also have high capacity; even Myanmar and India are near the top of the medium-capacity scale. Thailand, Sri Lanka, and the Philippines rank in the upper third of both indices (figure 5.8).

The Tropical American country with the highest threat to its biodiversity is El Salvador, followed by Costa Rica, Guatemala, Honduras, and Ecuador (figures 5.7 and 5.8). All of these, with the exception of Guatemala and, marginally, Honduras, rank in the highest third of the CRI (figure 5.8). Sisk et al. (1994:tables 4 and 5) identified four countries in Central America (Costa Rica, Guatemala, Honduras, Nicaragua) and two in South America (Colombia, Ecuador) as being areas of critical concern. There is very strong concordance between Sisk et al. and this study in their assessments for Central America (in this study Nicaragua is very nearly

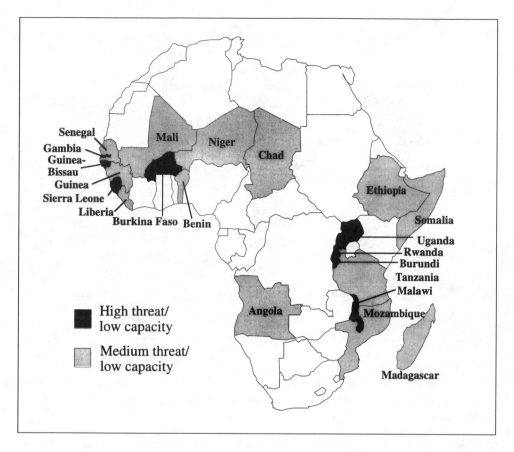

FIGURE 5.9

Potential areas of crisis and threat in sub-Sahara Africa. Low-capacity countries are particularly vulnerable to losing their biodiversity. Indeed, some of these countries have already lost a substantial portion of their habitats.

ranked as having high threat). With respect to South America, both analyses identify the biodiversity of Ecuador as being especially under threat. In this analysis, on the other hand, three other South American countries (Paraguay, Venezuela, and Mexico) rank higher than Colombia on the BTI (figures 5.7 and 5.8). Again, these differences arise from different measures of threat and the emphasis of Sisk et al. (1994) on the most diverse countries.

Needs and Priorities

The preceding discussion points to the fact that different perceptions of what constitutes threat can have obvious implications for any attempt at evaluating conservation need and identifying priorities to address those needs. This study has been predicated on the assumption that the capacity of a country is also an important element when undertaking these assessments. Both threat and action must, albeit in

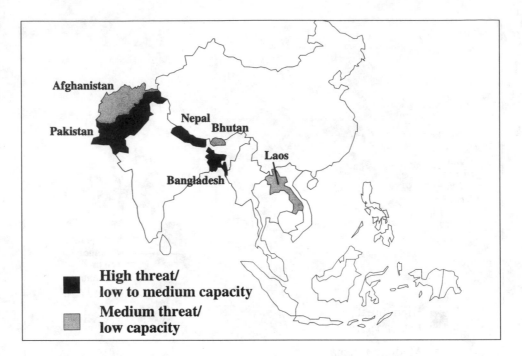

FIGURE 5.10

Potential areas of crisis and threat in South Asia. In addition to the countries shown here, three other high-capacity countries (Thailand, Philippines, and Sri Lanka) also have biodiversity under high threat.

complex ways, relate to capacity and be placed in that context. With respect to the 77 nations examined here, it must be remembered that none of these countries, all of which are normally classified as developing, would be judged to have high capacity when compared to the measures of capacity of the industrial countries. Thus it might be argued that all of these countries will need outside assistance in conserving and sustainably using their biodiversity.

Traditionally, predicting trends and setting priorities with respect to conservation have been based on relative species richness and perceived threat (figure 5.11). As diversity and threat are seen to decline, so does the assigned level of priority. Diversity in this sense might also incorporate some notion of the uniqueness of the biota, especially the degree of endemism. The results of Sisk et al. (1994) exemplify this approach.

Another way of thinking about the future prospects of a nation's biodiversity and the priority that might be given to its conservation involves consideration of the internal capacity of that nation to address its developmental and environmental problems. When this is done, the relationships among diversity, threat, and capacity become more complex (figure 5.12). It makes some intuitive sense that as threat and capacity increase, priority should also increase. Countries with little capacity might be expected to require more resources from the outside to have a given effect than would countries with higher internal capacity.

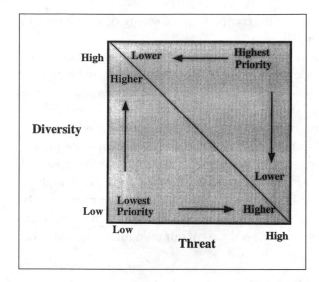

FIGURE 5.11

Possible assessments of priority
for conservation action in the
context of the relative amounts
of diversity and degree of threat.

Adding diversity to the equation expands the options for priority setting. How much diversity is currently remaining has always been relevant for establishing needs and setting priorities, which is why there has been a long-term emphasis on the species-rich nations. But should a country of moderate to high diversity and endemism that is also besieged by high threat and low capacity be judged to have a higher priority than a similarly diverse nation with medium or high capacity that is, in contrast, subject to medium or low threat? In the abstract, the chances of success clearly seem to be greater for the latter country. But what about the real world? The first country might be Madagascar, the second Botswana. In this case, aid agencies have decided to put much more into Madagascar than into Botswana (UNDP 1994). This is not unreasonable, not because of the characteristics of their respective biotas, but because the human need in Madagascar might be judged as

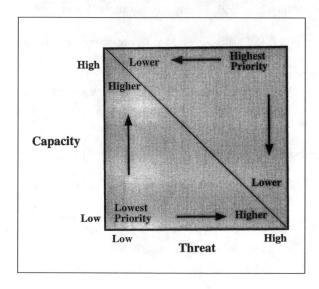

FIGURE 5.12

Possible assessments of priority
for conservation action in the
context of relative degrees of
threat and capacity.

being more acute. Yet it does point out the complexities of seeing priorities only in terms of diversity, threat, and capacity.

The difficulty is that even though conditions in most developing nations have improved over the last two decades as human development has risen, birth rates have fallen, and the commitment to conserve biodiversity has increased, the natural world of these countries is destined to face more trouble over the next decade or two without a significant turnaround in the economic and political relations between the industrial and developing worlds. Again, one cannot separate the future of biodiversity from the future of the people who depend on it. An essential component of this change is increased understanding on the part of policymakers in industrial countries that an isolationist worldview, which has unfortunately gained new life recently, especially in the United States, cannot work and that conserving biodiversity everywhere is an integral part of our own future security and well-being.

REFERENCES

Dinerstein, E., D. M. Olson, D. J. Graham, A. L. Webster, S. A. Primm, M. P. Bookbinder, and G. Ledec. 1995. *A conservation assessment of the terrestrial ecoregions of Latin America and the Caribbean*. Washington, D.C.: The World Bank.

Dinerstein, E. and E. D. Wikramanayake. 1993. Beyond "hotspots": How to prioritize investments to conserve biodiversity in the Indo-Pacific region. *Conservation Biology* 7: 53–65.

IUCN/WWF/UNEP. 1980. *World conservation strategy: Living resource conservation for sustainable development*. Gland, Switzerland: IUCN.

McNeely, J. A., K. R. Miller, W. V. Reid, R. A. Mittermeier, and T. B. Werner. 1990. *Conserving the world's biological diversity*. Gland, Switzerland: IUCN.

Moran, D., D. Pearce, and A. Wendelaar. 1996. Global biodiversity priorities: A cost-effectiveness index for investments. *Global Environmental Change* 6(2).

Sayer, J. A., C. S. Harcourt, and N. M. Collins, eds. 1992. *The conservation atlas of tropical forests: Africa*. New York: Simon & Schuster.

Sisk, T. D., A. E. Launer, K. R. Switky, and P. R. Ehrlich. 1994. Identifying extinction threats. *Bioscience* 44: 592–604.

Slud, P. 1976. Geographic and climatic relationships of avifaunas with special reference to comparative distribution in the Neotropics. *Smithsonian Contributions, Zoology* 212: 1–149.

United Nations Development Programme. 1991. *Human development report 1991*. New York: Oxford University Press.

United Nations Development Programme. 1994. *Human development report 1994*. New York: Oxford University Press.

World Bank. 1992. *World development report 1992. Development and the environment*. Washington, D.C.: World Bank.

World Resources Institute. 1994. *World resources 1994–1995*. New York: Oxford University Press.

WRI/IUCN/UNEP. 1992. *Global biodiversity strategy: Guidelines for actions to save, study, and use Earth's biotic wealth sustainably and equitably*. Washington, D.C.: WRI/IUCN/UNEP.

APPENDIX 5.1
Variables and Variable Ranks Used to Derive the SDI, BTI, and CRI for 20 Tropical American Countries

SDI	PLANT DIVERSITY	PLANT DIVERSITY RANK	BUTTERFLY DIVERSITY	BUTTERFLY DIVERSITY RANK	BIRD DIVERSITY	BIRD DIVERSITY RANK	MAMMAL DIVERSITY	MAMMAL DIVERSITY RANK	SDI
Argentina	9,372	11	267	3	717	13	285	13	40
Belize	12,894	14	392	7	368	4	173	5	30
Bolivia	17,367	15	842	16	1,042	15	303	15	61
Brazil	56,215	20	954	17	1,363	19	455	20	76
Chile	5,292	5	84	1	201	1	112	2	9
Colombia	51,220	4	1,276	20	1,425	20	379	17	61
Costa Rica	12,119	13	617	14	662	1	219	10	48
Ecuador	19,362	17	1,120	18	1,200	17	291	14	66
El Salvador	2,911	2	397	8	329	3	171	4	17
Guatemala	8,681	10	450	10	547	8	197	8	36
Guyana	6,409	7	368	6	642	10	227	11	34
Honduras	5,680	6	420	9	541	7	187	6.5	28.5
Mexico	26,071	19	471	12	776	14	446	19	64
Nicaragua	7,590	8	452	11	512	5	187	6.5	30.5
Panama	9,915	12	595	13	716	12	229	12	49
Paraguay	7,851	9	318	4	535	6	147	3	22
Peru	18,245	16	1,212	19	1,334	18	381	18	71
Suriname	5,018	3	355	5	602	9	207	9	26
Uruguay	2,278	1	174	2	227	2	75	1	6
Venezuela	21,073	18	757	15	1,151	16	311	16	65

(continued on next page)

APPENDIX 5.1 (continued from previous page)
Variables and Variable Ranks Used to Derive the SDI, BTI, and CRI for 20 Tropical American Countries

BTI	POPULATION PRESSURE INDEX	POPULATION PRESSURE RANK	PERCENTAGE OF LAND WITH MEDIUM/HIGH DISTURBANCE	DISTURBANCE RANK	PERCENTAGE CHANGE IN CROPLAND 1979–1989	CROPLAND RANK	ANNUAL PERCENTAGE OF FOREST LOST[a]	FOREST LOST RANK	BTI
Argentina	0.14	4	64	11.5	0	2	0.51	9	26.5
Belize	0.27	6	64	11.5	9	14	0.63	10	41.5
Bolivia	0.19	5	22	5	12.9	16	0.15	5	31
Brazil	0.34	9	33	7	23.1	18	0.37	7	41
Chile	0.32	8	44	9	3.9	8	0	1.5	26.5
Colombia	0.60	13	31	6	4.1	9	1.74	13	41
Costa Rica	1.51	17	89	17	4.5	11	3.72	19	64
Ecuador	0.85	14	53	10	9.4	15	2.10	15	54
El Salvador	7.59	20	100	19	1.1	5	2.54	17	61
Guatemala	2.76	19	74	15	7.9	13	1.78	14	61
Guyana	0.07	2	2	2	0.1	3	0.02	4	11
Honduras	1.57	18	67	13	3.7	7	2.50	16	54
Mexico	1.02	15	77	16	0.7	4	1.03	12	47
Nicaragua	1.07	16	69	14	2.1	6	2.71	18	54
Panama	0.59	12	1	1	16.7	17	0.76	11	41
Paraguay	0.30	7	16	4	26.7	19	3.93	20	50
Peru	0.39	10	40	8	6.1	12	0.36	6	36
Suriname	0.06	1	9	3	39.7	20	0.01	3	27
Uruguay	0.09	3	100	19	-9.5	1	0	1.5	24.5
Venezuela	0.52	11	100	19	4.3	10	0.40	8	48

(continued on next page)

APPENDIX 5.1 (*continued from previous page*)
Variables and Variable Ranks Used to Derive the SDI, BTI, and CRI for 20 Tropical American Countries

CRI	HDI: RANK-ORDER WORLD	HDI RANK	FERTILITY RATE 1990–1995	FERTILITY RATE RANK	PERCENTAGE LABOR FORCE IN AGRICULTURE	AGRICULTURAL LABOR FORCE RANK	CRI
Argentina	43	17	2.8	14.5	13	16.5	48
Belize	67	10	4.5	5	30	8	23
Bolivia	110	1	4.6	4	47	3	8
Brazil	60	12	2.8	14.5	25	11.5	36
Chile	38	19	2.7	17	19	15	51
Colombia	61	11	2.7	17	10	19	47
Costa Rica	40	18	3.1	10.5	25	11.5	40
Ecuador	77	8	3.6	9	33	7	24
El Salvador	94	4	4	7	11	18	29
Guatemala	103	2	5.4	1	50	1	4
Guyana	89	5	2.6	19	27	9.5	33.5
Honduras	100	3	4.9	3	38	5	11
Mexico	45	15	3.2	12	23	13	40
Nicaragua	85	6	5	2	46	4	12
Panama	54	14	2.9	13	27	9.5	36.5
Paraguay	73	9	4.3	6	48	2	17
Peru	78	7	3.6	8	35	6	21
Suriname	55	13	2.7	17	20	14	44
Uruguay	32	20	2.3	20	5	20	60
Venezuela	44	16	3.1	10.5	13	16.5	43

[a] Both WRI (1994) and World Bank (1992) show an increase in forest cover for both Chile and Uruguay. Because reliable data are not available on how much original forest is being lost, even while total forest cover may be increasing, the values for these two countries are set at zero.

APPENDIX 5.2
Variables and Variable Ranks Used to Derive the SDI, BTI, and CRI for 40 Sub-Sahara African Countries

SDI	PLANT DIVERSITY	PLANT DIVERSITY RANK	BUTTERFLY DIVERSITY	BUTTERFLY DIVERSITY RANK	BIRD DIVERSITY	BIRD DIVERSITY RANK	MAMMAL DIVERSITY	MAMMAL DIVERSITY RANK	SDI
Angola	5,185	28	346	29	872	35	269	33.5	125.5
Benin	2,201	11.5	314	25	630	18	209	22.5	77
Botswana	5,500	30	138	9	569	12	187	11	62
Burkina Faso	1,100	3	123	5	497	9	167	7	24
Burundi	2,500	14	312	24	633	20	197	14	72
Cameroon	8,260	36	577	39	848	34	311	37.5	146.5
Central African Rep.	3,602	21	359	31	668	26	266	31	109
Chad	1,600	6	165	10	496	8	200	17.5	41.5
Congo Republic	600	32	477	37	500	10	225	26	105
Equatorial Guinea	3,250	20	422	33	392	5	190	13	71
Ethiopia	6,603	34	236	17	836	33	269	33.5	117.5
Gabon	6,651	35	450	36	617	15	200	17.5	104
Gambia	974	1	109	1.5	489	7	116	2	11.5
Ghana	3,725	23	350	30	721	29	236	28	110
Guinea	3,000	16	279	21	529	11	200	17.5	65.5
Guinea-Bissau	1,000	2	109	1.5	376	3	125	3	9.5
Ivory Coast	3,660	22	345	28	683	28	224	25	103
Kenya	6,506	33	406	32	1,067	39	328	39	143
Lesotho	1,591	5	178	11	288	2	134	4	22
Liberia	2,200	10	298	23	590	13	169	8	54

(continued on next page)

APPENDIX 5.2 *(continued from previous page)*
Variables and Variable Ranks Used to Derive the SDI, BTL, and CRI for 40 Sub-Sahara African Countries

SDI	PLANT DIVERSITY	PLANT DIVERSITY RANK	BUTTERFLY DIVERSITY	BUTTERFLY DIVERSITY RANK	BIRD DIVERSITY	BIRD DIVERSITY RANK	MAMMAL DIVERSITY	MAMMAL DIVERSITY RANK	SDI
Madagascar	9,505	37	218	13	250	1	95	1	52
Malawi	3,765	24	263	19	630	18	208	21	82
Mali	1,741	7	128	6	647	24	198	15	52
Mozambique	5,692	31	230	16	666	25	230	27	99
Namibia	3,174	19	132	7.5	640	23	188	12	61.5
Niger	1,178	4	115	3	473	6	186	10	23
Nigeria	4,715	26	448	34	831	32	268	32	124
Rwanda	2,290	13	329	27	669	27	213	24	91
Senegal	2,086	8	121	4	625	16	171	9	37
Sierra Leone	2,090	9	275	20	614	14	154	5	48
Somalia	3,028	17	132	7.5	639	22	203	20	66.5
South Africa	23,420	40	220	14	774	31	249	30	115
Sudan	3,137	18	238	18	938	36	290	35	107
Swaziland	2,715	15	186	12	381	4	164	6	37
Tanzania	10,008	38	449	35	1,016	38	303	36	147
Togo	2,201	11.5	319	26	630	18	209	22.5	78
Uganda	5,406	29	491	38	989	37	311	37.5	141.5
Congo (Zaire)	11,007	39	705	40	1,086	40	422	40	159
Zambia	4,747	27	285	22	732	30	247	29	108
Zimbabwe	4,440	25	221	15	635	21	200	17.5	78.5

(continued on next page)

Variables and Variable Ranks Used to Derive the SDI, BTI, and CRI for 40 Sub-Sahara African Countries

BTI	POPULATION PRESSURE INDEX	POPULATION PRESSURE RANK	PERCENTAGE OF LAND WITH MEDIUM/HIGH DISTURBANCE	DISTURBANCE RANK	PERCENTAGE CHANGE IN CROPLAND 1979–1989	CROPLAND RANK	ANNUAL PERCENTAGE OF FOREST LOST	FOREST LOST RANK	BTI
Angola	0.20	6.5	47	10	0.5	6.5	1.90	36.5	59.5
Benin	1.38	27	88	26.5	3.9	23	0.68	17	93.5
Botswana	0.07	2	43	7	2	14	0.64	15	38
Burkina Faso	1.16	25	88	26.5	27.9	40	0.83	26	117.5
Burundi	6.67	39	100	36	3	16	0.87	28	119
Cameroon	0.85	21	84	24	1.2	11	0.49	10	66
Central African Rep.	0.13	5	54	11	3.4	19.5	0.25	4	39.5
Chad	0.10	3.5	46	9	1.7	13	0.83	25	50.5
Congo Republic	0.20	6.5	44	8	13.7	33	0.21	3	50.5
Equatorial Guinea	0.37	11	16	1	0	3.5	0.76	19	34.5
Ethiopia	1.24	26	98	30.5	0.4	5	0.39	8	69.5
Gabon	0.10	3.5	19	3	0.9	9	0.14	2	17.5
Gambia	2.19	34	100	36	14.5	34	1.18	30	134
Ghana	2.15	33	98	30.5	-2.6	1	1.38	32	96.5
Guinea	0.79	18	65	16.5	3.4	19.5	1.50	34	88
Guinea-Bissau	0.55	15	100	36	17.9	35	1.90	36.5	122.5
Ivory Coast	1.45	28	77	22	19	38	5.40	40	128
Kenya	1.66	31	57	14	6.5	27	1.47	33	105
Lesotho	1.82	32	100	36	12.8	31	0.72	18	117
Liberia	0.80	19	74	21	0.5	6.5	2	38	84.5

(continued on next page)

Variables and Variable Ranks Used to Derive the SDI, BTI, and CRI for 40 Sub-Sahara African Countries

BTI	POPULATION PRESSURE INDEX	POPULATION PRESSURE RANK	PERCENTAGE OF LAND WITH MEDIUM/HIGH DISTURBANCE	DISTURBANCE RANK	PERCENTAGE CHANGE IN CROPLAND 1979–1989	CROPLAND RANK	ANNUAL PERCENTAGE OF FOREST LOST	FOREST LOST RANK	BTI
Madagascar	0.65	17	85	25	3.3	18	1.37	31	91
Malawi	2.57	36	96	28	25.2	39	0.59	13	116
Mali	0.21	8.5	34	6	2.2	15	0.82	23.5	53
Mozambique	0.57	16	65	16.5	1.3	12	1.89	35	79.5
Namibia	0.06	1	25	4.5	0.8	8	0	1	14.5
Niger	0.21	8.5	25	4.5	4	25	0.81	22	60
Nigeria	2.92	38	97	29	6.2	26	0.77	20.5	113.5
Rwanda	9.94	40	100	36	12.9	32	2.15	39	147
Senegal	1.12	24	100	36	0	3.5	0.86	27	90.5
Sierra Leone	1.58	30	100	36	8.6	29	0.82	23.5	118.5
Somalia	0.38	12	68	18.5	3.9	23	0.35	7	60.5
South Africa	0.92	22	73	20	-0.6	2	0.47	9	53
Sudan	0.33	10	68	18.5	3.9	23	0.77	20.5	72
Swaziland	1.47	29	100	36	18.4	37	0.53	11	113
Tanzania	1.02	23	59	15	1	10	0.62	14	62
Togo	2.51	35	100	36	7.3	28	0.65	16	115
Uganda	2.74	37	55	12.5	18.3	36	1.17	29	114.5
Congo (Zaire)	0.50	14	55	12.5	3.5	21	0.30	5.5	53
Zambia	0.42	13	18	2	3.1	17	0.30	5.5	37.5
Zimbabwe	0.82	20	82	23	8.8	30	0.57	12	85

(continued on next page)

APPENDIX 5.2 *(continued from previous page)*
Variables and Variable Ranks Used to Derive the SDI, BTI, and CRI for 40 Sub-Sahara African Countries

CRI	HDI: RANK-ORDER WORLD	HDI RANK	FERTILITY RATE 1990–1995	FERTILITY RATE RANK	PERCENTAGE LABOR FORCE IN AGRICULTURE	AGRICULTURAL LABOR FORCE RANK	CRI
Angola	147	11	7.2	5	73	25	41
Benin	150	9	7.1	7	70	29.5	45.5
Botswana	95	39	5.1	37	28	38	114
Burkina Faso	154	6	6.5	19	87	4.5	29.5
Burundi	139	16	6.8	13	92	1	30
Cameroon	119	29	5.7	34	79	18	81
Central African Rep.	142	14	6.2	25	81	15.5	54.5
Chad	152	7	5.9	31.5	83	12	50.5
Congo Republic	115	32	6.3	23	62	33	88
Equatorial Guinea	137	18	5.9	31.5	77	20	69.5
Ethiopia	141	15	7	10	88	3	28
Gabon	97	38	5.3	35.5	75	22.5	96
Gambia	159	2	6.1	27	84	11	40
Ghana	121	28	6	29.5	59	34	91.5
Guinea	158	3	7	10	78	19	32
Guinea-Bissau	151	8	5.8	33	82	13	54
Ivory Coast	122	27	7.4	3	65	31.5	61.5
Kenya	113	33	6.3	23	81	15.5	71.5
Lesotho	107	35	4.7	39	23	39	113
Liberia	132	22	6.8	13	75	22.5	57.5

(continued on next page)

Variables and Variable Ranks Used to Derive the SDI, BTI, and CRI for 40 Sub-Sahara African Countries

CRI							
	HDI: RANK-ORDER WORLD	HDI RANK	FERTILITY RATE 1990–1995	FERTILITY RATE RANK	PERCENTAGE LABOR FORCE IN AGRICULTURE	AGRICULTURAL LABOR FORCE RANK	CRI
Madagascar	116	31	6.6	16.5	81	15.5	63
Malawi	138	17	7.6	2	87	4.5	23.5
Mali	156	4	7.1	7	85	8.5	19.5
Mozambique	146	12	6.5	19	85	8.5	39.5
Namibia	105	36	6	29.5	43	36	101.5
Niger	155	5	7.1	7	85	8.5	20.5
Nigeria	129	24	6.4	21	48	35	80
Rwanda	133	21	8.5	1	90	2	24
Senegal	135	19	6.1	27	81	15.5	61.5
Sierra Leone	160	1	6.5	19	70	29.5	49.5
Somalia	149	10	7	10	76	21	41
South Africa	57	40	4.1	40	13	40	120
Sudan	143	13	6.1	27	72	26	66
Swaziland	104	37	4.9	38	74	24	99
Tanzania	127	25	6.8	13	85	8.5	46.5
Togo	131	23	6.6	16.5	65	31.5	71
Uganda	134	20	7.3	4	86	6	30
Congo (Zaire)	124	26	6.7	15	71	27.5	68.5
Zambia	118	30	6.3	23	38	37	90
Zimbabwe	111	34	5.3	35.5	71	27.5	97

APPENDIX 5.3
Variables and Variable Ranks Used to Derive the SDI, BTI, and CRI for 17 South Asian Countries

SDI

	PLANT DIVERSITY	PLANT DIVERSITY RANK	BUTTERFLY DIVERSITY	BUTTERFLY DIVERSITY RANK	BIRD DIVERSITY	BIRD DIVERSITY RANK	MAMMAL DIVERSITY	MAMMAL DIVERSITY RANK	SDI
Afghanistan	4,000	2	106	1	456	6	132	3	12
Bangladesh	5,000	4	248	7	354	3	117	2	16
Bhutan	5,468	5	237	6	448	5	151	4	20
Cambodia	6,500	6	232	5	305	2	175	5	18
China	32,200	17	457	14	1,100	16	446	16	63
India	16,000	15	514	16	969	15	360	15	61
Indonesia	24,375	16	988	17	1,519	17	547	17	67
Laos	7,000	8.5	263	8	481	8	191	8	32.5
Malaysia	15,500	14	450	13	501	9	305	13	49
Myanmar	7,000	8.5	501	15	867	14	293	12	49.5
Nepal	6,973	7	198	4	629	12	185	7	30
Pakistan	4,950	3	114	3	476	7	180	6	19
Papua New Guinea	11,544	12	303	10	578	10	214	11	43
Philippines	8,931	10	339	11	395	4	195	9	34
Sri Lanka	3,314	1	110	2	221	1	99	1	5
Thailand	12,625	13	394	12	616	11	319	14	50
Vietnam	10,500	11	274	9	638	13	211	10	43

(continued on next page)

(continued on next page)

APPENDIX 5.3 (continued from previous page)
Variables and Variable Ranks Used to Derive the SDI, BTI, and CRI for 17 South Asian Countries

BTI	POPULATION PRESSURE INDEX	POPULATION PRESSURE RANK	PERCENTAGE OF LAND WITH MEDIUM/HIGH DISTURBANCE	DISTURBANCE RANK	PERCENTAGE CHANGE IN CROPLAND 1979–1989	CROPLAND RANK	ANNUAL PERCENTAGE OF FOREST LOST[a]	FOREST LOST RANK	BTI
Afghanistan	0.68	4	82	9	0	3	0	1	17
Bangladesh	18.57	17	100	16.5	2.1	9	0.65	10	40
Bhutan	0.30	2	71	5	8.5	14	0.45	9	30
Cambodia	1.11	5	78	7	0.4	5	0.79	12.5	29.5
China	1.59	10	68	4	-4	1	0.42	8	23
India	5.37	16	97	13.5	0.7	6	0.40	5.5	41
Indonesia	1.54	9	48	2	12.3	15	0.34	3	29
Laos	0.54	3	72	6	3.6	11	0.73	11	31
Malaysia	1.41	7	59	3	1.5	7.5	0.41	7	24.5
Myanmar	1.18	6	92	10.5	0.2	4	0.40	5.5	26
Nepal	3.51	11	79	8	14.3	16	3.57	17	52
Pakistan	4.69	14	95	12	4	12	0.37	4	42
Papua New Guinea	0.19	1	37	1	8.3	13	0.06	2	17
Philippines	5.10	15	97	13.5	2.5	10	1.01	14	52.5
Sri Lanka	4.02	12	100	16.5	1.5	7.5	1.34	15	51
Thailand	1.53	8	92	10.5	25.5	17	2.06	16	51.5
Vietnam	4.58	13	98	15	-2.9	2	0.79	12.5	42.5

APPENDIX 5.3 *(continued from previous page)*
Variables and Variable Ranks Used to Derive the SDI, BTI, and CRI for 17 South Asian Countries

CRI	HDI: RANK-ORDER WORLD	HDI RANK	FERTILITY RATE 1990–1995	FERTILITY RATE RANK	PERCENTAGE LABOR FORCE IN AGRICULTURE	AGRICULTURAL LABOR FORCE RANK	CRI
Afghanistan	157	1	6.9	1	61	11	13
Bangladesh	136	5	4.7	7	59	12	24
Bhutan	144	3	5.9	4	92	2	9
Cambodia	140	4	4.5	8	74	5	17
China	82	14	2.2	16.5	73	6	36.5
India	123	7	3.9	11	62	10	28
Indonesia	98	12	3.1	14	56	13	39
Laos	128	6	6.7	2	76	3.5	11.5
Malaysia	52	17	3.6	13	26	17	47
Myanmar	106	10	4.2	9	70	7	26
Nepal	145	2	5.5	5	93	1	8
Pakistan	120	8	6.2	3	47	15	26
Papua New Guinea	117	9	4.9	6	76	3.5	18.5
Philippines	84	13	3.9	11	45	16	40
Sri Lanka	75	15	2.5	15	49	14	44
Thailand	66	16	2.2	16.5	67	8.5	41
Vietnam	99	11	3.9	11	67	8.5	30.5

World Ranking (Relative Position) of 77 Countries for SDI, BTI, and CRI

	SDI	BTI	CRI
SOUTH ASIA			
Afghanistan	53	123.5	60
Bangladesh	74	200	117
Bhutan	88	177.5	52.5
Cambodia	92	121.5	99
China	281.5	138.5	166
India	268	141.5	133.5
Indonesia	301	156	164
Laos	124	118	70.5
Malaysia	211	140	197
Myanmar	237.5	134.5	136
Nepal	124	253	54
Pakistan	75	188	116
Papua New Guinea	172	84.5	109
Philippines	139	214	171
Sri Lanka	33	220	185
Thailand	216	250.5	180
Vietnam	176	178.5	147.5
SUB-SAHARA AFRICA			
Angola	198.5	111.5	47.5
Benin	129	178	53
Botswana	105	79.5	160.5
Burkina Faso	47	222	34.5
Burundi	123.5	227.5	36.5
Cameroon	247	135.5	95
Central African Republic	175	76.5	62
Chad	74.5	100	57
Congo Republic	174	103	108.5
Equatorial Guinea	115	73.5	79
Ethiopia	186.5	140.5	33
Gabon	174.5	39.5	121
Gambia	28.5	253.5	44
Ghana	177	185	113
Guinea	110.5	162	34
Guinea-Bissau	22.5	232	61
Ivory Coast	166	242.5	77.5
Kenya	235	195	84.5
Lesotho	35	228.5	162
Liberia	91	158	67.5
Madagascar	82	174	76
Malawi	133.5	224	29.5

(continued on next page)

World Ranking (Relative Position) of 77 Countries for SDI, BTI, and CRI

	SDI	BTI	CRI
Mali	96	104	23.5
Mozambique	162	147.5	45.5
Namibia	103.5	30.7	135.5
Niger	49	113.5	24.5
Nigeria	199	217.5	102.5
Rwanda	149	279	31
Senegal	68.5	173	71.5
Sierra Leone	86	230	58
Somalia	111.5	113.5	46
South Africa	202	110	196
Sudan	169	138	78
Swaziland	57	220.5	124.5
Tanzania	244.5	120	56.5
Togo	131	220	87
Uganda	233	212.5	37
Congo (Zaire)	274	99	82
Zambia	173	75.5	120.5
Zimbabwe	126.5	170.5	118
TROPICAL AMERICA			
Argentina	198	73	215.5
Belize	141	135.5	176.5
Bolivia	270.5	88.5	146.5
Brazil	301	121.5	202
Chile	42	80	218
Colombia	300	150	213
Costa Rica	225	229	206.5
Ecuador	278	188.5	182
El Salvador	91.5	238.5	186
Guatemala	170.5	222.5	137
Guyana	185	20.5	192.5
Honduras	144	199	153.5
Mexico	268.5	154.5	203.5
Nicaragua	159	181.5	155
Panama	228	138.5	200.5
Paraguay	127	173	166.5
Peru	291	103	180
Suriname	150	85.5	210
Uruguay	32	78.5	229
Venezuela	280	164.5	211

Data used to determine relative positions on each index can be found in appendices 5.1–5.3.

CHAPTER 6

BIODIVERSITY, AGRICULTURAL PRODUCTIVITY, AND PEOPLE

John Burnett

Humans continue to modify their natural environment to better satisfy their needs.

—*Anonymous*

Of all human activities, agriculture has had the greatest adverse effect on natural biodiversity. Land clearance before cropping destroys or changes the existing flora and fauna above and below ground. Clearance, cropping, and extensive grazing result in wind-induced soil erosion, followed by greater runoff and sedimentation, thus increasing loss of soil fertility and biodiversity over a much wider area than that cultivated. Palynological and historical records show this clearly. Once-fertile regions have been made barren; arborescent communities have been replaced by a mixture of mainly herbaceous, weedy, and cultivated crops and stock with an associated new fauna. Only rarely, in temperate regions, have the resulting herb-rich communities shown an increase in biodiversity over that which has been replaced. Despite the loss of vast tracts of the natural environment, in many ancient and medieval agricultural systems there was a quasisymbiotic balance between crops, stock, and land use. This balance preserved localized fertility, or even increased it in some cultivated areas, and some natural biodiversity was retained.

The increased urbanization of human populations in the postmedieval period and their rising growth rates coincided with an equally rapid growth in understanding of the natural world. For an inherently exploitive species such as *Homo sapiens*, the consequential adaptive strategy was to try to manipulate the agricultural environment, both to increase its productivity and to lessen the burden of labor. The development of mechanized agriculture in Europe in the late eighteenth century and of fertilizers and fungicides at the end of the nineteenth century were natural outcomes. The tempo of change increased. The application of genetics to breeding in the first half of the twentieth century was followed by the development of effective pesticides and herbicides. Objectives became increasingly clear: to achieve the highest possible yields and optimal adaptation of crops and stock in defined, manipulated conditions. In the second half of this century, maximum biological and economic efficiency in producing crops and stock have almost been achieved through agricultural industrialization, resulting in standardized outputs. These initially European tendencies, rapidly adopted and expanded in North America, have been copied worldwide. But it has become apparent that the technology involved in the development of modern, uniform varieties and breeds and the manner of their adoption result in the unavoidable loss of *their* biodiversity. Thus there has been a double loss of biodiversity as a result of the pursuit of increased agricultural productivity. Today and in the predictable future, a world that is suffering great losses in its natural biodiversity faces major losses in the biodiversity of its cultivated crops and domesticated stock.

The objective of this chapter is to explore these aspects of agriculture and consider whether their effects can be ameliorated.

EFFECTS OF AGRICULTURAL ACTIVITIES ON NATURAL BIODIVERSITY

Decline in natural biodiversity coupled with associated environmental degradation correlated with increased agricultural activity is detectable in palynological records since at least 5,000 yrbp, for example, in southwest Asia (Bender 1975). Global statistics for the last hundred years show an unabated change in the balance of ecosystems (James et al. 1992; table 6.1), especially the continuing loss in forest and woodland and increase in cropped and grazed areas. These trends have continued to 1995 so that over 1 billion ha of woodland and forest have been lost over the last 100 years.

Only 3.2 billion hectares of land worldwide are potentially suitable for arable cultivation, although only 10 to 15 percent are without some inherent physical or chemical constraints. Remedial measures are therefore necessary in most cropped soils to uphold water supply, improve drainage, restore or replace mineral deficiencies or excesses, or improve and preserve texture. Such measures also restore some of the lost biodiversity of the soil, although its composition is likely to differ from its preagricultural pattern. One and a half billion hectares are now under cultivation. Available reserves of land potentially usable for agriculture in nonindustrial countries are unevenly distributed and much of this land is of poorer soil quality than that cultivated already (table 6.2).

TABLE 6.1

Gross Global Changes in Land Use 1892–1995

CATEGORY	ACTUAL CHANGE	PERCENTAGE CHANGE
Forest/woodland	−1.10	+21.2
Arable/permanent crops	+0.64	+74.4
Permanent pasture	+2.01	+133.3

Based on James et al. (1992).

TABLE 6.2

Land in Arable Use and Reserves in Nonindustrial Regions Predicted to 2000

REGION	IN USE 1982–1984 (MILLION HA)	EXPANSION TO 2000 (MILLION HA ARABLE LAND)	RESERVES AFTER 2000 (MILLION HA)
Near East / North Africa	92	0	3
Sub-Saharan Africa	201	33	582
93 Developing countries[a]	768	83	1,291[b]
Asia (excluding China)	280	14	48
Latin America	195	36	658[b]
Total or % of total	1,536	10.8%	168.1%

Based on Alexandratos (1988).
a. †These are listed in Alexandratos (1988). Note that these estimates are based on a different basis from data given in table 6.1.
b. In these areas the available reserves are heavily concentrated in Zaire and Brazil, respectively.

One-third of this land is in areas of good rainfall, but two-thirds, located largely in Asia, will require increased irrigation to become usable for agriculture. Agricultural expansion almost anywhere will involve loss of further tropical forest (currently estimated at 17 million ha per year) and inevitably will be accompanied by more large losses in natural biodiversity and soil fertility (Alexandratos 1988). Thus even the possibility of restoring natural biodiversity eventually will be compromised, if not completely eliminated.

Associated with changes in agricultural use, the loss of soil globally is estimated to be 0.7 percent per year, or about 23×10^9 metric tons per year (Brown 1988). Over the last 15 years the per capita loss of soil has increased from 4.0 to 4.8 metric tons per year—that is, by 20 percent annually. The best available evidence suggests that despite remedial measures, 25 percent of existing arable and grazing land is adversely affected by wind and water erosion, salinization, and desertification caused directly and indirectly by human activities (Oldeman et al. 1990). Indeed, environmental degradation leading to loss of biodiversity can have social causes arising from population and other human pressures (Delpeuch 1994).

Tables 6.3 and 6.4 show the devastating effects of erosion in tropical regions and how it can be reduced, provided there is enough plant cover (Repetto 1986; Magrath and Arens 1987).

Another serious concern is the ever-increasing annual consumption of water for agriculture and the uncertainty about its availability and quality. For example, in 1980 (the last year for which reliable data are available), agriculture accounted

TABLE 6.3

Annual Relative Rate of Erosion for Differently Vegetated Areas in West Africa Compared with That from Equal Areas of Bare Ground

COVER	RELATIVE VALUES TO BARE GROUND AT 1
Cotton	0.500
Groundnuts	0.400–0.800
Maize, sorghum, millet	0.300–0.900
Forage and cover crops, first year	0.300–0.800
Forage and cover crops, second year	0.100
Cassava, first year, and yams	0.200–0.800
Palms, coffee, cocoa, and cover crops	0.100–0.300
Intensive rice, second cycle	0.100–0.200
Rapid cover crops	0.100
Ungrazed savanna/grassland	0.010
Dense forest/thick straw mulch	0.001

Based on Repetto (1986).

for almost 70 percent of all water consumed and almost 89 percent of the total irretrievable loss of water. Projections suggest little change in these figures (Ayibotele and Falkenmark 1992; Shiklomanov 1990).

The overall demand for water by agriculture and the high percentage irretrievably lost are exacerbated by the increasing contamination of groundwater and excessive demands on irreplaceable available water. Groundwater contamination has led to legislation, particularly in the United Kingdom, restricting the use of nitrogen fertilizers within "nitrogen-vulnerable zones" resulting from previous excessive use of fertilizers. Excessive demand, through irrigation in water-limited regions, has worsened existing water shortages (e.g., in irrigated maize-growing regions of the United States).

Estimates of freshwater available to support natural vegetation and, therefore, biodiversity, agriculture, or any other land use, suffer from two major areas of uncertainty (Ayibotele and Falkenmark 1992):

- Lack of data on available freshwater and its freedom from contamination, especially in the dry and humid tropics and the subtropics
- The hydrologic features that determine the world's freshwater resources

Despite the uncertainty, it seems probable that available freshwater stocks are steadily declining. This decline will eventually result in widespread loss of biodiversity.

The loss of covering flora and its associated fauna as a consequence of converting natural communities to agriculture has been described often. But at least as serious, if not more so in the long term, are consequential changes in the soil pop-

TABLE 6.4

On-Site Losses of Soil in Relation to Vegetation Cover in Java

LAND USE	AREA (MILLION HA)	TOTAL SOIL LOSS (MILLION METRIC TONS)	SOIL LOSS (METRIC TONS/HA)
Forest	2.4	14	5.8
Degraded forest	0.4	35	87.2
Natural wetlands	0.1		
Rain-fed cropland ("tegal")	5.3	737	183.3
Wet riceland ("sawah")	4.6	2	0.5

Based on Magrath and Arens (1987).

ulations (bacterial, fungal, algal, and invertebrate). Such changes are poorly documented, and taxonomic and biological ignorance of them is so great that they are largely unknown except in a very general way. The majority of such life forms remain to be described and the number of practitioners able to describe them is inadequate. This is especially so in the tropics, where "the numbers of personnel involved in taxonomic work with invertebrates and micro-organisms based in the tropics are insignificant in relation to the magnitude of the task before them" (Hawksworth 1993:41).

Table 6.5 provides some general data about changes in soil fauna when woodland is converted to agricultural use. However, because of their undeniable importance, more detailed data are given in tables 6.6 and 6.7 for soil fungi as a specific example of the adverse effects of agriculture on soil biodiversity. Soil fungi play a major role in regulating the balance and cycling of soil nutrients. It is now clear that mycorrhizal fungi play a major role, both indirectly and directly, in these

TABLE 6.5

Consequences for Soil Biodiversity of Conversion to Pasture and Wheat on Former Beechwood Soil (abundance in millions of organisms/acre)

ORGANISM TYPE	BEECHWOOD	GRASSLAND	WHEAT	WHEAT AND MANURE
Earthworms	0.72	0.84	0.60	2.60
Potworms	2.16	7.60	0.20	1.10
Gastropods	0.42	0.05	—	0.04
Millipedes/centipedes	1.04	1.80	1.80	4.50
Mites/springtails	17.85	57.00	30.20	47.10
Diptera/elaterid larvae	0.99	13.40	4.70	25.30
Other insects/Arachnida	1.91	12.20	0.62	3.57

TABLE 6.6

Qualitative Changes in Types of Soil Fungi in Relation to Agricultural Management of North Temperate Forests and Temperate Seminatural Grassland

NORTH TEMPERATE FORESTS				
MANAGEMENT	BIOLOGICAL FUNGAL TYPES			
	ECTOMYCORRHIZA	SOIL SAPROTROPHS	WOOD SAPROTROPHS	TREE PARASITES
None	d	i	I	0
Clear felling	d	d	D	D
Litter removal	I	D	d	0
Lime added	i	0	0	i
Nitrogen added	D	d	0	0/?

TEMPERATE SEMINATURAL GRASSLANDS			
MANAGEMENT	BIOLOGICAL FUNGAL TYPES		
	MYCORRHIZA	SOIL SAPROTROPHS ON POOR SUBSTRATES	SOIL SAPROTROPHS ON RICH SUBSTRATES
Light grazing	0	0	0
Intensive grazing	d	d	i
Added slurry	D	D	I
Artificial fertilizer	D	D	i
Removal of topsoil	0	D	D
Plowing/resowing	D	D	D

Based on Arnolds (1990).
Key: 0, no effect; i, small increase; d, small decrease; **I**, large increase; **D**, large decrease; 0/?, variable effect.

processes. Thus they affect the success of the surface flora and therefore its fauna. As Read has written, "There seems little doubt that mycorrhizal fungi play a central role in the nutrient dynamics of the heathland, forest and grassland ecosystems of the world" (Read 1990:123). In western Europe over the last 50 years, especially over the last 25 years, dramatic changes in soil fungi, correlated with land management and probably aerial pollution, have been detected. Whether comparable changes occur in the nonindustrial world has not been studied—there is almost no one to study it! But it would be surprising if similar changes are not occurring there, or will not occur in the future. Tables 6.6 and 6.7 set out some summarized results (Arnolds 1990).

The data in Table 6.6 are only qualitative, but they show clearly that agricultural operations—clear felling, plowing and resowing, heavy grazing or adding fertilizer, especially with a high nitrogen content—all decrease markedly the mycorrhizal fungi and often the soil saprotrophs. High nitrogen concentrations are

TABLE 6.7

Changes in Species Composition and Abundance of Above-Ground Sporocarps Between 1972 and 1989 in Three 1,000-m² Replicate Plots in Mature Oak Forest, NE Netherlands

FUNGAL CLASS	AVERAGE SPECIES/PLOT			AVERAGE MAXIMUM SPOROCARPS/1,000 M²		
	1972–1973	1976–1979	1988–1989	1972–1973	1976–1979	1988–1989
Mycorrhiza	37	32	12	4,720	1,110	370
Soil saprotrophs	21	33	22	2,430	920	350
Wood saprotrophs/ parasites	15	15	26	180	330	680

Based on Arnolds (1990).

known to inhibit the growth of mycorrhizal fungi (Dörflet and Braun 1980). Clear felling of woodlands markedly decreases wood saprotrophs, crucial to decay and hence to carbon and nutrient recycling, and an increase in tree parasites reduces the potential for healthy regeneration. Some quantitative data are available (table 6.7). A marked decline in the number and abundance of the sporocarps of mycorrhizal fungi is obvious over the period 1972–1989 in the Dutch oakwood sampled. This decline was not associated with differences in any noticeable climatic variable, but aerial pollution had increased appreciably over the period. Such changes are long-term ones. Total macromycete records for the Netherlands over the period 1900–1989 showed that mycorrhizal fungi accounted for 45 to 48 percent of the total macromycete record up to 1969. After that, they decreased to 35 percent over 1970–1979 and to only 26 percent in the decade 1980–1989. The decline is associated mainly with ectomycorrhizal fungi of conifers. Over the same period marked decreases in the numbers and abundance of saprotrophic species of grasslands and heaths appeared to be correlated with changing agricultural practices, most notably increased use of artificial fertilizers (Arnolds 1988). Reduced endomycorrhizal infection and potential for infection is also correlated with disturbance of the soil and the continual use of nitrogenous fertilizers on crops (Harley and Smith 1983; Miller and McGonigle 1992).

CHANGES IN AND LOSS OF CROP AND STOCK BIODIVERSITY

At least 12,000 wild plants have been used by humankind over the millennia and virtually all were domesticated to some extent. Today, however, almost 90 percent of the world's food needs are supplied by 20 crops. Only about 20 to 30 species of animals have been domesticated; just over half of these have been used in some way in agriculture. From the outset, selection (of crops and stock, site, and season) has been the basic technique used in agriculture. Selection implies retention of the plants or animals that best meet the requirements of the farmer; the rest are large-ly rejected. Inevitably, over the millennia there has been a continuous overall ero-sion of the biological range and genetic variability of farmed plants and animals. This trend has seen partial reversals, but in today's high-yield cultivated varieties (HYCVs) and modern stock it has reached its limit.

Much of this erosion of biodiversity has gone unnoticed until recently. How did it come about and what is the magnitude of the largely inadvertent changes involved?

From the origins of agriculture, about 10,000 yrbp, to the middle of the eigh-teenth century crops and stock were selected by eye and experience as the best adapted to produce the best yields for each locality. In parallel, cropping practices developed to select and maintain the best sites for optimum yields. Unfortunately, the limited destruction of existing flora on primitive, small crop sites was extend-ed greatly in Europe, North Africa, and Asia as a consequence of clearance by fire and the early domestication of sheep and goats. Fire damage is often reversible, and wildlife recovery is often possible. However, grazing and browsing accounted far more for the early, often irreversible destruction of the existing vegetation and subsequent erosion of the environment than did arable farming. Overgrazing by domesticated livestock has continued to the present and it is still a predominant factor in biodiversity loss as a consequence of environmental erosion, as in the Sahel region. Globally, overgrazing is said to account for 90 percent of the deserti-fication of rangelands (Mabbutt 1984).

As agriculture developed, selective breeding was based almost entirely on phenotypic selection within small localities. Inevitably, it involved an appreciable degree of inbreeding and resulted in an enormous number of locally adapted Lan-draces, or "folk" varieties, and breeds. These improved, locally adapted forms showed some degree of phenotypic uniformity, which was a selection criterion, but they were still highly variable and remarkably diverse genetically. More inten-sive selection was practiced on crops in the nineteenth century by Vilmorin in France, Rimpau in Germany, and Shirreff in Scotland. The first and last of these se-lected individually superior plants, with good agronomic and yield characters, from many different Landraces of wheat, oats, and barley and propagated them. Because these are self-pollinating crops, the new varieties were highly inbred, so genetic variability was lost. With rye, however, Rimpau chose small numbers of su-perior individuals (usually about 20), interbred them, and selected from the prog-eny. Because rye is outbreeding, he maintained vigor together with improved yield and some degree of genetic diversity. However, the small numbers within which selection was practiced reduced the genetic diversity of the crop. These procedures were adopted and widely copied at that time before their underlying genetic basis was understood.

Although controlled hybridization had been used with plants since about 1750, scientific breeding did not begin until the start of this century. This greatly improved the effectiveness of selection. Within 10 to 15 years it resulted in an increased understanding of quantitative characters, the recognition that resistance to disease and pests had a genetic basis, and the application of scientific cross- and inbreeding, coupled with the use of heterosis. Yields began to increase steadily. New, often more inbred varieties and more specialized stock, such as beef or dairy breeds, were used in preference to the older Landraces and general-purpose breeds. In the period 1920–1940 yields increased steadily, if not spectacularly. The introduction of more bureaucratic intervention, such as government lists of recommended varieties, and the more effective spread of official recommendations and information by the agricultural press and radio increased the rate at which less favored varieties were discarded. Discarded varieties gradually began to be lost completely, a tendency that accelerated in Europe and North America during World War II and immediately afterward. For example, before 1930 all the wheat grown in Greece was made up of local Landraces. Some 25 percent of these were replaced by improved varieties in the decade 1930–40, over half (58 percent) by 1950, and 75 percent by 1960; in 1970 only about 5 percent of all the wheat in Greece represented the old Landraces (Frankel 1973).

As genetic diversity within crops was slowly being reduced, the need for new, genetically determined attributes, most notably resistance and yield-affecting genes, led to the use of wider sources of plant material. In some cases the attributes sought were derived from the old Landraces. In addition, inspired by the work of Vavilov carried out between 1923 and 1933, completely new sources of genetic material were sought in crop-related wild species or "primitive" races occurring in the so-called centers of origin (or diversity)—often centers of early cultivation (Vavilov 1951). Two benefits accrued. First, knowledge of the range of biodiversity of a crop's progenitors greatly increased understanding of its genetic diversity; second, if these newly discovered forms were used for breeding with the crop, its genetic diversity was increased.

Despite such broadening of the genetic basis of many crops, the application of improved selective and breeding techniques for increased genetic and physiological uniformity, coupled with other developments in agriculture and the progress toward industrial crop production, opposed this trend. Developing technology, therefore, led more often to a narrowing of the genetic basis of many crops and breeds. This tendency reached its climax for crops in the HYCVs, developed first in the industrialized countries and later for the developing world. This geographic extension was a response to the rapid growth in the populations of these countries and their consequent need for more food.

Postwar internationalism, symbolized by the establishment of the Food and Agriculture Organization (FAO), was coupled with a recognition that techniques that had led to self-sufficiency, or even to food surpluses in the industrial world, could be applied to the problem of feeding other countries. The practical response was, in essence, the application worldwide of the full gamut of industrial agricul-

TABLE 6.8

Adoption of Modern HYCVs of Cereals in India 1965–1984

	PERCENTAGE HECTARAGE IN HYCVS				
	WHEAT	RICE	SORGHUM	PEARL MILLET	MAIZE
1965	—	—	—	—	—
1970	35.5	14.9	4.6	8.9	7.9
1975	65.8	31.5	12.2	25.0	18.8
1980	72.3	45.4	21.1	31.3	26.8
1984	82.9	56.9	32.5	49.3	35.5

From Lipton and Longhurst (1989).

tural technologies to the major food crops and, to a lesser extent, to stock. This so-called green revolution resulted in huge increases in yields and productivity of the major food crops in developing countries.

Tables 6.8 and 6.9 illustrate the spread and yields of these new HYCVs in developing countries. In India the adoption of HYCVs of wheat over the 20-year period 1965–1984 was highly successful, leading to self-sufficiency and even the capacity to export surplus grain. Within 7 years of their adoption, over 50 percent of the wheat-growing area in India was planted with HYCVs. This figure is now well over 80 percent, and 50 percent of the rice and pearl millet areas are now sown with HYCVs (table 6.9). In other developing countries these proportions have been exceeded, but on average, almost half the area of wheat, rice, and maize in developing countries is sown with HYCVs. Landraces have declined dramatically and even been totally replaced by HYCVs.

The yields of HYCVs per hectare far exceed those of former cultivated varieties. However, in many countries these gains have done no more than keep pace with the food requirements of growing populations. For example, in Asia the total yield of rice increased from 292 million tons in 1971 to 479 million tons in 1990, a 64 percent increase, yet because of the annual population increase of about 2 percent, yield per capita increased only slightly. To maintain a constant yield per capita, the yield of rice will have to grow by about 2.6 percent per year at least to the year 2000. It is uncertain whether such a rate of increase can be sustained, even if achieved, for in many cases the yields of the HYCVs appear to have leveled out (Hammond 1994). This uncertainty increases the probability of agriculture being extended into new areas with the loss of more natural vegetation and biodiversity.

There is no question that HYCVs have promoted agricultural productivity and economic gain worldwide and reduced the risk of famine, and saving lives in many developing countries, but there are some exceptions. For instance, the contribution of HYCVs, with the exception of hybrid maize, to the food problems of sub-Saharan Africa has been small (Grigg 1993). A more serious, less immedi-

TABLE 6.9

Areas of Wheat and Rice in Developing Regions, 1982–1983, and of Maize, 1983–1986

	WHEAT			RICE			MAIZE		
	TOTAL (MILLION HA)	HYCVS (MILLION HA)	HYCVS (%)	TOTAL (MILLION HA)	HYCVS (MILLION HA)	HYCVS (%)	TOTAL (MILLION HA)	HYCVS (MILLION HA)	HYCVS (%)
Noncommunist Asia[a]	25.4	32.1	79.2	36.4	81.1	44.9	15.7	44.1	35.6
Communist Asia[b]	8.9	29.1	30.6	33.4	41.2	81.0	19.2	27.0	71.1
Near East[c]	7.6	24.8	30.6	0.1	1.2	8.3	2.4	5.1	46.6
Africa[d]	0.5	1.0	50.0	0.2	4.3	4.7	14.9	29.0	51.4
South America	8.3	10.7	77.6	2.5	7.6	32.9	27.3	50.5	54.1
Total	50.7	97.7	51.9	72.6	135.4	53.6	79.5	155.7	51.1

From Lipton and Longhurst (1989).

a. Excluding Taiwan and West Asia.

b. Excluding North Korea and underestimating short CVs in China.

c. North Africa, West Asia, and Afghanistan.

d. Excluding North Africa, including Sudan; excluding South African Republic.

ately visible disadvantage of HYCVs is that in addition to leading to serious reductions in crop biodiversity, their cultivation is causing environmental problems in some regions. Some of these problems are direct, such as the consequences of high fertilizer use; others are indirect. In particular, the social impact
of industrial agriculture on subsistence farmers in some developing countries has
triggered a chain of consequences terminating in land degradation. Therefore, in
both industrialized and developing countries it is questionable how far the original biodiversity of crop plants can be recovered, how far current HYCVs can be
regarded as sustainable crops, and how long the associated husbandry practices
can continued unchanged.

The data given earlier for the loss of Landraces of wheat in Greece and those
implicit in tables 6.8 and 6.9 starkly illustrate the consequences of replacing other
varieties with HYVCs. These data do not reveal the small number of HYCVs

TABLE 6.10

Recent Crop and Varietal Uniformity in the Netherlands and the United States

CROP UNIFORMITY IN THE NETHERLANDS, 1992		
CROP	% ACREAGE SOWN	CUMULATIVE %
Fodder maize	26	
Potato	21	47
Winter wheat	17	64
Sugar beet	16	80
All other crops	20	100

VARIETAL UNIFORMITY IN THE NETHERLANDS, 1989			
CROP	% ACREAGE SOWN WITH LEADING VARIETIES (CVS)		
	MAIN CV	FIRST TWO CVS (2D)	FIRST THREE CVS (3D)
Winter wheat	61	73 (12)	79 (6)
Spring wheat	94	98 (4)	99 (1)
Potato	78	82 (4)	84 (2)
Fodder maize	21	37 (16)	53 (16)
Forage peas	45	83 (38)	95 (12)

VARIETAL UNIFORMITY IN THE UNITED STATES, 1986		
CROP	MAJOR VARIETIES	ACREAGE (%)
Wheat	9	50
Maize	6	71
Potato	4	72
Pea	2	96

Adapted from Vellve (1992).

grown. Tables 6.10A and 6.10B (Vellve 1992) provide information on crop and varietal uniformity in the Netherlands, as an example of the situation in a modern
and agriculturally progressive European country, and Table 6.10C provides similar information about the recent situation in the United States. Some action has
been taken to promote the genetic diversity of U.S. crops since their uniformity
was recognized, but it is unclear how much their basis has been widened.

A few favored varieties often cover immense hectarages; for example, the most
favored spring wheat in the Netherlands in 1989 accounted for 94 percent of the
total spring wheat hectarage (table 6.10B). Similar situations occurred in developing countries. In India, for example, it is estimated that over 30,000 different varieties of rice have been grown this century, but in just 15 years after the adoption of
the HYCVs, only 10 varieties may cover as much as 75 percent of the total rice
acreage in India (Jain 1982). There are similar large hectarages in other crops. For
example in the 1980s IR36, the most widely cultivated rice variety known, covered
11 million hectares in India alone and was also grown in some regions of Africa.

Such uniformity greatly decreases crop diversity, but the actual situation is
often worse than appears, for the favored HYCVs are themselves remarkably uniform in their germplasm. Table 6.11 sets out the origins of the germplasm associated with three genetically determined attributes used recently by European breed-

TABLE 6.11

Germplasm Used by European Breeders in Barley and Onions

CROP AND SOURCE OF GERMPLASM	ATTRIBUTE AS % OF TOTAL GERMPLASM		
	DISEASE RESISTANCE	STRESS TOLERANCE	YIELD INCREASE
Barley			
Advanced HYCVs	68	63	96
Landraces	22	28	4
Weedy relatives	4	3	—
Wild species	6	6	—
Breeder's material	59	71	85
Gene bank material	41	28	15
Onion			
Advanced HYCVs	46	59	82
Landraces	31	28	9
Weedy relatives	5	6	6
Wild species	18	6	3
Breeder's material	80	91	84
Gene bank material	20	9	16

Adapted from Vellve (1992).

ers in developing modern HYCVs of barley and onion, respectively (Vellve 1992). For instance, in barley, 68 percent, 63 percent, and 96 percent of the incorporated germplasm relating to disease resistance, stress tolerance, and yield increase, respectively, was derived from other HYCVs, themselves having much of their germplasm in common. Only 32 percent, 37 percent, and 4 percent of the germplasm used was derived from the genetically more diverse Landraces, weedy relatives, or wild species for these three attributes, respectively. In other words, for barley about 76 percent and for onions about 62 percent of the germplasm for these attributes was derived from closely related varieties. This high degree of genetic uniformity is borne out by the related information on source of origin, where it can be seen that the majority of material used in the breeding programs was the breeder's own material. This is likely to be more genetically uniform than that derived from gene banks, where widely collected material is stored; 72 percent and 85 percent for barley and onion, respectively, was derived from breeders, against only 28 percent and 15 percent from gene banks. Data such as these are typical of most current HYCVs. Their innate genetic variability is immensely reduced when compared with that exhibited by the large numbers of Landraces or their earliest cultivated progenitors.

The high degree of genetic uniformity and the vast hectarages over which many are planted enable HYCVs to be planted, sprayed, and harvested with minimal fixed costs. All these factors render them particularly susceptible to diseases if their inbred genetic resistance is broken down. For ease of breeding, single-gene resistance to specific races of pests and diseases is often incorporated into HYCVs. Unfortunately, once a new pathogenic form arises, the whole hectarage of the HYCV is at risk. The ineffective resistance gene can be replaced by one or more different genes, but most of the rest of the genotype remains unchanged. This strategy can prolong the period of resistance, but sooner or later the newly introduced genes are overcome. Hence the productive life of a modern HYCV is 5 to 10 years at the most and can be appreciably shorter. The replacement of specific gene resistance by general or durable resistance with a polygenic basis, characteristically found in many of the old Landraces, is much more difficult, takes longer, and is often associated with some unavoidable loss in yield. Therefore, it has not been favored by breeders of HYCVs. Change is desirable but rendered more difficult because much of the available genetic resistance of this nature has been lost with the loss of the old Landraces.

Further problems with the HYCVs are caused by the environmental impacts of associated crop husbandry practices. First, most modern wheat and rice varieties have a high demand for fertilizers and, because of their uniformity, are designed to be harvested industrially. The continual incorporation of added fertilizers, the effects of mechanical harvesting with heavy machines, and the exposure of large hectarages of the soil surface at harvest often deplete soil fertility and promote its erosion. This is especially true for many of the more fragile soils in developing countries, less so for those of the north temperate zone, where the majority of industrial farming is located. As a consequence, soil fertility is impaired. Of course, these problems are also well known in industrial countries, particularly in high soil-nitrogen areas. Such soil degeneration reduces biodiversity.

The effectiveness of modern pesticides and the problem of finding genes conferring resistance to each pest have led to widespread reliance on pesticide spraying for their protection. Ever since *Silent Spring* (Carson 1963), the devastation that pesticides can cause on a wide variety of animal and microbial life, in addition to the pest concerned, has been common knowledge. However, industrialized countries still consume most pesticides—57 percent of the world market in 1991—but 23 percent is consumed in the Far East and 8 percent in Latin America. Since 1970 their use has increased virtually linearly and more rapidly than has agricultural production (Hammond 1994). The concomitant reduction in animal biodiversity inevitably spreads far beyond the sites of application through runoff and through food chains. The damage to wildlife is worse than it need be. Pesticides banned in industrialized countries because of their known harmful effects on wildlife are still sold to developing countries. A further disadvantage is that pesticide resistance often develops frequently and rapidly. Therefore, the percentage crop loss from pest damage since they were introduced on a widespread commercial scale 50 years ago has not declined measurably. Biodiversity loss through pesticides has continued unabated over the same period.

It is pertinent to point to two economic features that underlie the use of HYCVs other than the apparent advantages of industrialized agriculture. The first is the monopoly position of the small number of global agricultural suppliers. In particular, seed production and marketing are increasingly becoming the monopoly of about 15 multinational companies already responsible for producing fertilizers, fungicides, herbicides, and pesticides. They offer a complete service—seed and the necessary chemicals to sustain and protect the crops—which tends to perpetuate the use of these manufacturer's products, including those that have adverse effects on soil or wildlife. Second, vast international aid subsidies to adopt HYCVs are offered to developing countries by international organizations. Naturally, governments are eager to accept.

Static yields and subsidized government promotion of HYCVs coupled with high population increases have only one consequence in developing countries: the agricultural area is extended even more rapidly into adjoining areas and further natural vegetation and biodiversity loss occurs. Table 6.12 provides some data concerning recent high rates of increase in arable and grazed land in developing countries to accommodate the growth of HYCVs. These should be compared with the data already provided on the availability of additional cultivable land (see table 6.2).

The social problems that the HYCVs have brought to developing countries, which have also affected to a far lesser degree the industrial countries, are well documented and will not be described here. They have had two important consequences for the maintenance of natural and agricultural biodiversity.

TABLE 6.12

Examples of at Least 20% Conversion to Agriculture and at Least 10% Conversion to Pasture from Natural Vegetation in Developing Countries

COUNTRY	% TO CROPS		COUNTRY	% TO PASTURE	
	1960–1980 (10-YR AVERAGE)	1980–1990		1960–1980 (10-YR AVERAGE)	1980–1990
Paraguay	71.2 (35.6)	26.7	Ecuador	61.5 (30.8)	29.2
Niger	32.4 (16.2)	4.0	Costa Rica	34.1 (17.1)	15.6
Mongolia	31.9 (16.0)	17.0	Thailand	32.1 (16.1)	29.7
Brazil	22.7 (11.4)	23.1	Philippines	26.2 (13.1)	24.0
Ivory Coast	22.4 (11.2)	19.0	Paraguay	26.0 (13.0)	**33.5**
Uganda	21.4 (10.7)	18.3	Vietnam	14.0 (7.0)	**18.3**
Guyana	21.3 (10.7)	0.1	Nicaragua	11.8 (5.9)	**10.7**
Burkina Faso	19.4 (9.7)	**27.9**			

Compiled from annual reports of the World Resources Institute, Washington, D.C.
Bold figures indicate that the 10-yr rate of conversion is increasing.

- Because of the often high cost of HYCV seed, necessary fertilizers and pesticides, and the scale of production, many poor farmers have been driven from the land so that a great deal of knowledge of older farming techniques concerning mixed cropping and varieties has been lost. There has thus been a loss of actual agricultural and quasinatural biodiversity, exemplified by mixed cropping systems, and of "folk" varieties.
- The social upheaval associated with the introduction of HYCVs has increased the number of itinerant, landless people whose demands on available land—to the extent that they can find it—far exceeds its carrying capacity and leads to environmental erosion of all kinds. There is thus a general tendency for biodiversity to be lost (for example through the use of standing timber for firewood).

Figure 6.1 summarizes many of the consequences, all ultimately inimical to the maintenance of biodiversity, that these social changes have provoked. There are clearly many reasons why agricultural systems designed for modern HYCVs are unlikely to be sustainable, not least because HYCVs are highly adapted to specific, human-devised conditions and are dependent on an extremely narrow genetic base. Is change possible?

IMPROVING THE BIODIVERSITY OF CROPS AND STOCK

Unless the loss of genetic diversity within crops and stock is halted, it will become increasingly impossible to restore their biodiversity. The approach initiated

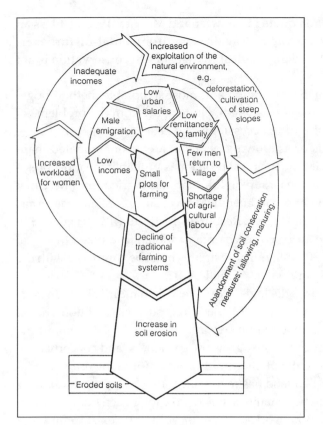

FIGURE 6.1

Social causes of environmental degradation caused by unequal distribution of land in Asia.

After Delpeuch (1994) and Catholic Institute for International Relations.

by Vavilov in the 1920s to find new breeding material has been replaced by the recognition of a parallel necessity to conserve the biodiversity so discovered. The immense importance of conserving genetic resources as opposed to simply exploiting them was revealed largely by the pioneering work of Otto Frankel and Erna Bennett. As a result, genetic conservation of crops was a major theme in the International Biological Programme (IBP) between 1964 and 1974 (Frankel and Bennett 1970; Frankel and Hawkes 1975). Since then there has been an immense national and international expansion of collecting and conserving material of old Landraces and varieties, early crop progenitors, and related wild species.

The techniques available are varied but essentially fall into two classes: in situ and ex situ conservation. Most of the old centers of crop diversity and development are located in developing countries and the biological conservation of such areas appeared at one time to be an effective way of conserving crop biodiversity. However, in practice this is often impossible, either because of the environmental degradation that has occurred in such areas or because the old varieties have been ousted by HYCVs. A particularly striking example is given by Carlos Ochoa from northern Peru. He found 45 different Landraces of potato in two areas in 1945, none of which could be found in the early 1970s, having been replaced entirely by a new HYCV, *Renacimiento* (Rebirth), "one of the varieties I bred . . . for this country" (Ochoa quoted in Fowler and Mooney 1990). Unless an area or site is especially well protected against environmental disturbance, human interference, or unavoidable ecological change, in situ conservation is not likely to be successful.

Ex situ conservation is the most common and most acceptable method of genetic conservation but it too has disadvantages. Quite apart from the problems of locating, collecting, and cataloguing accessions, the cost of maintaining gene banks, whether as viable seeds or as growing plants in an uncontaminated manner, has proved to be very great. The financial problem is probably the most pressing. The budget of the Consultative Group on International Agricultural Research in 1995 was about $280 million and had been held level since 1987 in real terms. From this, the International Plant Genetic Resources Institute (IPGRI) spent $16 million on ex situ conservation and on its own organization. It estimated that total global expenditure was $20 to 30 million. These figures must be compared with the estimated requirement of $300 million for AGENDA 21.

Even with adequate financing, the logistics are very demanding. First, there is the problem of actually collecting material. Expeditions such as the international potato collecting expeditions, although effective, are very costly. Throughout the world there are many subsistence farmers, growers, gardeners, and concerned individuals who have continued to maintain the old varieties for many different reasons. For example, a farmer in Scotland has maintained, for her own interest, over 200 (originally 400) potato varieties, many of which have long been out of cultivation. In some countries national networks, official or voluntary, exist to find and maintain similar collections of other crop plants. These must be extended and IPGRI intends to establish a global register of individuals and their holdings. Harnessing individual interest is a key to discovering and maintaining the scattered Landraces and old varieties. Great opportunities exist in developing countries, where women are often the farm workers. It is they who possess the knowledge and often retain a few examples of the old varieties. Actions such as increasing the numbers of female farm advisers, usually fewer than 5 percent of the total at present, could improve information exchange and plant collection.

Once accessions are available, a fundamental issue is what and how many different samples should be retained in a collection. One of the largest collections of varieties of any crop is that of rice. The International Rice Research Institute at Los Banos in the Philippines holds about 85,000 accessions (wild species, related species, and varieties)—a collection duplicated in the United States and, in part, elsewhere. It is still not clear how adequately this immense and costly collection represents the genetic diversity of the rice plant, but it probably represents no more than 50 percent. Indeed, it is not clear what size of collection adequately represents a crop or even a species. Table 6.13 gives some indication of the numbers of holdings of crops and timbers in ex situ collections, and table 6.14 indicates the estimated percentage variability stored for some crops in these holdings.

TABLE 6.13

Global Ex Situ Gene Bank Holdings, 1994

CROP	NO. OF HOLDINGS	% OF TOTAL
BY TYPE OF CROP		
Cereals	2,067,400	46.8
Food legumes	718,350	16.3
Forages	425,350	9.6
Vegetables	336,600	7.6
Roots and tubers	179,850	4.0
Fruit	174,850	4.0
Oil crops	89,750	2.0
Fiber crops	70,300	1.6
Beverages	42,900	1.0
All other, including timber	310,800	7.1
BY SOURCE		
Wild species	213,700	16.0
Landraces	579,200	43.0
Breeders' stocks	287,700	22.0
Commercial varieties	247,200	19.0

Unpublished data from FAO meeting, Rome, 1994; used with permission.

Only the holdings of wheat, maize, potato, and tomato approach a complete representation of a crop's genetic diversity. It should be noted that even these figures are only estimates; there is no satisfactory method for determining the adequacy of representation of total variability. Problems of this nature become even more acute when, to allow for the constraints imposed on collections by finance, space, equipment, and labor, core collections have been proposed. A core collection has been defined as "a limited set of accessions derived from an existing collection, chosen to represent the genetic spectrum in that collection" (Hodgkin et al. 1995). The danger with core collections is that they will represent only the range seen to be of immediate, or immediately foreseeable, practical utility.

Maintenance of gene banks is another unending and complex task. The major problem of maintenance was highlighted at the Keystone Conference in 1991, where it was said, "In some cases we may be losing as much diversity in the genebanks as we are in the fields." It is too soon to be certain that DNA banks will provide a simpler, equally effective form of conservation collection (Adams and Adams 1992).

Breed conservation is achieved almost entirely by ex situ collections maintained by voluntary associations such as the British Rare Breed Society, the nu-

TABLE 6.14

Estimated Variability, as Percentage of Total Variability, in Gene Bank Holdings

CULTIVATED VARIETIES				
	PERCENTAGE VARIABILITY REPRESENTED			
	90–95%	50%	<50%	10% OR LESS
	Wheat (75%)	Phaseolus bean	Millet	Other crops
	Maize (50%)	Sweet potato	Cassava	
	Potato (70%)	Cucurbits		
	Tomato (95%)	Rice		

WILD RELATIVES				
	70%	60%	10% OR LESS	
	Potato	Wheat	Rice	Lupin
			Sorghum	Lentil
			Millet	Yam
			Cowpea	Cassava
			Phaseolus bean	
			Sweet potato	

Unpublished estimates from various sources.

merous individual breed societies around the world, or government agricultural services, as in India. Although the loss in biodiversity among domestic animals is not as great as with crops, many are at risk, as shown in table 6.15. The region where most breeds are at risk is Europe and the former USSR. Here, at least 355 breeds of an estimated 1,200 in the region are at risk of extinction (Loftus 1993).

The use of greater genetic diversity in breeding is inevitably a long-term problem. Now that the narrow genetic base of modern varieties has been recognized, there are clear signs that attempts will be made to broaden it. The practical issue is that with the perpetual threat of world food shortage, yields cannot be allowed to fall dramatically, so any change must be phased in slowly. A further encouraging change is the wider range of crops, including more root crops, that the international institutions are now trying to improve through breeding, but without the constraints such as high fertilizer requirements. Green manuring, which is environmentally more acceptable, has been found to be effective for several tropical crops. It may involve growing additional crops if suitable ones for manuring are not available. This coupled with a change from monoculture to mixed cropping can often restore soil fertility effectively and maintain reasonable food supplies. As a result, crop biodiversity is achieved and the restoration of natural biodiversity encouraged. Equally desirable is the replacement of massive pesticide use by integrated pest management (IPM). This is already being accomplished in indus-

TABLE 6.15

Breeds of Domesticated Animals at Risk of Extinction

SPECIES	BREEDS RECORDED	RECORDED WITH POPULATION DATA	AT RISK[a]
Ass	69	32	6
Buffalo	70	30	1
Cattle	845	544	120
Goat	330	148	39
Horse	360	204	99
Pig	368	200	69
Sheep	880	521	108
Total	2,922[b]	1,679	442

Based on Loftus (1993).
a. Estimated from breeds for which population data are available (about 40% of all breeds).
b. It is estimated that there are about 4,000 breeds in the world (about 75% have been recorded).

trial countries, most successfully with protected crops, but is beginning to be effective in field situations. In the developing world the use of IPM to control infestations of the Indonesian rice crop by brown leaf hoppers stands as a model application. Over a 15-year period, using HYCVs and the associated agricultural technology, Indonesia became self-sufficient for rice in 1980, having originally been a major importer. However, in that year it accounted for 20 percent of the world use of pesticides. By 1986 pesticide-resistant leaf hoppers had become a major threat to the crop and a switch was made to IPM, using natural predators such as spiders and wasps. The leaf hoppers were controlled, yields of rice were restored, and rice paddy fish, eliminated by pesticides, were restored. Leaf hoppers have not recurred as a major pest, although in 1990 the white-stem rice borer began to increase. It too was eliminated by an appropriate IPM technique. The restoration of natural biodiversity as a result of using IPM rather than pesticides was dramatic.

This remarkable success was achieved through an effective educational training program to the level of the subsistence farmer. It is clear that in developing countries such programs are an essential element if change is to be achieved. Most importantly, the greater involvement of women in agricultural education, training, and decision-making is essential. In Latin America women do 20 percent of all agricultural work, in South and Southeast Asia 40 percent, and in sub-Saharan Africa over 60 percent, most of it actually working on the land. There is evidence that they will initiate change. In Kenya, for example, women's organizations are largely responsible for carrying out biological terracing, planting up the unplowed strips with perennial crops, elephant grass (*Pennisetum* spp), and fruit trees. This has not only reduced erosion in over half of the 1.1 million farms at an average cost of $28 per farm, but has increased agricultural biodiversity and improved crop yields (Wenner 1979). Similarly, in the Sudano-Sahara, women are largely responsible for the agroforestry program that is helping to stem soil erosion.

At present, changes that will increase the genetic diversity of high-yield crop plants and stock and reduce the environmental damage resulting from their associated agricultural systems are coming slowly. In general, they affect only small, local areas compared with the total devoted to agriculture worldwide. The continuing large requirements for food must be met by agricultural systems based on monocultures of HYCVs for several decades at the very least. Gradual amelioration, such as smaller hectarages of individual HYCVs, replacement of pesticides by IPM, and more multicropping and agroforestry, is the best that can be expected in the immediate future. Once the goal of maintaining biodiversity, whether of crops or stock, is recognized widely as a requirement for an effective agricultural system, change in technological thinking will follow. The best way to bring about such change is through the involvement of people at all levels in industrial and developing countries. Inevitably, much biodiversity will continue to be lost and many organisms will become extinct, but change in agricultural practice is now in the air.

Fifty years ago, it seemed to be a proper and morally legitimate goal of agricultural scientists to improve agricultural productivity and food yields at any cost if it would save and maintain human life, a goal that was also economically attractive to their business communities. Now a new morality is gaining ground that recognizes the importance of maintaining the diversity of life as well as ensuring the continuation of human society. It can be seen that it is potentially practicable as well as necessary. The best and first place to ensure that it does occur is in the industrial countries because only they have the capabilities to bring about technical and economic change of this magnitude.

For all humankind, perhaps nothing better expresses the approach now needed than the wise words of that visionary scientist, discoverer, and conserver of agricultural biodiversity, Nikolai Vavilov: "It is better to display excessive concern now than to destroy all that has been created by Nature over thousands and millions of years."

REFERENCES

Adams, R. P. and J. E. Adams. 1992. *Conservation of plant genes: DNA banking and in vitro biotechnology.* San Diego: Academic Press.
Alexandratos, N. 1988. *World agriculture toward 2000.* Rome and London: Food and Agriculture Organization and Belhaven Press.
Arnolds, E. 1988. The changing macromycete flora in the Netherlands. *Transactions of the British Mycological Society* 90: 391–406.

Arnolds, E. 1990. Mycologists and nature conservation. In L. Hawksworth, ed., *Frontiers in mycology*, 243–264. Wallingford, U.K.: CAB International.

Ayibotele, N. B. and M. Falkenmark. 1992. Freshwater resources. In J. C. I. Dooge et al., eds., *An agenda for science for environment and development into the 21st century*, 187–203. Cambridge, U.K.: Cambridge University Press.

Bender, B. 1975. *Farming in prehistory: From hunter-gatherer to food-producer*. London: John Baker.

Brown, L. R. 1988. *The changing world food prospect: The nineties and beyond*. Washington, D.C.: Worldwatch Institute.

Carson, R. 1963. *Silent spring*. London: Hamish Hamilton.

Delpeuch, B. 1994. *Seed and surplus: An illustrated guide to the world food system*. London and Norwich: Catholic Institute for International Relations and Farmer's Link.

Dörflet, H. and U. Braun. 1980. Untersuchungen zur Bioindikation durch Pilze in der Dubener Heide (DDR). In R. Schubert and J. Schuh, eds., *Bioindikation 4:* 15–20. Halle, Germany: Saale.

Fowler, C. and P. Mooney. 1990. *The threatened gene. Food, politics, and the loss of genetic diversity*. Cambridge, U.K.: Lutterworth.

Frankel, O. H. 1973. *Survey of crop genetic resources in their centres of diversity*. Rome: FAO/IBP.

Frankel, O. H. and E. Bennett. 1970. *Genetic resources in plants : Their exploration and conservation*. Oxford, U.K.: Blackwell Scientific.

Frankel, O. H. and Hawkes. 1975. *Genetic resources for today and tomorrow*. Cambridge, U.K.: Cambridge University Press.

Grigg, D. 1993. *The world food problem*. Oxford, U.K.: Blackwell.

Hammond, A. L. 1994. *World resources 1994 – 95*. New York: Oxford University Press.

Harley, J. L. and S. E. Smith. 1983. *Mycorrhizal symbiosis*. New York: Academic Press.

Hawksworth, D. L. 1993. *Biodiversity and biosystematic priorities: Microorganisms and invertebrates*. Wallingford, U.K.: CAB International.

Hodgkin, T., A. H. D. Brown, T. J. L. Van Hintum, and E. A. V. Morales. 1995. *Core collections of plant genetic resources*. Chichester, U.K.: Wiley.

Jain, H. K. 1982. Plant breeders' rights and genetic resources. *Indian Journal of Genetics* 42: 122–131.

James, C., D. Norse, B. J. Skinner, and Q. Zhao. 1992. Agriculture, land use and degradation. In J. C. I. Dooge et al., eds., *An agenda of science for environment and development into the 21st century*, 79–89. Cambridge, U.K.: Cambridge University Press.

Lipton, M. and R. Longhurst. 1989. *New seeds and poor people*. London: Unwin Hyman.

Loftus, R. 1993. World Watch list for domestic animal diversity released by FAO and UNEP provides "early warning system." *Diversity* 9: 34–36.

Mabbutt, J. A. 1984. A new global assessment of the status and trends of desertification. *Environmental Conservation* 11: 100–113.

Magrath, W. B. and P. Arens. 1987. *The costs of soil erosion on Java: A natural accounting resource approach*. Washington, D.C.: World Resources Institute.

Miller, M. H. and T. P. McGonigle. 1992. Soil disturbance and the effectiveness of arbuscular mycorrhizas in an agricultural ecosystem. In D. J. Read et al., eds., *Mycorrhizas in ecosystems*, 156–163. Wallingford, U.K.: CAB International.

Oldeman, L. R., R. T. A. Hakkeling, and W. G. Sombroek. 1990. *Global assessment of soil degradation*. Nairobi: International Soil Reference and Information Centre/United Nations Environment Programme.

Read, D. J. 1990. Mycorrhizas in ecosystems. In D. L. Hawksworth, ed., *Frontiers in mycology*, 101–130. Wallingford, U.K.: CAB International.

Repetto, R. 1986. *Economic policy reform for natural resource conservation*. Washington, D.C.: World Resources Institute.

Shiklomanov, I. A. 1990. *The world's water resources. International symposium to commemorate the 25 years of IHD/IHP, Paris 15– 17 March 1990*. Paris: UNESCO.

Vavilov, N. I. 1951. *The origin, variation and immunity and breeding of cultivated plants* (trans. K. S. Chester). Waltham, Mass.: Chronica Botanica.

Vellve, R. 1992. *Saving the seed: Genetic diversity and European agriculture*. Barcelona and London: Grain & Earthscan Publications.

Wenner, G. 1979. *An outline of soil conservation in Kenya*. Nairobi: Ministry of Agriculture.

CHAPTER 7

THE IMPLICATIONS OF BIODIVERSITY LOSS FOR HUMAN HEALTH

Francesca T. Grifo and Eric Chivian

Habitat degradation, overexploitation, introduced species, pollution and contamination, and global climate change, driven by an ever-growing human population and greatly increased consumption levels, are the primary factors behind the loss of biodiversity. This loss can be characterized as a two-step process consisting of ecosystem disruption and the subsequent extinction of species. Others in this volume have characterized the causes, extent, and speed of this devastating loss of species. It is the objective of this chapter to examine one set of specific consequences of this loss: the profound implications it has had and will continue to have for human health.

Ecosystem disruption has a variety of consequences for human health. The quality and quantity of freshwater and available food may be adversely affected. In addition, damage to ecosystems can change the equilibria between hosts and parasites and between predators and prey. Parasites may switch to humans when their usual hosts become rare, accidentally introduced disease vectors and pathogens may thrive, and predators on both pathogens and vectors may be reduced in numbers or even go locally or globally extinct. All of these may lead to the emergence of rare or unknown diseases or the resurgence of diseases previously controlled. Global warming and its consequences may also be increased by ecosystem disruption. Finally, living in a world devoid of the beauty and tranquility of diverse, intact ecosystems has profound effects on our mental health.

The loss of species means we lose the raw materials for existing and new drugs to alleviate human suffering and premature death. This is true both for the products of pharmaceutical companies and for traditional medicine, which is still an important source of medical care for 80 percent of the world's population (Farnsworth et al. 1985). We lose the models through which we learn about human physiology, the prerequisite for our understanding of health and illness. We also lose organisms related to the pathogens and vectors of disease, whose study allows us to make predictions about the pathogenicity, epidemiology, and genetic mechanisms of the causal agents and carriers of disease. Finally, we lose indicators of the ability of ecosystems to support life of all kinds, including human life.

ECOSYSTEM DISRUPTION

Water Quality and Quantity

The massive perturbations of watersheds and wetlands and manipulations of rivers and streams have resulted in the extinction of many species and the loss of key ecosystem services once provided by healthy habitats. Although decreases in water supply result from many complex causes, massive global deforestation of watersheds and riparian forests has crippled the ability of ecosystems to naturally purify and regulate the flow of water and has altered hydrologic cycles both locally and on a broader scale. Therefore, we observe flooding and drought where formerly forested regions boasted stable streams and rainfall, and massive soil erosion and agricultural runoff into formerly clean water. The health consequences include increased transmission of waterborne diseases and decreased water supply for drinking and for agriculture. Flooding, contaminated water, and drought, whatever their causes, promote diarrhea, cholera, dysentery, and a host of other infectious diseases. A 1993 report by the Natural Resources Defense Council (NRDC) concluded that nearly 1 million Americans contract diseases every year from contaminated water and 900 die as a result. The NRDC noted 250,000 violations of federal drinking water laws nationwide, and 83 percent of the nation's water systems went virtually unmonitored by state and federal agencies. In 1993, Milwaukee experienced the worst outbreak of waterborne disease ever reported in the United States (Komar and Speilman 1994). An estimated 403,000 people became ill with cryptosporidiosis. Runoff from dairy farms into Lake Michigan, the source of Milwaukee's water supply, was suspected as a source of the pathogen.

Food Supply

The relationships among ecosystem disturbance, agriculture, food production, malnutrition, and human health are many-layered and complex. Agriculture on almost any scale entails the transformation of ecosystems, and the resulting loss of species also has profound consequences for agriculture. Agricultural expansion, consisting largely of massive-scale monocultures designed to produce cash crops for export, has certainly served to feed more people (Pearse 1980). It has also resulted in the abandonment of traditional farming methods that, on more modest scales, incorporated diverse crops, used fewer chemicals, and were meant to feed a village. The loss of genetic diversity of crop plants and their relatives, the loss of pollinators and natural predators on potential plant pests (worth billions of dollars each year), and increased soil erosion are additional negative consequences of changes in agricultural methods and philosophies. The deforestation of watersheds also affects the quality and quantity of water available for growing crops. Water limits agriculture in some of the most populated countries in the world, including China, India, and the former USSR. Conversely, flooding caused by upstream deforestation is a threat to agriculture in many developing countries, including Bangladesh, China, and Brazil.

Although the proportion of undernourished people in the world has almost halved since midcentury, the absolute numbers have edged upward. The Food and Agricultural Organization (1992) estimated that about one-fifth of the people in the developing world are seriously undernourished, including about 200 million children, who are therefore destined for stunted growth, vulnerability to infectious disease, and perhaps intellectual impairment. Each year, 7 to 11 million children die from hunger and hunger-associated infectious diseases. Although malnutrition is caused primarily by social and political inequities, it is exacerbated by the effects of biodiversity loss, which, ironically, is ultimately driven by the same inequities.

Disease Ecology

Biodiversity plays a significant role in controlling pests, pathogens, and human parasites. When ecosystems are disrupted, important host organisms may become rare, causing pathogens to switch to humans instead. Accidentally introduced disease vectors and pathogens may thrive in disturbed ecosystems where many native organisms have become rare or extinct. Predators on both pathogens and disease vectors may be reduced in numbers or even go extinct, leading to the emergence of "new" diseases. The following examples illustrate some of the complex connections between biodiversity and human disease.

ARGENTINE HEMORRHAGIC FEVER

Since 1958, more than 20,000 Argentineans have been infected with Argentine hemorrhagic fever, from the Junin virus (Morse and Schluederberg 1990). Because of changing agricultural practices after World War II, the central Argentinean pampas—whose biologically diverse grasslands had for centuries supported wild fauna, weeds, and grasses—was widely cleared to plant corn. Heavy herbicide use to control these native grasses and weeds resulted in the widespread growth of other grasses that could live under the shade of the corn and tolerate the herbicides. Thriving on the seeds of these newly dominant grasses and with its natural predators eliminated when the pampas vegetation was destroyed, a field mouse (*Calomys musculinus*), which carries the Junin virus, flourished in the cornfields. As mouse numbers grew, farmers were exposed to mouse feces when they reaped the

corn. The fever spread, and as native grasslands were steadily replaced by monocultures, the incidence of the disease increased dramatically in 1990, affecting people across 40,000 square miles.

MALARIA

The worst of the vector-borne diseases, malaria (caused by the *Plasmodium*
parasite) is endemic in 91 countries, putting about 40 percent of the world's population at risk. Up to 500 million cases occur every year, 90 percent of them in
Africa. There are up to 2.7 million deaths annually. In Amazonian Brazil, malaria
was largely controlled in the 1960s, but epidemic outbreaks occurred in the 1980s
because of massive settlement and ecological disruption in the Amazon basin.
Over half a million cases were reported in 1988 (Kingman 1989). The outbreaks
were caused partly by the large influx of people with little or no immunity to
malaria and partly by rainforest disturbance that allowed malarial mosquitoes
(*Anopheles* spp.), which typically fed on other hosts in the jungle, to come into contact with and bite people. Additionally, road construction, runoff from land clearing, and open mining in the forests left pools of water standing everywhere
(Walsh et al. 1993), ideal conditions for unlimited breeding of malarial mosquitoes
(*A. darlingi*).

In Honduras, malaria increased from 20,000 cases in 1987 to over 90,000 in
1993. The earlier expansion of cattle grazing and sugarcane plantations had raised
local temperatures, making it too hot for malarial mosquitoes, and the incidence
dropped. However, many people who then had little or no immunity to the disease
migrated to the northern forested regions and encountered malarial vectors in the
rainforest. Furthermore, because of the heavy use of pesticides in the converted
agricultural areas, these mosquitoes were largely resistant. Malaria increased dramatically (Almendares et al. 1993).

SCHISTOSOMIASIS

Schistosomiasis flourished in Senegal because of similar ecosystem distur-
bances. In 1985 the Diama Dam, built on the Senegal River, diverted the river's
water through irrigation canals, which transformed desert soils into fields of sugarcane, mint, potatoes, and rice. As the hydrology of the river was altered, water
level fluctuations abated and stopped the daily input of saltwater from the ocean.
The dam, along with the new canals, provided a steady and fairly calm supply of
freshwater. This caused populations of freshwater snails, which carry the *Schisto-
soma* fluke, to explode in the canals and river, causing an epidemic of schistosomiasis among the local people (World Watch 1996).

KYASANUR FOREST DISEASE

The Kyasanur forest disease was the first tropical tick-borne arbovirus to be
discovered (Dobson et al. 1997). Carried by *Haemaphysalis spinigera* tick vectors in
the tropical forests of Mysore in southern India, the ticks multiplied when rain-

forest cutting and the subsequent introduction of sheep and cattle led to the invasion of the cleared areas by a thick brush species, *Lantana camara.* This brush provided habitat for small mammals, and the cleared rainforest patches were used for grazing sheep and cattle. Tick density increased because adult ticks could feed on the cattle and sheep, and the immature ticks, which infect humans, had the small mammals as reservoirs. These abundant nymphs could then attack people coming into the clearings to gather fuel or produce. First noted in 1957, the disease peaked in 1983 with 1,555 cases and 180 deaths recorded in that year alone (Bannerjee 1988).

SLEEPING SICKNESS

Spread by the tsetse fly (which carries *Trypanosoma* parasites), sleeping sickness is prevalent in 36 countries of sub-Saharan Africa, putting 55 million people at risk (World Health Organization 1994). A rinderpest epidemic in East Africa wiped out many species of wild game, causing a large reduction in the tsetse flies' preferred hosts. The flies switched to feeding on humans, causing a massive outbreak of sleeping sickness throughout East Africa (Simon 1962).

ENCEPHALITIS

Introduced insect species often fare better than native species in degraded habitats and contribute to the rise of infectious diseases. The Asian tiger mosquito (*Aedes albopictus*) accidentally arrived in the United States from Japan in 1985, stowed away in container shipments of used, water-filled tires. This mosquito can multiply rapidly and take over from native mosquitoes in converted habitats, breeding mainly in small pools of water that collect in discarded tires, cans, wrappers, and logged tree stumps. It has already spread to 21 states and is well established in the southeastern United States, as far west as Texas and north to Chicago (Chivian 1993). An aggressive species, it attacks more hosts than any other mosquito in the world, including many mammals, birds, and reptiles, and can thus vector diseases from one species to another and to humans. Among these diseases are various forms of encephalitis, including the La Crosse variety, which infects chipmunks and squirrels. In Florida, the Asian tiger mosquito has already been found to transmit the virus causing eastern equine encephalitis (EEE), a rare but deadly brain infection, and could cause EEE epidemics in other areas where the virus is already present in wild populations. It is also able to transmit yellow fever and dengue fever.

LEISHMANIASIS

Leishmaniasis is spread by sandflies, which transmit the causative *Leishmania* parasite (a protozoan). Forest rodents carry the protozoan and are the usual reservoir hosts for sandflies, but when the rainforest is cleared and new villages are introduced, these rodents are displaced. The flies then turn to biting humans in the absence of sufficient numbers of their preferred rodent hosts. Leishmaniasis affects over 12 million people globally in its various forms; its spread is further accelerated by road building, dam construction, mining, and other development programs that bring more people into contact with the sandflies (Chivian 1993).

BOLIVIAN HEMORRHAGIC FEVER

In eastern Bolivia in 1962, Bolivian hemorrhagic fever (caused by the Machupo virus) surfaced as a result of deforestation and predator removal. Following a social revolution in the early 1950s, villagers had neither money nor food, so they cleared dense jungle patches to grow corn and other vegetables. This disrupted the natural habitat of the field mice (genus *Calomys*), which are the host for the Machupo virus. The corn itself provided the mice with an excellent food source. Furthermore, all the village cats had been killed from massive DDT spraying to control malarial mosquitoes. The mice populations flourished, overrunning settlements and spreading the virus in their urine and feces, killing up to 20 percent of the villagers. The epidemic abated when cats were reintroduced to the area (Garrett 1994).

RIFT VALLEY FEVER

Additional examples include Rift Valley fever, a mosquito-borne virus that until recently was primarily a disease of sheep and cattle. In 1977, a major outbreak of human disease in Egypt followed the completion of the Aswan Dam, which flooded 800,000 ha of land, much of which remained in puddles after the dam stabilized the water table. This standing water provided ideal breeding sites for virus-infected mosquitoes, enabling them to thrive and spread the disease to local people and cattle. Over 200,000 people became sick and 598 died (Wilson 1994).

YELLOW FEVER

Yellow fever persists in Western Africa, despite a vaccine, because of rainforest encroachment (Morse and Schluederberg 1990). Yellow fever was originally transmitted in a jungle cycle from monkey to monkey via the mosquito *Aedes africanus*. However, when logging was introduced, humans became part of the disease life cycle. When they returned home from the forest, these people spread the disease from person to person in an urban cycle via *A. bromiliae* and *A. aegypti,* which thrive in disturbed, urban environments (Haddow et al. 1947).

Mental Health

Circumstantial evidence abounds for the unique relationship between humanity and nature. We spend substantial resources on travel and on cultivating nature in our yards, windowboxes, and homes. Nature is abundantly depicted in art, literature, and song, as well as in the fabrics we wear and use to decorate our surroundings. It seems that we all are profoundly moved by aspects of the natural world. A significant amount of research has linked our intuitive sense of the restorative powers of nature to our mental and physical health. Verderber has shown that the quality of the view from the window was an important element in the recovery of patients in the physical medicine and rehabilitation wards of six hospitals (Verderber 1986; Verderber and Reuman 1987). Ulrich (1984) demonstrated that the content of the view is important in hospital patients' recovery from surgery, with natural views contributing to faster recovery. In prisons, Moore (1981) and West (1986) found that inmates with views of nature sought health care much less often than inmates whose views were of other inmates or urban settings. Cimprich (1993) found that women recovering from breast cancer, who were asked to design a plan of restorative activities during the 3 months following surgery, most often selected activities that involved the natural environment. In addition, those keeping to their plans exhibited a consistent improvement not observed in a control group. In a study of college students, Tennessen and Cimprich (1995) found that natural views from dormitories were associated with better ability to direct their attention.

Mental fatigue is more pervasive than simple burnout at the end of a work day. A mentally fatigued person is far more likely to commit what accident investigators call "human error" (Broadbent et al. 1982). Mentally fatigued people are also less likely to help someone in need, more likely to react aggressively, and less tolerant and less sensitive (Cohen and Spacapan 1978; Donnerstein and Wilson 1976; Cohen and Lezak 1977). Based on their work and that of many colleagues, Kaplan and Kaplan (1995) provide a synthesis of the aspects of nature that enable it to alleviate mental fatigue. Their results suggest that natural environments are particularly rich in the characteristics necessary for restorative experiences.

SPECIES LOSS

Raw Materials for Pharmaceuticals and Traditional Medicines

There are three major reasons to continue searching for unique natural compounds:

• We are currently dependent on biodiversity-derived pharmaceuticals and alternative medical therapies. The World Health Organization reports that 80 percent of the world's people depend on non-Western medicine that is heavily based on biodiversity (Farnsworth et al. 1985). Examination of the 150 most prescribed pharmaceuticals in the United States in 1993 reveals that 57 percent are in some way derived from nature (Grifo et al. 1997). From 1959 to 1980, 25 percent of all the drugs dispensed by U.S. pharmacies came from plants (Farnsworth 1990). These include amoxicillin, cefaclor, codeine, digoxin, erythromycin, penicillin, and warfarin. Vincristine and vinblastine, the most effective medicines against childhood leukemia, are derived from the rosy periwinkle; Taxol, used to treat breast and ovarian cancer, comes from the bark of the Pacific yew trees.

- New diseases, such as certain cancers, heart disease, and AIDS, are constantly emerging. Even without newly discovered viruses, human physiology is in flux as our diet changes, our environment changes, and we live longer.

- There has been a surge in antibiotic- and disinfectant-resistant pathogens. Hundreds of bacteria, viruses, and parasites are now resistant to dozens of drugs. Some examples that have received recent public attention include cholera, staph and strep infections, tuberculosis, herpes and cytomegalovirus, salmonella, shigella, and cryptosporidiosis.

Medical Models

The unique biology of a wide range of terrestrial and aquatic organisms provides medical models that enable us to understand human physiology and disease. Much of what we know about human physiology and genetics has come from animal models. The familiar models are mice, rats, rabbits, many species of monkeys, drosophila, and, of course, the guinea pig. But many more obscure animals turn out to mimic the behavior of certain diseases in humans. The armadillo is the only animal to acquire leprosy, the horseshoe crab has one of the largest and most accessible optic nerves for the study of human vision (Cohen 1978), and cone snails produce a wide variety of toxins that are important in the study of pain and other conditions (Olivera et al. 1990).

Sharks, which are being lost in record numbers, rarely develop tumors or infections, presumably an adaptive advantage that evolved over their 400 million years of species survival. Scientists at the Massachusetts Institute of Technology recently isolated a substance from one shark species that strongly inhibits the growth of new blood vessels toward solid tissue tumors, thus preventing their growth (Lee and Langer 1983). In another species of sharks, researchers discovered a compound that has demonstrated potent efficacy against a variety of bacteria, fungi, and parasites. As more and more infectious diseases grow resistant to antibiotics, our best hope may lie with these creatures and the knowledge we gain from them.

Bears, which are endangered in many parts of the world, are the only vertebrate that does not lose bone mass after sustained periods of immobility (Floyd et al. 1990). Understanding how this is possible could lead to a cure or treatment for osteoporosis, which costs the U.S. economy $10 billion dollars annually in direct health care costs and lost productivity (National Osteoporosis Foundation 1993).

Understanding how bears are able to recycle their wastes during hibernation without urinating might enable us to find effective treatment for patients now totally reliant on dialysis (Nelson 1987). Currently, renal failure costs the U.S. economy $7 billion annually (New York Times 1993).

Organisms related to the pathogens and vectors of disease allow us to make predictions about the pathogenicity, epidemiology, and genetic mechanisms of the causal agents and carriers of disease. Their loss eliminates these possibilities.

Thermostable enzymes from thermophilic bacteria have greatly enhanced our ability to manipulate DNA, including such techniques as polymerase chain reaction and sequencing of genetic material (Lovejoy 1997). Current and future pharmaceutical production, gene therapy, diagnostic tools, identification of newly emergent microbes, and basic research all depend on these processes. In addition, new techniques may be uncovered based on new microbes or other species yet to be discovered.

Indicator Species

The world today is a series of accidents waiting to happen in terms of global climate change and toxic chemicals and other contaminants of our air, water, and food. Many nonhuman species react to these perturbations at much lower thresholds than humans, and their reactions serve as warnings that allow us to mitigate their effects before humans are affected. Examples include the severely deformed tern chicks discovered in 1970 on Great Gull Island (a research station of the American Museum of Natural History) that called attention to high levels of PCBs in Long Island Sound and the thinning of eggshells of many birds of prey from DDT ingestion. Many mysterious dieoffs among marine mammals and amphibians are warnings we are just beginning to understand. Loss of biodiversity equals the loss of these "miner's canaries" that we depend on to tell us of imminent environmental problems.

CONCLUSIONS

No matter how skilled we become at chemistry, or biotechnology, we will never outdo mother nature. The most sophisticated pharmaceutical firms in the world still find entirely new classes of compounds by studying biodiversity. We know that if ecosystems continue to be disrupted and species lost at current rates, we will pay a monumental price in the potential medicines we lose, diseases we fail to understand, or new pathogens we face. However, we are profoundly limited in our ability to continue to use biodiversity as both a source of raw materials and a place to learn physiology, biochemistry, natural products chemistry, epidemiology, and much more by our own ignorance of the identity, relationships, and natural history of most of the world's biodiversity. This is because we are faced with a global shortage of systematists trained, employed, and supported to do this work. All our natural history collections require additional resources to serve the informational needs of society; however, the few collections in developing countries (where most of the world's biodiversity is located) are especially likely to be too small, understaffed, or poorly maintained.

REFERENCES

Almendares, J., M. Sierra, P. K. Anderson, and P. R. Epstein. 1993. Critical regions: A profile of Honduras. *Lancet* 342: 1400–1402.

Bannerjee, J. 1988. Kyasanur forest disease. In T. P. Monath, ed. *The arboviruses: Epidemiology and ecology*. Boca Raton, Fla.: CRC Press.

Broadbent, D. F., P. F. Cooper, P. FitzGerald, and K. R. Parkes. 1982. The Cognitive Failures Questionnaire (CFQ) and its correlates. *British Journal of Clinical Psychology* 21: 1–16.

Chivian, E. 1993. Species extinction and biodiversity loss: The implications for human health. In E. Chivian, M. McCally, H. Hu, and A. Haines, eds. *Critical condition: Human health and the environment*. Cambridge, Mass.: MIT Press.

Cimprich, B. 1993. Development of an intervention to restore attention in cancer patients. *Cancer Nursing* 16(2): 83–92.

Cohen, E., ed. 1978. *Biomedical applications of the horseshoe crab (Limulidae): Proceedings of a symposium, Marine Biological Laboratory, Woods Hole, Massachusetts, October 1978*. New York: Alan R. Liss.

Cohen, S. and A. Lezak. 1977. Noise and inattentiveness to social cues. *Environment and Behavior* 9: 559–572.

Cohen, S. and S. Spacapan. 1978. The aftereffects of stress: An attentional interpretation. *Environmental Psychology and Nonverbal Behavior* 3: 43–57.

Dobson, A., M. S. Campbell, and J. Bell. 1997. Fatal synergisms: Interactions between infectious diseases, human population growth, and loss of biodiversity. In F. T. Grifo and J. Rosenthal, eds. *Biodiversity and human health*. Washington, D.C.: Island Press.

Donnerstein, E. and D. W. Wilson. 1976. Effects of noise and perceived control on ongoing and subsequent aggressive behavior. *Journal of Personality and Social Psychology* 34: 774–781.

Farnsworth, N. R. 1990. The role of ethnopharmacology in drug development. In *Bioactive compounds from plants*. Ciba Foundation Symposium. New York: Wiley.

Farnsworth, N. R., O. Akerele, A. S. Bingel, D. D. Soejarto, and Z. Guo. 1985. Medicinal plants in therapy. *Bulletin of the World Health Organization* 63(6): 965–981.

Floyd, T., R. A. Nelson, and G. F. Wynne. 1990. Calcium and bone metabolic homeostasis in active and denning black bears. *Clinical Orthopaedics and Related Research* 255: 301–309.

Food and Agriculture Organization. 1992. World declaration and plan of action for nutrition. International Conference on Nutrition, Rome, December 1992.

Garrett, L. 1994. *The coming plague: Newly emerging diseases in a world out of balance*. New York: Farrar, Straus & Giroux.

Grifo, F. T., D. Newman, A. S. Fairfield, B. Bhattacharya, and J. T. Grupenhoff. 1997. The origins of prescription drugs. In F. T. Grifo and J. Rosenthal, eds. *Biodiversity and human health.* Washington, D.C.: Island Press.

Haddow, A. J., K. C. Smithburn, A. F. Mahaffy, and J. C. Bugher. 1947. Monkeys in relation to yellow fever in Bwamba County, Uganda. *Transactions of the Royal Society of Tropical Medicine and Hygiene* 40: 677–700.

Kaplan, R. and S. Kaplan. 1995. *The experience of nature.* Ann Arbor: University of Michigan Press.

Kingman, S. 1989. Malaria runs riot on Brazil's wild frontier. *New Scientist* August 12: 24–25.

Komar, N. and Speilman, A. 1994. Emergence of eastern encephalitis in Massachusetts. In M. E. Wilson, R. Levins, and A. Spielman, eds. *Disease in evolution: Global changes and emergence of infectious diseases.* New York: New York Academy of Sciences.

Lee, A. and R. Langer. 1983. Shark cartilage contains inhibitors of tumor angiogenesis. *Science* 221: 1185–1187.

Lovejoy, T. E. 1997. Foreword. In F. T. Grifo and J. Rosenthal, eds. *Biodiversity and human health.* Washington, D.C.: Island Press.

Moore, E. O. 1981. A prison's environment's effect on health care demands. *Journal of Environmental Systems* 3: 1–6.

Morse, S. and A. Schluederberg. 1990. Emerging viruses: The evolution of viruses and viral diseases. *Journal of Infectious Diseases* 162: 1–7.

National Osteoporosis Foundation. 1993. *Fast facts on osteoporosis.* Washington, D.C.: NOF.

Natural Resources Defense Council. 1993. *Think before you drink: The failure of the nation's drinking water system to protect public health.* Washington, D.C.: NRDC.

Nelson, R. A. 1987. Black bears and polar bears: Still metabolic marvels. *Mayo Clinic Proceedings* 62: 850–853.

New York Times. 1993, November 4. A bleak U.S. report on kidney-failure patients.

Olivera, B. M., J. Rivier, C. Clark, C. A. Ramilo, G. P. Corpuz, F. C. Abogadie, E. E. Mena, S. R. Woodward, D. R. Hillyard, and J. J. Cruz. 1990. Diversity of *Conus* neuropeptides. *Science* 249: 257–263.

Pearse, A. 1980. *Seeds of plenty , seeds of want: Social and economic implications of the green revolution.* Oxford, U.K.: Oxford University Press.

Simon, N. 1962. *Between the sunlight and the thunder: The wildlife of Kenya.* London: Collins.

Tennessen, C. M. and B. Cimprich. 1995. Views to nature: Effect on attention. *Journal of Environmental Psychology* 15: 77–85.

Ulrich, R. S. 1984. View through a window may influence recovery from surgery. *Science* 224: 420–421.

Verderber, S. 1986. Dimensions of a person—window transaction in the hospital environment. *Environment and Behavior* 18: 450–466.

Verderber, S. and D. Reuman. 1987. Windows, views, and health status in hospital therapeutic environment. *Environment and Behavior* 18: 450–466.

Walsh, J. F., D. H. Molyneux, and M. H. Birley. 1993. Deforestation: Effects on vector-borne disease. *Parasitology* 106: 55–75.

West, M. J. 1986. *Landscape views and stress responses in the prison environment.* Unpublished master's thesis. Seattle: University of Washington.

Wilson, M. L. 1994. Rift valley fever virus ecology and the epidemiology of disease emergence. In M. E. Wilson, R. Levins, and A. Spielman, eds. *Disease in evolution: Global changes and emergence of infectious diseases*. New York: New York Academy of Sciences.

World Health Organization. 1994. Alarming increase in sleeping sickness: Appeal for international solidarity. Press release, October 7.

World Watch. 1996. Infecting ourselves. Paper #129. Washington, D.C.: Worldwatch Institute.

CHAPTER 8

BIODIVERSITY LOSS AND ITS IMPLICATIONS FOR SECURITY AND ARMED CONFLICT

Arthur H. Westing

The worldwide numbers of humans and livestock continue to increase at essentially exponential rates in a world of finite size. In 1850, humans and their livestock represented perhaps 5 percent of total terrestrial animal biomass, a century later this value represented just over 10 percent and currently it is somewhat more than 25 percent (Westing 1981:180). Ten years from now it is sure to be in the neighborhood of 30 percent. This increase in human and livestock biomass occurs at the expense of wildlife biomass, a loss that is measurable in both quantitative and qualitative terms—that is, both in loss of numbers of individuals within a species and in loss of numbers of species (Groombridge 1992). In other words, these losses in biodiversity result largely from an arrogation of nature by the ever-expanding human population (Morris 1995), an expansion that has been aptly likened to a biospheric pathology (Hern 1990). Indeed, human demands on the environment have been growing even more rapidly than population increases suggest, as indicated by the even more rapid increases in productive and consumptive activities.

Technical advances—agricultural, industrial, and other improvements—that expand the global carrying capacity for the expanding human population are now increasing at almost comparable rates. But these technical advances have not been sufficiently innovative or rapid to alleviate the growing human destitution and squalor that represent the present lot of most people on Earth. Even if human population growth were somehow stemmed, the need to mitigate human poverty would continue to undermine efforts to prevent losses in biodiversity.

The questions being addressed here are the extent to which this grave and seemingly intractable predicament affects security issues, including those involving armed conflict, and the policy recommendations that flow from that analysis. This requires a brief exposition of what is subsumed under the notion of security. What is not addressed here is the impact that the military sector of society has on biodiversity, whether in times of war or peace, whether direct or indirect, or whether for better or worse. These issues are covered elsewhere (e.g., Westing 1980, 1985, 1990a).

THE CONCEPT OF SECURITY

The peoples of the world have sorted themselves into a number of jealously sovereign nations, about 192 at present. One of the main functions of these political entities is (or should be) to provide security for their citizens. That obligation of national governments has traditionally been seen as referring to security from foreign attack, thus falling within the aegis of the nation's military sector. Indeed, in the industrialized world military expenditures typically consume 10 percent of a nation's total central government expenditures and in the nonindustrialized world perhaps 12 percent, although in many instances not all of those military expenditures are devoted to reducing vulnerability to external military threats (ACDA 1996:table I; Westing 1988).

Especially since World War II there has been a growing concern over basic human rights, which found eloquent and detailed expression in 1948 in the Universal Declaration of Human Rights (UNGA 1948). But the human rights espoused by governments in this and other international instruments—especially the right to life, security of person, a system of justice and law, and a standard of living adequate for health and well-being—cannot exist in a vacuum. That is, the promulgation of these human rights implies obligations on the part of the national governments that have proclaimed them.

Thus the notion of national security—and its corollary of government obligations—has been expanded from the traditional narrow emphasis on the provision of security from potential foreign enemies to a more comprehensive one that includes a considerable assortment of interdependent civil and other social rights. Of particular importance in the present context, comprehensive security must embrace the environmental prerequisites to the achievement of those social rights (Westing 1986, 1989, 1992a, 1993e). Indeed, the social and environmental ingredients of comprehensive security are thoroughly enmeshed in a web of interactions; none can survive and flourish in the long term unless they all do.

The environmental prerequisites to basic human rights—which can be operationally grouped under the heading "environmental security"—fall into two major categories. One component of environmental security is sane resource use, based on use or harvesting at levels and with procedures that either maintain or restore optimal resource services or stocks. To satisfy this component, renewable resources must be used strictly on the principle of sustained use or sustained discard and nonrenewable resources strictly on the principles of efficiency and frugality.

The second component of environmental security is environmental protection, based on protection from medically unacceptable environmental pollution or other disruption, protection from war and similar vandalism, and—for special areas—

protection from all permanent human intrusions. All aspects of environmental security contribute to the conservation of biodiversity. However, the protection of
special areas from all permanent human intrusions—which, in the aggregate,
should amount to perhaps 10 to 12 percent or even more of each nation's area, ap-
propriately distributed across the various habitat types, both terrestrial (land plus
freshwater) and marine—provides the centerpiece for such conservation.

LOSS OF BIODIVERSITY AS A CAUSE OF INSECURITY

The security status of most of the approximately 192 countries in the world is
found to be thoroughly inadequate when measured against the necessarily expanded notion of the term *security*. Indeed, such comprehensive security is actually deteriorating for a large (at least 90) and growing number of countries, especially in Africa but also elsewhere (Westing 1993d:308). A crucial factor in the
declining security of the many countries thus afflicted is the exploitation of renewable natural resources beyond their sustainable yield or service, leading to soil
degradation, freshwater shortages, overloaded disposal sinks, continuing encroachment on agriculturally marginal lands, and so forth.

The resulting losses in biodiversity per se seem to have little immediate negative impact on the security of these countries; indeed, in some instances there is an
associated immediate (although perhaps only transitory) gain in security as wild
or semiwild habitats are taken over for human exploitation. On the other hand, the
accelerating losses in biodiversity occurring in these many countries—and thus
throughout much of the global biosphere—contribute subtly to the erosion of their
long-term environmental security (McNeely et al. 1990; Reid and Miller 1989).
Such local or regional contributions to environmental insecurity can lead to the
generation of environmental refugees, with destabilizing social ramifications at
both their locus of origin and sites of destination (Westing 1992b, 1994b). Ultimately such environmental insecurity compromises the security of all the nations
in the world.

LOSS OF BIODIVERSITY AS A CAUSE OF ARMED CONFLICT

Finally, it is of interest to explore briefly whether a causal relationship can be dis-
cerned between biodiversity loss and armed conflict. To begin with, little appears
to have changed in modern times—if ever—in the overall conflictual or bellicose
nature of *Homo sapiens*. Nations continue to settle through violent (deadly) action
some modest fraction of their numerous inevitable conflicts, whether such armed
conflicts are with neighboring countries or with groups inside their own borders.
Perhaps three dozen or so conflicts, each with annual fatality rates of at least 1,000,
are always in progress somewhere (Westing 1994b:table II). With resort to violence

the continuing norm for the ultimate settlement of conflict between peoples, clans, and nations, some fraction of these conflicts derives at least in part from environmental issues.

In recent decades the frequency of interstate (external) armed conflict has been declining as the frequency of intrastate (internal) armed conflict has been increasing, especially in the nonindustrialized world (Sollenberg and Wallensteen 1997). Thus no obvious positive correlation exists between environmental decline and interstate armed conflict. On the other hand, the rising incidence of internal armed conflicts coincides with increasingly intense environmental constraints in the nonindustrialized world. However, careful study has, so far at least, revealed no discernible upward trend in the frequency of environmentally motivated armed conflicts (Bächler et al. 1993; Homer-Dixon 1994–1995; Smith 1994; Swain 1993). And with biodiversity loss largely a second-order effect of environmental (habitat) deterioration and arrogation, it becomes especially difficult to suggest that biodiversity loss per se has become a cause of armed conflict.

POLICY RECOMMENDATIONS

Policy responses to the issues raised herein can be suggested on a number of levels. To begin with, because the underlying cause of the unequal conflict between humankind and the rest of the living world is gross human overpopulation, enormous efforts must be made to bring human numbers into balance with the availability of sustainable resources (Westing 1981, 1990b).

Next, because biodiversity will continue to decline unless at least 10 to 12 percent of all major habitat types is placed under appropriate protection, Herculean efforts must be made throughout the global biosphere—both within the nations of the world and within areas beyond national jurisdiction—to achieve that goal (Westing 1990b:116). Global cooperation will be of paramount importance, with special emphasis on transfers of relevant resources (both tangible and intangible) from the industrialized to the nonindustrialized world (Glowka et al. 1994; Imber 1994; UN 1992). And because a substantial fraction (perhaps one-third) of the areas in great need of such protection overlap national boundaries, special attention must be devoted to them (Westing 1993a). Although the creation and protection of transfrontier reserves (refuges) presents a diplomatic as well as administrative challenge, such efforts simultaneously provide an important and welcome opportunity for building political confidence and security among neighboring countries (Westing 1993b).

The maintenance of traditional national security will continue to rest largely on a country's military forces (these to be structured in a strictly nonprovocative, defensive manner) and is a topic in itself. Suffice it to note here that the size of the military sector can be reduced to the extent that regional and global political confidence and security are enhanced, an area in which environmental measures can be especially useful. The possibility of transfrontier cooperation in biodiversity protection has already been noted; the joint management of shared freshwater resources is another valuable candidate for such confidence-building collaboration. Environmental pollution problems also lend themselves extremely well to regional and global cooperation, both informal and formal.

The various approaches to enhancing security in its properly expanded sense are largely beyond the scope of this chapter. But two points must be made. First, suitable education at all levels—primary, secondary, and tertiary—available equally to both sexes, must be greatly strengthened throughout the world in order to develop socially and environmentally relevant cultural norms as the basis for the needed political changes (Brock-Utne 1988; Kennedy 1993:339–343; Westing 1993c, 1996). Second, a huge stumbling block to the enhancement of comprehensive national security is all too often overlooked: the lack of real interest by many governments in the welfare of most of their citizens. Indeed, despotic rule (Westing 1994b:table I) and government corruption (Theobold 1990:76–106) are the hallmarks of many nonindustrialized countries suffering from severe environmental insecurity. In the quest for security, it becomes truly necessary to address the interlocking issues associated with government accountability and the rule of law.

A word is in order regarding armed conflicts. Some armed conflicts do derive in part from environmental issues; these issues can be addressed directly to the extent that they are identifiable. But the causes of armed conflict are almost always multifaceted and extremely difficult to sort out. Even though environmental factors may well be imbedded within the panoply of proximate and ultimate causes of an armed conflict, efforts to develop and apply means of nonviolent conflict resolution should emphasize non–cause-specific approaches.

CONCLUSION

In recent decades, the earth has been subjected to conditions that are leading in most countries to ever-greater losses in biodiversity, despite concerted efforts to the contrary (Glowka et al. 1994; McNeely et al. 1990; Norse 1993; UN 1992; Westing 1994a; WRI et al. 1992). The frequency of interstate armed conflicts has been declining during that same time frame, while the frequency of internal armed conflicts, by contrast, has been rising. However, it has not yet been possible to link in a causal sense the noted trends in biodiversity loss with those in armed conflict. Thus it is safe to suggest that biodiversity loss per se does not make a country more vulnerable to attack by an external power (the traditional measure of national security), nor is there an indication that such loss leads a country to aggressive action against another. Whether biodiversity loss per se has been contributing to the increased frequency of internal armed conflicts has not been demonstrated.

As soon as one recognizes that national security depends on the satisfaction of a panoply of factors, it becomes evident that losses in biodiversity contribute to the long-term deterioration of such security. Environmental security cannot be achieved or maintained in the long run without adequate protection of biodiversity; with deficiencies in environmental security, it becomes impossible to attain or sustain an optimal level of social security.

The poignant question arises as to whether in preparing for the twenty-first century via the expanded approach to security that such preparation implies (Kennedy 1993:14, 128–131), it will be possible to cope with the increasing need for economic development in ways that do not continue to be coupled with losses in biodiversity. It seems that the only durable hope to stem the tide of biodiversity loss is greatly strengthened regional and global cooperation and a widespread commitment to environmental education that instills an environmental ethic conducive to a respect for nature, population controls, frugal lifestyles, equitable distribution of resources, and pacific approaches to conflict resolution.

ACKNOWLEDGMENT

The author is pleased to acknowledge suggestions from Carol E. Westing.

REFERENCES

ACDA. 1996 (24th ed.). *World military expenditures and arms transfers 1995*. Washington, D.C.: U.S. Arms Control & Disarmament Agency.

Bächler, G., V. Böge, S., Klötzli, and S. Libiszewski. 1993. *Umweltzerstörung: Krieg oder Kooperation?: Ökologische Konflikte im internationalen System und Möglichkeiten der friedlichen Bearbeitung*. Münster, Germany: Agenda Verlag.

Brock-Utne, B. 1988. Formal education as a force in shaping cultural norms relating to war and the environment. In A. H. Westing, ed., *Cultural norms, war and the environment*, 83–100. Oxford, U.K.: Oxford University Press.

Glowka, L., F. Burhenne-Guilmin, and H. Synge. 1994. *A Guide to the Convention on Biological Diversity*. Gland, Switzerland: World Conservation Union (IUCN), Environmental Policy & Law Paper no. 30.

Groombridge, B., ed. 1992. *Global biodiversity: Status of the earth's living resources*. London: Chapman & Hall.

Hern, W. M. 1990. Why are there so many of us?: Description and diagnosis of a planetary ecopathological process. *Population & Environment* 12: 9–39.

Homer-Dixon, T. F. 1994–1995. Environmental scarcities and violent conflict. *International Security, Cambridge* 19(1): 5–40.

Imber, M. F. 1994. *Environment, security and UN reform*. New York: St. Martin's Press.

Kennedy, P. 1993. *Preparing for the twenty-first century*. New York: Random House.

McNeely, J. A., K. R. Miller, W. V. Reid, R. A. Mittermeier, and T. B. Werner. 1990. *Conserving the world's biological diversity*. Gland, Switzerland: IUCN.

Morris, D. W. 1995. Earth's peeling veneer of life. *Nature* 373: 25.

Norse, E. A., ed. 1993. *Global marine biological diversity: A strategy for building conservation into decision making.* Washington, D.C.: Island Press.

Reid, W. V. and K. R. Miller. 1989. *Keeping options alive: The scientific basis for conserving biodiversity.* Washington, D.C.: World Resources Institute.

Smith, D. 1994. Dynamics of contemporary conflict: Consequences for development strategies. In N. Graeger and D. Smith, eds., *Environment, poverty, conflict,* 47–89. Oslo: International Peace Research Institute, PRIO Report no. 2/92.

Sollenberg, M. and P. Wallensteen. 1997. Major armed conflicts. *SIPRI Yearbook* 1997: 17–30.

Swain, A. 1993. *Environment and conflict: Analysing the developing world.* Uppsala: Uppsala University, Department of Peace & Conflict Research, Report no. 37.

Theobold, R. 1990. *Corruption, development and underdevelopment.* London: Macmillan.

UN. 1992. *Convention on biological diversity , Rio de Janeiro, 5 June 1992; entry into force, 29 December 1993 ,* UNTS no. 30619. New York: United Nations.

UNGA. 1948. *Universal declaration of human rights.* New York: UN General Assembly, Resolution no. 217(III)A, September 10, 1948.

Westing, A. H. 1980. *Warfare in a fragile world: Military impact on the human environment.* London: Taylor & Francis.

Westing, A. H. 1981. A world in balance. *Environmental Conservation* 8: 177–183.

Westing, A. H., ed. 1985. *Explosive remnants of war: Mitigating the environmental effects.* London: Taylor & Francis.

Westing, A. H. 1986. An expanded concept of international security. In A. H. Westing, ed., *Global resources and international conflict: environmental factors in strategic policy and action,* 183–200. Oxford, U.K.: Oxford University Press.

Westing, A. H. 1988. Military sector vis-à-vis the environment. *Journal of Peace Research* 25: 257–264.

Westing, A. H. 1989. Comprehensive human security and ecological realities. *Environmental Conservation* 16: 295.

Westing, A. H., ed. 1990a. *Environmental hazards of war: Releasing dangerous forces in an industrialized world.* London: Sage.

Westing, A. H. 1990b. Our place in nature: Reflections on the global carrying capacity for humans. In N. Polunin and J. H. Burnett, eds., *Maintenance of the biosphere,* 109–120. Edinburgh: Edinburgh University Press.

Westing, A. H. 1992a. Environmental dimensions of maritime security. In J. Goldblat, ed., *Maritime security: The building of confidence,* 91–102. Geneva: UN Institute for Disarmament Research, Document no. UNIDIR/92/89.

Westing, A. H. 1992b. Environmental refugees: A growing category of displaced persons. *Environmental Conservation* 19: 201–207.

Westing, A. H. 1993a. Biodiversity and the challenge of national borders. *Environmental Conservation* 20: 5–6.

Westing, A. H. 1993b. Building confidence with transfrontier reserves: The global potential. In A. H. Westing, ed., *Transfrontier reserves for peace and nature: A contribution to human security,* 1–15. Nairobi: UN Environment Programme.

Westing, A. H. 1993c. Global need for environmental education. *Environment* 35(7): 4–5, 45.

Westing, A. H. 1993d. Human instability and the release of dangerous forces. In N. Polunin and J. Burnett, eds., *Surviving with the biosphere,* 307–319. Edinburgh: Edinburgh University Press.

Westing, A. H. 1993e. Human rights and the environment. *Environmental Conservation* 20: 99–100.

Westing, A. H., ed. 1994a. Biodiversity in the context of science and society. Middletown Springs: Vermont Academy of Arts & Sciences, Occasional Paper no. 27.

Westing, A. H. 1994b. Population, desertification, and migration. *Environmental Conservation* 21: 110–114, 109.

Westing, A. H. 1996. Core values for sustainable development. *Environmental Conservation* 23: 218–225.

WRI, IUCN & UNEP. 1992. *Global biodiversity strategy: Guidelines for actions to save, study, and use Earth's biotic wealth sustainably and equitably*. Washington, D.C.: World Resources Institute, World Conservation Union [IUCN], and United Nations Environment Programme.

CHAPTER 9

THE ECONOMIC CONSEQUENCES OF BIODIVERSITY LOSS

Dominic Moran and David Pearce

The components of biodiversity are recognized to have a number of values, including intrinsic, economic, cultural, and aesthetic. Biodiversity has economic value in that some of its components are scarce and are capable of generating human well-being. In assessing the economic consequences of loss, an important distinction should be made at the outset between losses due to the erosion of diversity per se and losses resulting from the depletion of more broadly defined categories of biological resources such as forests and wetlands. In the latter case, economics is providing useful approaches for valuing individual welfare change in assessing the costs and benefits of alternative development or conservation options. Precise definition and quantification of the former losses remain elusive, although the issue of what we are losing and its translation into terms of human well-being is currently a common focus between some economists and conservation biologists.

Of equal importance from an equity perspective is the global distribution of the consequences of biodiversity loss or, conversely, the distribution of benefits from its preservation. Conservation decisions come at a price. Any emphasis on the conservation of resources in tropical, predominantly developing countries gives rise to several dilemmas related to the spatial distribution of conservation costs and benefits and compensation issues designed to reconcile conflicting priorities in some of the world's poorest but most species-rich nations. The valuation of biodiversity is central to the issue, and this chapter outlines the economic approach to evaluating the consequences of biodiversity loss and potential responses to the current biodiversity crisis.

ECONOMICS, VALUE, AND BIODIVERSITY

How do economists determine the economic consequences of biodiversity loss? Economic valuation looks at the value people actually place on biodiversity, as opposed to the value they would place on biodiversity if they were perfectly informed or maximizing some other metaethical objective unrelated to their own welfare.

Value is typically inferred from what people are observed to pay in actual markets or from their willingness to pay (WTP), as revealed by questionnaires. The most obvious manifestation of this WTP is through established markets. Thus the price of the Malagasy periwinkle plant as part of a prospecting arrangement provides one—albeit unconventional—gauge of human preferences for that plant, using money as a unit of value. Logically, if the periwinkle ceases to exist because of extinction, the same WTP reflected in the price measures some part of the resulting loss of well-being.

But conventional markets are not always necessary to determine economic value. Indeed, the fact that many biological resources are unpriced has forced environmental economics to refine innovative methods to investigate individual preferences for states of the environment. For reasons to be discussed, the use of such methods—in particular those that ask people directly to value highly unfamiliar goods—may place cognitive bounds on the meaningful use of some valuation techniques. The development of methods to investigate the value of complex goods is nevertheless a growth area in economics, such information being much sought after by environmental agencies.

Economic valuation of nonmarket goods is controversial. The equation of consequences with a money-equivalent loss of well-being, for example, may strike some as a peculiarly anthropocentric view of the world. Economics makes no apologies for such an approach, as economic values are necessarily those held by people, and individual maximization of well-being is consistent with any number of ethically based motives often supposed by critics to be neglected in monetary evaluation. However, it is important to appreciate the claims made for, and the reason to support, economic valuation over other approaches. Money simply translates unobserved well-being or utility into an observable unit of account that is the most convenient metric of comparison of gains and losses across space and time. If money is objectionable, then some other universal unit of well-being may serve the same function just as well, providing it makes such comparisons possible. Thus if it were possible to infer or elicit a value for bald eagles in Wisconsin today (and it is), the numeration of value should be the same metric for an exercise to value, say, raccoons in Florida tomorrow, thereby facilitating the comparison of costs and benefits over respective conservation programs. It is difficult to overemphasize the importance for efficient resource allocation of the ability to make such tradeoffs. Unfortunately the oft-cited charge of "pricing everything and valuing nothing," which arises from a confusion over the interpretation of economic value, inhibits the adoption of valuation without proposing alternative decision criteria and by even denying that tradeoffs are necessary. The use of a monetary numeration does not suggest that such comparisons are always necessary, inevitable, or mutually exclusive. Nor does it claim to be completely defined to include intrinsic or spiritual values, which are often invoked in the context of species preservation and our duty to future generations. Putting an economic value on biological resources is important because conservation tradeoffs represent very real choices for allocating scarce resources. When constraints bind, spending on items perceived as providing little or zero return will be minimal. A low default value for biodiversity and its services compared to the very real returns from conversion through deforestation or agriculture is historically reflected in land use decisions the world over. Other value systems may have a lot to say about the rights and wrongs of this, but by any objective assessment their practical role in changing the behavior precipitating the current global biodiversity crisis has been limited. On the other hand, valuing an asset creates a rationale for conservation. The "use it or lose it" dictum is one, albeit extreme, interpretation of this approach. A more sympathetic interpretation, which we hope to convey, simply seeks equal treatment for biological resources in development decisions.

TOTAL ECONOMIC VALUATION

Economic values can be categorized according to their underlying motives. Common components of total economic value that motivate observed or elicited WTP are dictated by current direct and indirect use, expected future use, and existence value (figure 9.1). Various methods can be used to determine the magnitude of these values. Current use can be direct and indirect.

Direct Uses

Even in the most developed societies, direct use of naturally occurring products demonstrates the fundamental contribution of biodiversity to well-being and the satisfaction of immediate needs. Direct use valuation tends to be straightforward, as many natural foods and fibers are traded in conventional markets. Even where market prices do not currently exist, as is the case for some nontimber forest products and traditional medicines, prices may be inferred from the value of close substitutes. Direct use does not have to be consumptive and therefore mutually exclusive of other values. For example, we may view an elephant in a wildlife reserve without precluding further use. But a problem with relying on direct use as the mainstay of value is that, whereas many people may want to "use" elephants, the demand for use of, say, any of the 10^9 microorganisms in a typical gram of soil is likely to be nonexistent. Similarly, there has been much euphoria over the potential for direct uses of nontimber forest products as economic arguments for maintaining biodiverse environments in pristine shape (Peters et al. 1989). On methodological grounds such conclusions have been shown to be premature. Nontimber forest products and hunting and gathering can represent very important values in

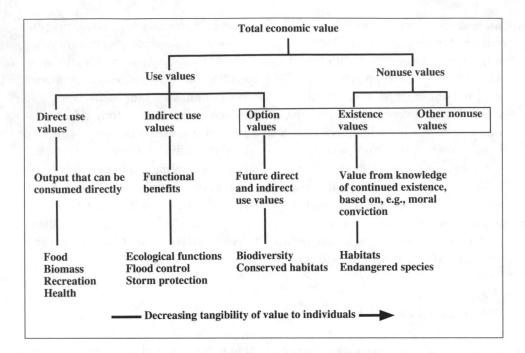

FIGURE 9.1

Categories of economic values attributed to environmental assets.

specific locations, but the findings of studies that have scrutinized the economics of such values have pointed to the need for more robust economic arguments beyond such direct uses.

Indirect Uses

Even if direct uses are avoidable, most people benefit indirectly from biological resources. Indirect uses are related to the functional benefits of species and ecosystems. Thus forests provide hydrologic benefits (flood control) and act as oxygen sources and carbon sinks. Similarly, mangroves serve as natural storm defenses, breeding grounds for aquatic species that are directly used, and natural water purification systems. On current knowledge the routine valuation of some of these functions may provide some of the most powerful reasons for maintaining functioning ecosystems. In fact, the realization of these values may be a sufficient conservation rationale without having to stray into the more contentious area of nonuse valuation.

The measurement of indirect values is more subtle because these functions are rarely bought and sold in regular markets. Full appreciation of the extent of indirect benefits generally requires cooperation between economists and natural scientists to establish the role of functions and the extent of impacts resulting from their loss. Once impacts can be related to market goods or some expression of WTP for resulting damages, an economic relationship is established. The degree of detective work required to attach economic values to the loss of these functions varies in complexity. Thus the relationship between deforestation, soil erosion, sediment loading in dams, and reduced electricity generation may be straightforward compared to the assignment of a benefit for global warming damage avoided by the preservation of carbon biomass of a protected forest. In both cases, economics has devised methodologies to assign values that are now routinely included in the economic assessment of environmental projects. The inclusion of such values is proving highly significant for the economic justification of biodiversity-related projects in which the biodiversity element itself remains elusive to available techniques.

Option Value

The concept of option value is at the heart of the issue of diversity and the maintenance of the widest portfolio of natural assets. In the face of potentially irreversible change and under present uncertainty, rational risk averse agents may reveal an economic value for the option not to foreclose on access to the resource in question, whether this be a set of protected areas or the diversity of the genetic portfolio. This WTP arises over uncertainty related to personal future use (direct use, indirect use, and nonuse) and is akin to an insurance premium or a risk premium.[1]

The cost of trading off current use and future choice represents the closest approximation of an economic value for the preservation of diversity per se. Such behavior is often implicit in collective decisions, where conservation is done with a view to the potential value of resources in the event that relevant use-related information is generated before resource depletion. This quasi–option value associated with potentially useful information is sometimes evoked as a rationale for safe minimum standards and the adoption of a precautionary approach to conservation.

Although it represents the value most closely associated with the popular perception of biodiversity, empirical assessment of biodiversity-related option values turns out to be extremely difficult. The exact definition of diversity is uncertain, and in relation to its demand, we can only suppose that on the basis of risk aversion, nonsatiated individuals want more rather than less. In order to say something about the economic consequences of reduced diversity, some diversity attribute must be translated into a value-of-diversity objective (Weitzman 1992), to be maximized subject to known financial constraints of global conservation expenditure. But how different facets of diversity may be given money equivalents is far from clear, and the development of a unified theory of diversity (particularly covering numerous species) is a formidable challenge for natural and social scientists.

Recent interest in conservation priorities has given renewed impetus to the search for a value-of-diversity function; in the first instance the surrogate of value is typically a cladistic distance criteria. The priorities issue has produced an array of measures for biological diversity covering measures representing functional di-

versity (Walker 1992) and measures using different levels of phylogenetic information to represent genetic or phenotypic diversity (e.g. Vane-Wright et al. 1991;
Williams and Humphries 1996; Weitzman 1992; Solow et al. 1993; Faith 1994). Faith
(1992) and Weitzman (1992) linked phylogenetic diversity to the concept of option
value, thereby providing a common basis of inquiry between conservation biologists and economists. Williams and Humphries (1996) elaborated on the links between option value and different possibilities for currencies of diversity, currency
models, diversity measures, and surrogates. Several such measures have also been
the basis of innovative area-selection algorithms maximizing taxonomic diversity
in a minimum area. As yet, data restrictions limit such methods to consideration
on a species-by-species basis, which is somewhat short of the assessment of the
economic value corresponding to the degradation of whole ecosystems. As a result, option value estimation has used the somewhat less discriminating approach
of direct survey-based elicitation, an issue to which we now turn.

Nonuse Value

Many people derive pleasure from knowing the world is a diverse place. We are
psychologically diminished by the irreversible extinction of species we never see
and cherish the continued existence of the greatest variety of life on Earth, irrespective of use motives. The exact motives of this existence value are uncertain, but
they may range from pure self-interest (perhaps motivated by optional use)
through various forms of altruism to rights-based and other ethical views. Equally uncertain (given the remoteness of these motives from the validation of everyday market transactions) is the question of whether such motives systematically
obey any axioms of individual preferences, such as the dictates of economic theory, or whether they translate into real economic commitments. The latter issue is of
particular interest because it seems likely that, for many charismatic species, the
aggregate global WTP for existence would be enormous. The presence or absence
of this potentially latent demand is therefore of considerable interest for a number
of reasons, not least in the optimal allocation of conservation resources in a
cost–benefit framework. For example, if significant existence values related to the
conservation of Amazon rainforests can be validated, then their realization may go
some way toward easing current international funding constraints.

Careful scrutiny of existence values for reasons related to the issue of conservation has recently increased in the United States, where federal legislation admits
them as compensable entities in law. In other words, damages due to oil spills,
water pollution, or other impacts that diminish existence values may be recover-
able from guilty parties (Ward and Duffield 1992). Needless to say, the stakes involved in multi–million-dollar lawsuits have injected considerable urgency into
inquiries into existence values.

Whether manifestations of sympathy, "warm glows" or other somehow "im-
pure" preferences, donations of money, time, or the purchase of wildlife-related

merchandise appear to put the existence of some form of nonuse value beyond dispute. But donations are typically cause-specific and, for the most part, other species have no such economic value representation. The question then arises whether surrogate or hypothetical markets can be created to value the existence of unfamiliar resources or elicit unobserved option value. From recent experience in valuing nonmarket goods, the response appears to be positive. Using a method known as contingent valuation (CV), economists have attempted to simulate marketlike tradeoffs for nonmarket goods using sophisticated questionnaires and experiments. Applications of constructed market methods are growing steadily, with interesting applications over a range of species and even complex ecosystem functions.

Not surprisingly, asking people what they think about biodiversity is controversial on several counts. First, do people know anything about what they are being asked to value? Second, what is to stop biased or strategic responses? Third, in the absence of marketlike benchmarks for many goods now being valued using CV, can responses ever be tested? If they cannot, what is the justification of using them in resource allocation decisions?

The answers to some of these questions lie in a growing literature dedicated to reducing biases and validating stated preference methods. The resulting methodological challenges are taking CV practitioners well beyond the traditional domain of economics and into fields such as cognitive psychology, where the robustness of consumer sovereignty and stable preferences underlying the enterprise of economic valuation are put to severe tests. In the process, increased use of nonmarket methods has also fueled debate both within and outside the economics profession over the ability to establish reliable values for environmental goods. Inevitably, criticism has centered on the displacement of an ethical stance on the inviolability of nature by instrumentalist consumer sovereignty. Proponents of valuation have not been slow to reclaim some of the moral high ground by pointing to the inevitable consequences for decision-making when nature is inviolable and to the lamentable consequences of the inaction implicit in a rights-based view of the world.

DEMONSTRATING ECONOMIC VALUE

The difficulties involved in measuring the full value of biological resources point to one of the causes of the current biodiversity crisis. The values that can be demonstrated begin to put a price on the consequences of biodiversity loss. Yet it is certain that much remains beyond the boundaries of economic analysis. For example, we know little about system resilience, the existence and nature of damage thresholds and system discontinuities. It is feasible that the transgression of some catastrophic threshold, for example by infraspecific genetic concentration in agriculture, may visit severe food shortages on large populations. At present we can

TABLE 9.1

Comparing Local and Global Conservation Values of Forest Resources (US$/ha, present values at 8%)

	MEXICO (PEARCE ET AL. 1993)	COSTA RICA (WORLD BANK 1992; CARBON VALUES ADJUSTED)	INDONESIA (WORLD BANK 1994; CARBON VALUES ADJUSTED)	MALAYSIA (WORLD BANK 1991)	PENINSULAR MALAYSIA (KUMARI 1994)
Timber	—	1,240	1,000–	4,075	1,024
Nontimber products	775	—	38–125	325–1,238	96–487
Carbon storage	630–3,400	3,046	1,827–3,654	1,015–2,709	2,449
Pharmaceutical	1–90	2	—	—	1–103
Ecotourism/recreation	8	209	—	—	13–305
Watershed protection	<1	—	—	—	—
Option value	80	—	—	—	—
Nonuse value	15	—	—	—	—

Adapted from Kumari (1994) but with additional material and some changed conversions. All values are present values at 8% discount rate, but carbon values are at 3% discount rate. Uniform damage estimates of $20.3 tons carbon have been used, so original carbon damage estimates in the World Bank studies have been reestimated.

do little more than speculate while arresting erosion by the most complete assessment of the full costs of development options.

Following the typology of figure 9.1, table 9.1 offers some orders of magnitude for the economic values of forest resources. Information from existing studies is patchy simply because of the paucity of good studies and the methodological difficulties encountered in applying those that have been carried out. Furthermore, table 9.1 reports orders of magnitude for whole countries, but per hectare present values[2] vary from site to site. Several observations are noteworthy. Option value as calculated in one study is based on a stylized model for pharmaceuticals that is not strictly consistent with the description of option value given previously (Adger et al. 1995). The summed sustainable nondestructive use and nonuse values (that is, uses most consistent with preservation) are generally higher than timber values associated with destructive logging. Contrary to ethnobotanical folklore, there is little convincing evidence for potential pharmaceutical megavalues, although the secrecy surrounding prospecting arrangements does not make such information easy to validate. On the other hand, given current scientific understanding, the certain carbon storage values are clearly dominant, giving rise to several questions about the rationale for forest conservation and methods to capture an essentially global value.

Table 9.2 reports CV results in several studies focused on species and habitat valuation. Clearly, if such values are typical over wider nonusing populations, the orders of magnitude for welfare change related to conservation merit further attention.

CAPTURING ENVIRONMENTAL VALUES

The rationale for biodiversity conservation can be greatly enhanced by emphasis on the range of nonmarket values, particularly the often extensive global values. The issue of measuring global value introduces the question of how easily such benefits may be captured and harnessed as conservation incentives in conserving countries. The internalization of these benefits is an important caveat when the supply of global value is derived from poorer countries that are only marginal beneficiaries of their own efforts. To make any difference in behavior, economic values must translate into real cash transfers for agents making conservation conversion decisions.

The presence of a spatial mismatch between the costs of conservation, often seen to fall on the resource-poor south, and the benefits accruing to a vociferous conservation-minded north is a geopolitical conundrum acknowledged by the Convention on Biological Diversity and subsequent deliberations over financing and north–south transfers. Mechanisms to capture the latent demand for conservation goods have arisen through spontaneous informal and formal deals, ranging from carbon offset agreements and debt-for-nature swaps to pharmaceutical prospecting deals. An important aspect of such deals is the possibility to devise dy-

TABLE 9.2

Preference Valuations for Endangered Species and Prized Habitats (1990 US$ per year per person)

SPECIES		
Norway	Brown bear, wolf, and wolverine	15.0
United States	Bald eagle	12.4
	Emerald shiner	4.5
	Grizzly bear	18.5
	Bighorn sheep	8.6
	Whooping crane	1.2
	Blue whale	9.3
	Bottlenose dolphin	7.0
	California sea otter	8.1
	Northern elephant seal	8.1
	Humpback whales[a]	40–48 (without information)
		49–64 (with information)
HABITAT		
United States	Grand Canyon (visibility)	27.0
	Colorado wilderness	9.3–21.2
Australia	Nadgee Nature Reserve	28.1
New South Wales	Kakadu Conservation	40.0 (minor damage)
	Zone, Northern Territory[b]	93.0 (major damage)
United Kingdom	Nature reserves[c]	40.0 ("experts" only)
Norway	Conservation of rivers against hydroelectric development	59.0–107.0

From Pearce (1993).
a. Respondents divided into two groups, one of which was given video information.
b. Two scenarios of mining development damage were given to respondents.
c. Survey of informed "expert" individuals only.

namic incentives that will minimize compliance and monitoring costs. Evidently the objectives of such deals are not always related to biological diversity; cross-country carbon offset agreements for forest conservation and management, for example, clearly serve climate-change obligations first and foremost. But if such deals are consistent with forest conservation, then they are to be encouraged.

Fortunately, in the case of carbon sequestration, a further economic justification is available. Table 9.3 shows various estimates of land values in Amazonia compared to the equivalent area values that can be calculated for the carbon sequestration function derived by multiplying estimated biomass of standing forest

TABLE 9.3

Conservation Versus Development

EXPECTED VALUE OF AMAZON FOREST LAND IN AGRICULTURE		VALUE OF AMAZON FOREST LAND IN CARBON SEQUESTRATION		
LOCATION	VALUE ($/HA)	VALUE ($/HA)	VALUE NET OF CARBON IN PASTURE	VALUATION BASIS
Para Paragomina area near Belem, Brazilia	300	7,200	4,950	Carbon tax in Sweden and Netherlands
Rondonia, BR364, south of Porto Velho	150	1,168	803	Nordhaus's "medium" estimate of marginal damage from global warming
Rondonia, BR364, west of Porto Velho	50	976	671	Carbon tax in Finland
Rondonia, Guajara Mirim area	15	560	385	Carbon equivalent of one-cent-a-gallon gas tax
Para Transamazonia Altamira area	2.5	288	198	Nordhaus's "low" estimate of marginal damage from global warming

From Schneider (1994).

by the estimated damage cost per ton of carbon that would be emitted if the forest were lost. Because deforested land will give way to crop production, these damage figures are shown gross and net of the subsequent sequestration by pasture biomass. The example shows a significant disparity between the value of standing forest as a carbon sink relative to alternative production.[3] From a global perspective, welfare is enhanced by carbon sequestration rather than clearing and agricultural production. But agricultural returns accrue to the converter, whereas conservation benefits inherent in reduced global warming are a global benefit. Thus there is apparently little incentive for a land owner to conserve carbon biomass. Assuming that no cheaper technological fix is available, therefore, an efficient global trade might involve buying out agricultural uses with a payment, possibly using a land rental agreement, at least equal to the maximum agricultural return and up to a maximum of the lower of the value of sequestered carbon or the cost of an alternative technological fix. The results of such a deal would leave land dwellers at least as well off with the payment as with the alternative of clearing forests for agricultural production, while possibly providing a cost-effective method for carbon sequestration.

From a global perspective, and abstracting from transaction costs, the fact that these options are available but not exploited represents a very real welfare cost. If the transaction costs can be circumvented, the potential to implement such deals is facilitated. Forest conversion limits the scope for such deals and conservation options for forest diversity.

CONCLUSION

Economics has two essentially related roles in the current biodiversity crisis. As a social science it is concerned with analyzing the motives and patterns of self-interested behavior of individual agents as they maximize welfare, subject to natural and artificial constraints. Such self-interested behavior, itself a fundamental cause of biodiversity loss, also reveals the values placed on biodiversity, which are equivalent to the welfare losses incurred when biodiversity disappears.

Because current unprecedented rates of biodiversity loss are human-induced, a second objective, not fully addressed here, involves the development of improved incentives for conservation (see Pearce and Moran 1994). More importantly, can the identified economic values be effectively harnessed for use in benefit–cost analysis or damage litigation, or as the basis of cross-country financial transfers?

Economic consequences are instrumentalist in that they are related to the human welfare effects of biodiversity loss. Individuals maximize well-being as they see it; this is not inconsistent with a variety of conservation-related motives such as sympathy, altruism, and rights-based ethics. Intrinsic values, on the other hand, cannot by definition be assigned. Although it is possible to have sympathy with the assignment of intrinsic rights, such standpoints are associated with their own costs and distributional effects and do not obviate tradeoff behavior. On the other hand, any denial of the need to price scarce resources simply tilts the balance against the unpriced.

A more interesting critique of the instrumentalist approach is related to our limited scientific understanding of the complexity of biological diversity. There are large, potentially enormous stakes involved in the loss of biodiversity, but the unknown simply cannot be valued. Similarly, while acknowledging (and occasionally attempting to model) the nature of system discontinuities, the presence of uncertainty handicaps the range of values that can be factored into the analysis. But uncertainty does not necessarily invalidate the contribution of economic valuation. Instead, it may be interpreted as an argument for an improved scientific understanding of biodiversity. In the meantime, the costs of precautions and the adoption of safe minimum standards provide benchmarks for the economic assessment of competing conservation strategies as they arise.

We have sounded a precautionary note about the values dominating the economic valuation of biological resources. Specifically, the dependence on carbon values as a catchall conservation rationale gives rise to a "good news–bad news" paradox associated with the corroboration of the science of global warming. Biodiversity is valuable in its own right, and there are reasons to be optimistic about the development of scientific and economic methods that say as much. Meanwhile, valuing the services we can identify buys time for improved understanding. Even with this approach, a final and probably the most vital caveat relates to the need to translate these benefits into definite returns so that biodiversity is regarded as an asset rather than a liability.

NOTES

1. The analogy to financial market portfolio diversification trading off yield for reduced risk should be clear.

2. Present values are an accounting convention to reduce an intertemporal stream of costs or benefits to a base year to facilitate comparison of investment alternatives.

3. Note that the current market price can be interpreted as a present value equivalent of the annual returns to land.

REFERENCES

Adger, W. N., K. Brown, R. Cervigni, and D. Moran. 1995. Estimating the total economic value of forests in Mexico. *Ambio* 245: 286–296.

Faith, D. 1992. Conservation evaluation and phylogenetic diversity. *Biological Conservation* 61: 1–10.

Faith, D. 1994. Phylogenetic pattern and the quantification of organismal biodiversity. *Philosophical Transactions of the Royal Society of London, Series B* 345: 45–58.

Kumari, K. 1994. An environmental and economic assessment of forestry management options: A case study in Peninsular Malaysia. In K. Kumari, *Sustainable forest management in Malaysia*, Ph.D. Thesis, University of East Anglia, U.K.

Pearce, D. W. 1993. *Economic values and the natural world.* London: Earthscan.

Pearce, D. W., N. Adger, K. Brown, R. Cervigni, and D. Moran. 1993. *Mexico forestry and conservation sector review: Substudy of economic valuation of forests.* Centre for Social and Economic Research on the Global Environment (CSERGE) for World Bank Latin America and Caribbean Country Department.

Pearce, D. W. and D. Moran. 1994. *The economic value of biodiversity.* London: Earthscan and Island Press.

Peters, C., A. Gentry, and R. Mendelsohn. 1989. Valuation of an Amazonian rainforest. *Nature* 339: 655–656.

Schneider, R. 1994. *Government and the economy on the Amazon frontier.* Latin America and the Caribbean Technical Department Report no. 34. Washington, D.C.: World Bank.

Solow, A., S. Polasky, and J. Broadus. 1993. On the measurement of biological diversity. *Journal of Environmental Economics and Management* 24: 60–68.

Vane-Wright, R. I., C. J. Humphries, and P. H. Williams. 1991. What to protect: Systematics and the agony of choice. *Biological Conservation* 55: 235–254.

Walker, B. H. 1992. Biodiversity and ecological redundancy. *Conservation Biology* 6: 18–23.

Ward, K. M. and J. W. Duffield. 1992. *Natural resource damages: Law and economics*. New York: Wiley.

Weitzman, M. L. 1992. On diversity. *Quarterly Journal of Economics* 57: 363–405.

Williams, P., D. Gibbons, C. Margules, T. Rebelo, C. Humphries, and R. Pressey. 1995. A comparison of richness hotspots, rarity hotspots, and complementarity areas for conserving diversity of British birds. *Conservation Biology* 10: 155–174.

Williams, P. and C. J. Humphries. (1996). Comparing character diversity among biotas. In K.J. Gaston, ed., *Biodiversity: A biology of numbers and difference*. Boston: Blackwell Scientific.

World Bank. 1991. *Malaysia: Forestry subsector study*. Report 9775-MA. Washington, D.C.: World Bank.

World Bank. 1992. *Costa Rica: Forestry sector review*. Draft report no. 11516-CR. Washington, D.C.: World Bank.

World Bank. 1994. *Environment and development*. Washington, D.C.: World Bank.

PERSPECTIVE

BIODIVERSITY IN AGRICULTURAL AND FORESTRY SYSTEMS

David Pimentel

More than 99 percent of the world's food comes from agricultural land and less than 1 percent comes from the oceans and other aquatic systems. The cultured plant and animal species of agriculture and forestry provide the basic food, fiber, and shelter to support human society and for this reason contribute several trillion dollars to the world economy. Yet the continued viability of agriculture and forestry also depends on most of the estimated 10 million natural species to provide the genetic and other resources to improve production.

Although many gaps in biodiversity science exist, augmented research and an appreciation of the importance of preserving biodiversity will improve agriculture and forestry as well as our quality of life. These gaps and the potential benefits for improving the sustainability of agricultural and forestry systems are discussed here.

BIOLOGICAL CONTROL OF PESTS

An estimated 67,000 different pest species attack agricultural crops worldwide, destroying more than 40 percent of all potential agricultural production valued at about $250 billion. This occurs despite the annual application of pesticides costing $26 billion.

Each pest or potential pest has an estimated 10 to 15 natural enemies attacking it, and some pests such as the gypsy moth have as many as 100 natural enemies. In fact, approximately 99 percent of potential pest species are kept under adequate control by natural enemies and the presence of resistance in their host plants. Unfortunately, the value of these natural enemies often does not become clear until after they are killed by insecticides. For example, when citrus trees were sprayed in California with dieldrin and endrin insecticides, the purple scale pest population increased 50 to 130 times because several natural enemies of the purple scale were eliminated by the pesticides.

In addition to helping control pests in agriculture, natural enemies also are effective in protecting forests. Bird predation is estimated to prevent $440 million ($1.70/ha) in insect damage to U.S. forests each year. In addition, other natural enemies and beneficial insect pathogens provide $3/ha/yr in forest benefits. A conservative estimate of the benefits of natural enemies to forests is about $10 billion worldwide each year.

Greater knowledge of the biodiversity of natural enemies would help reduce the reliance on pesticides and the associated environmental problems they cause. The potential for augmenting biological control of pests to increase the sustainability of agriculture and forestry is enormous.

HOST-PLANT RESISTANCE IN PEST CONTROL

Host-plant resistance (HPR) has become a prime control method in pest management. Using resistant genes in agricultural crops is economically and environmentally beneficial because HPR reduces the need for pesticides. In addition, the danger of eliminating natural enemies is avoided. At present, 75 to 100 percent of the crops planted in agriculture have some host-plant resistance traits. For example, breeding wheat for resistance to the Hessian fly pest saves American agriculture about $250 million each year.

Resistant genes have been identified and are now available for transfer to control all major cereal pathogens, making chemical control methods no longer necessary for many grain crops. A study involving reducing fungicide spray to suppress early and late blights on potato crops found that only about one-half the fungicide was needed when HPR was integrated into the control scheme. This provided a savings of about $7 million per year to farmers.

In the United States, host-plant resistance prevents about 40 percent of the pest losses in crops, equaling a saving of about $80 billion worldwide each year. The benefit–cost ratio for the use of HPR in U.S. agriculture is estimated to be $300 for every $1 spent for research and development. Host-plant resistance also prevents serious losses to insect pests and plant pathogens in world forests. Estimates are that this resistance prevents losses valued yearly at $10 billion worldwide.

The economic benefits of using HPR in pest management have been proven. Preserving biodiversity is essential to continued productivity and crop protection because the natural sources of resistant genes are found in diverse wild plant species. Eventually, when pests evolve tolerance to resistance factors in the crops, new forms of genetic resistance must be obtained from natural ecosystems in order to replace the old resistant factors. This fact reinforces the importance of preserving biodiversity.

POLLINATION

Plant pollination by animal species is vital in both agricultural and natural ecosystems. As much as one-third of the world's human food relies on insect pollination of plants. Worldwide, more than 400 species of crops valued at $200 billion require insect pollination.

There are about 20,000 species of bees capable of pollinating blossoms. Although honeybees in the United States are major pollinators of crops, about two-thirds of all pollination on cropland is by wild native bees. Some pollinators are keystone species. For example, euglossine bees are the primary pollinators of Brazil nuts and orchids, and one wasp species pollinates nearly all of the 800 species of figs found throughout the world. Unfortunately, pesticides kill or substantially reduce pollinator populations. For example, in California, where pesticides are used extensively in agriculture, most if not all of the wild bees have become extinct. This has forced fruit and vegetable farmers to rent honeybees to pollinate their crops. Economic losses also occur because many wild bee species are more effective pollinators than rented honeybees.

NITROGEN FIXATION

Nitrogen is vital to plants and animals and is often a prime limiting factor in the functioning of all natural ecosystems. Without biological nitrogen fixation, many natural biota and agricultural ecosystems would cease to exist. Approximately 150 million tons of nitrogen valued at $83 billion are fixed by natural organisms in world agricultural systems each year. Genetic engineering combined with the use of natural biodiversity has been reported to increase nitrogen fixation in beans tenfold, to 550 kg/ha/yr, and has significantly increased bean production. This was achieved without the addition of commercial nitrogen.

In addition to agricultural systems, biological nitrogen fixation occurs in forest ecosystems. For example, leguminous trees such as *Acacia* fix nitrogen at rates comparable to that of leguminous crops, thereby increasing the productivity of the soil.

Through the use of biodiversity science and genetic engineering, it may be possible to introduce nitrogen fixation capabilities into world grain crops. This will save about $100 billion in fertilizer costs each year, reduce nitrate pollution of ground and surface waters, and substantially cut the drain on the world's fossil energy supplies. When science and technology achieve this goal, agriculture will become more sustainable and environmentally sound.

PERENNIAL CEREAL GRAINS

The dominant grain crops grown worldwide are rice, wheat, corn, millet, barley, and rye. These and other grains provide 80 percent of the food consumed by humans. Because many are not cold tolerant, most of these crops are planted as annuals. Even in the tropics they are planted annually because of existing seasonal rainfall and weed problems. The impacts of this system on the environment are significant. For instance, spring tilling to prepare the soil for planting causes soil erosion and rapid water runoff from agricultural land. U.S. agricultural practices are responsible for 64 percent of the total pollution entering streams and 57 percent of the pollutants entering lakes.

In contrast, perennial grains could be grown and harvested continuously for a period of 4 to 5 years without replanting. A perennial corn genotype has been collected from the wild. If this wild corn type could be engineered as an economic perennial, the savings in diesel fuel alone could reach $300 million per year because energy would not have to be expended to sow the corn each year. Also, the savings in reduced erosion could total nearly $4 billion annually. Worldwide, the potential economic benefits of a perennial grain system could reach $300 billion per year in reduced fuel and soil erosion costs.

The future development of perennial grains and other desirable polycultures will depend on the availability of a diverse group of species in nature and their subsequent selective introduction into agricultural production.

GENETIC RESOURCES INCREASE CROP YIELDS

Approximately 15,000 species of the 350,000 known plant species have been consumed by humans. In the United States, nearly 6,000 crop species have been introduced for food, forage, and ornamentals. Crop yields have increased two- to fourfold throughout the world since 1945. This increase has resulted primarily because of the development of high-yield varieties, including hybrid varieties such as corn. From 20 to 40 percent of the increase in crop yields per hectare has resulted from improvements provided by biodiversity and genetic breeding. Assuming that the contribution of genetic resources to increased crop yields is 30 percent, the estimate is that genetic breeding accounts for about $175 billion per year worldwide in benefits.

Biodiversity science and genetic breeding in crops will continue to be essential in increasing food supplies to help feed the rapidly growing world population.

CONCLUSION

Maintaining a productive agricultural and forestry ecosystems depends on the availability of a healthy biodiversity. Implementing strategies that ensure the availability of diverse species of plants and animals is therefore essential. Without this gene and species pool to use in agriculture and forestry, greater crop and forest yields and quality improvement will be impossible to achieve.

PART III

BIODIVERSITY SCIENCE AND POLICY FORMULATION

CHAPTER 10

SAVING BIODIVERSITY AND SAVING THE BIOSPHERE

Norman Myers

There is a growing need for us to expand our policy purview to enhance biodiversity conservation. It is becoming clear that the predominant mode of protecting biodiversity—setting aside more parks and reserves—is falling further short of meeting overall conservation aims.

The present network of protected areas safeguards only a limited proportion of biodiversity at risk, and most such areas are proving incapable of preserving more than a modicum of their biodiversity in the long run. More importantly, protected areas are increasingly subject to threats that even the best-managed areas cannot resist. These threats are well established, especially the multitudes of landless and destitute people—a problem that will grow for a long time to come (as many as one-third of developing country parks and reserves are already being overtaken by agricultural encroachment). The new threats also stem from more distant and diffuse problems, notably atmospheric pollution in the form of acid rain, ozone layer depletion, and global warming. These threats can be countered only by remedial measures that achieve much more than protection of specific local wildland environments. The measures include efforts to slow, halt, and reverse problems such as desertification, deforestation, overuse of water supplies, acid rain, and global warming. They also include population planning, poverty relief, and a host of other development activities.

In short, we have reached a point where we can save biodiversity only by saving the biosphere. Hence, we can best save biodiversity by doing many other things that we should be doing for many other reasons. In turn, this implies that conservation planners should expand their policy approach to include action responses in lands far outside the main loci of biodiversity concentrations. Fortunately, then, there is much complementarity between our policy responses to the biodiversity problem and our responses to other problems of the biosphere.

BACKGROUND

The protected areas movement has a remarkable record. The number of areas now totals over 25,000 and the aggregate expanse amounts to 5 percent of Earth's land surface (Groombridge 1992; McNeely et al. 1994). Many of these areas are in the tropical developing world, especially in the biome that features greatest biotic richness and experiences greatest depletion: the tropical forests. Especially heartening is the recent upsurge in establishment of new areas, an additional 5 million km^2 or so since 1970. For all this, the protected areas movement deserves emphatic credit.

At the same time, we should recognize that the present protected expanse is too limited. For purposes of ecological comprehensiveness, it should be doubled at least. Moreover, many areas are too small to do their jobs adequately. Many are not sited in the best places; 2.2 million km^2 are in the Nearctic, twice as much as in the biotically richest zone on Earth, the Neotropics.

Worse, many protected areas are not as well safeguarded as they might be, notably in developing countries, which generally are coextensive with the tropics. These areas are subject to poaching, logging, grazing, and other forms of illegal resource exploitation; as many as 95 percent of all tropical parks and reserves are thought to suffer from some degree of poaching (Groombridge 1992; McNeely 1990). Many are subject to extensive encroachment by small-scale cultivators (Brown and Pearce 1994; Myers 1995a). Whatever the form of unwanted human incursion, the general result is a decline in the ecological integrity of protected areas, whereupon their conservation value is reduced. Worse, these derogations of protected status seem to be increasing, largely in response to pressures from multitudes of impoverished peoples—a familiar factor in developing countries (otherwise, they would not be "developing"). We have no good figures on how many protected areas are thus depleted, nor do we have a sound idea of how significant the damage is. But the depletion appears to be widespread already and it is becoming steadily more pervasive.

What of the future? Can we hope it will be a simple extension of the past, a case of "the same as before, only more so and better—that is, a moderate success story for protected areas? Or should we anticipate there could be some sharp divergences from the past, with unusual nonlinear changes? In particular, will there be a superscale increase in the pressures on protected areas, with a similarly superscale increase in their depletion? Or will there be a new approach, an imaginative and innovative approach that constructively addresses the rising tide of depletive pressures and, rather than trying to build higher walls against the incoming tide, seeks to divert the tide? Perhaps we should go so far as to accept the creative piece of speculation (McNeely 1990) that proposes there could eventually be no protected areas at all, on the grounds that either they will have been overrun by land-hungry people of the developing world and by resource-hungry consumers of the developed world or there will be no need for protected areas because humankind will have devised ways to manage all landscapes and ecosystems in a sufficiently rational way that the need for wildlands will be met.

In this chapter I examine the prospect for protected areas in a world that is going to become more crowded and more subject to economic activity. The chapter asserts that, to paraphrase Wordsworth, a leading English celebrant of nature, the world is too much with us already and will be far more so throughout the foreseeable future. This highlights the urgency of the question we should increasingly ask about protected areas—"Protected from what?"—in light of what surely lies ahead. That is to say, what sort of world, how alien if not antagonistic a world, are they to be safeguarded from in a future that will surely be very different from the past?

We are in a watershed phase of the human enterprise, and hence of the biosphere's security. Changes are overtaking the world with such variety and such speed, and they are interacting with such mutually reinforcing impact, that the period ahead will be different from anything we have known (Myers 1990). True, crystal-ball gazers have proclaimed new eras aplenty in the past, and in retrospect we have not noticed much difference in the most significant areas. But this time it will be different in multiple and unprecedented ways.

Consider the profound departures postulated by the World Commission on Environment and Development (1987). We must anticipate that within little more than a single human generation there will be twice as many people, seeking three times as much food and fiber, needing four times as much energy, and engaging in five to ten times as much economic activity. In other words, we shall be trying to build several more worlds on top of the one we already have, and the present one is proving too much for the survival of many protected areas.

This chapter describes some of the pressures that are gathering around protected areas and will put a progressive squeeze on the areas' survival if we do not anticipate their full scope and extent. In particular, the chapter considers the interactive processes that, through their multiple linkages, are likely to exert the greatest pressures of all.

THE KEY QUESTION OF LINKAGES

We are all too familiar with the myriad constraints that lie ahead—in principle at least. But do we know how they might work out in practice? Have we considered the diverse interactions between, say, population growth, economic development, and resource exploitation, and how they will jointly affect the outlook for protected areas? Have we taken systematic account of the way in which individual sectors (e.g., population growth) interconnect through their proliferate relationships? In short, have we taken on board the key question of linkages, with all that means for a realistic appraisal of what lies ahead? (For some exploratory assessment of the linkage question, see Myers 1991; Ehrlich and Ehrlich 1990, 1991; Ramphal 1992.)

To gain an insight into how all the critical factors will work out when interacting with each other—that is, to focus more sharply on the crucial question of linkages—consider a simple equation that demonstrates the environmental impacts of population growth and other factors:

$$I = PAT$$

In this equation, I stands for the impacts, P for population growth, A for economic welfare, and T for the technology needed to produce the economic welfare (Ehrlich and Ehrlich 1990). Note that the three items on the right-hand side of the equation, P, A, and T, are not merely additive. They interact in multiplicative fashion, each one compounding the others' impacts. This basic equation demonstrates why developing nations, with large populations but limited economic advancement, can have a vast impact on the environment (and hence on the survival outlook for protected areas), if only because the effect of P on A and T is so large. Also significant is that the developing world's population is projected to expand at least two and a half times before it attains zero growth in about 100 years.

The equation makes clear that developed nations generate population impacts when the A and T multipliers for each person are exceptionally large. If the almost 100 million people of Eastern Europe and the 170 million of the western Soviet Union achieve economic advancement to match that of Western Europe, then even though their population growth will probably remain small, their environmental impact will be large indeed, not only for these regions but, through ozone layer depletion and global warming, for the rest of the world.

This analysis contrasts with the approach that views prominent environmental sectors in isolation from other factors. By way of comparison, demographers are well known for their projections made in apparent disregard for intersectoral relationships. Of course demographers can protest that they are making only projections, not predictions or forecasts; that is, they are merely extrapolating from earlier demographic data. All the same, they might well ask whether it is realistic to keep on offering their super-significant assessments of the future (e.g., a projection of almost twice as many people within five decades) while doing it in an environmental vacuum.

After all, the projection presupposes that potential parents of the future—that is, parents already born—will find sufficient food resources to sustain themselves long enough to produce children in their turn. What if it turns out that they have overloaded their life support systems (croplands and the like) to an extent that they can no longer sustain themselves in the manner of the past? What if their progressive pressures start to trigger ecological discontinuities that lead to large-scale breakdown of life support systems? What if multitudes of impoverished people eventually (or soon) find themselves in such desperate straits that they feel driven to encroach on whatever unoccupied areas remain available, including protected areas (Jacobson 1988; Peters and Neunschwander 1988; Schumann and Partridge 1989)? For reasons set out in the rest of this chapter, we could shortly witness a fast-growing demand on the part of land-hungry peoples for more space to support themselves, exerting much greater pressure on protected areas (Myers 1995a).

The record of the past 10 years is hardly reassuring, with famines and mass starvation in many parts of sub-Saharan Africa and declining diets in several parts of the Indian subcontinent and several Andean nations. So is it still appropriate to assert that sub-Saharan Africa will increase its present population of almost 600 million people to 2.5 billion by the time the region reaches zero population growth? This is a much larger rate and scale of growth than is projected for any other part of the world, even though the region suffers far greater agricultural constraints, dietary deprivation, and associated forms of poverty than any other. Plainly, and humanitarian considerations apart, we should bear in mind that there is no greater threat to protected areas than the destitute person who disregards the boundaries of the best-protected park if he or she feels there is no other place to gain a livelihood.

Indeed, the region demonstrates that for a variety of reasons (not all of its own making), it is less capable than any other region of maintaining even a modicum of economic advancement (for an overview, see Myers and Kent 1995). In fact, per capita food production has been declining for at least two decades, and the perilous situation is not likely to be reversed within a decade. Of course, our first and overriding concern for the region must be human survival. But the prospect has ominous implications for protected areas. As we have noted, the principal threat to protected areas comes from multitudes of impoverished people practicing rudimentary agriculture. An underdeveloped region will be unable to modernize its agriculture, so the small-scale farmer will continue to practice extensive rather than intensive agriculture. It is likely that farmer will need 10 ha or more of inefficiently farmed land to support a family, rather than 3 ha or less of efficiently farmed land. So even if sub-Saharan Africa does not eventually reach as high a population as projected, it will surely have vast numbers of impoverished people spreading out across every last portion of landscape that can support them, including many landscapes that should remain as wildlands.

In other words, the biggest threat of all for protected areas is lack of development. Although it is often easy to demonstrate that development will not succeed in the long run without environmental protection, it is not always so simple to demonstrate the converse. But based on my quarter century of residence in the region, I have no doubt that an undeveloped Africa will become a barren Africa, ultimately bereft of its wildlands as well as any other sustainably productive lands. In Asia and Latin America, too, we should anticipate that the future is not likely to be a simple extension of the past, especially in terms of such a basic requisite as food. The green revolution has enabled growth in grain production to keep ahead of growth in human numbers for most of the past several decades. But a number of covert costs have been built into the process in the form of overloading of crop-land soils leading to erosion, depletion of natural nutrients, and salinization, for example. These costs, though unnoticed through disregard of the linkage factor, are already levying a price in terms of cropland productivity. In Pakistan, at least 32,000 km² of irrigated lands (20 percent of the total) and in India 200,000 km² (36 percent) are now so salinized that they have lost much of their productivity. Yet these two countries have often been ranked among the prime exponents of green revolution agriculture.

Worldwide the environmental underpinnings of agriculture are becoming so severely depleted that many apparent advances in agricultural output are being critically cut back. According to a recent analysis (Brown et al. 1990), soil erosion now leads to an annual loss in grain output that is roughly estimated at 9 million metric tons; salinization and waterlogging of irrigated lands, 1 million metric tons; and loss of soil organic matter, shortening of shifting-cultivator cycles, and soil compaction, 2 million metric tons. This makes a total from these forms of land degradation of 12 million metric tons. In addition, there are various types of dam-age to growing crops: air pollution costs 1 million metric tons of grain output each year and flooding, acid rain, and increased UV-B radiation cost another million metric tons.

Thus the overall total from all forms of environmental degradation amounts to 14 million metric tons of grain output per year. This total is to be compared with gains from increased investments in irrigation, fertilizer, and other inputs, worth 29 million metric tons per year. In other words, environmental stresses and strains are causing the loss of almost half of all gains from technology-based advances in agriculture. Note a linkage to population growth: we need an additional 28 million metric tons of grain output each year just to feed the extra mouths (let alone to meet demands of economic advancement and nutritional improvement). Whereas the net gain in grain output is worth less than 1.0 percent per year, population growth is 1.5 percent.

The faltering status of agriculture has momentous implications for protected areas. As indicated earlier, the principal threat to wildlands lies with poverty, ex-pressed through manifold repercussions for land use patterns and trends. In addi-tion to the agricultural connection, there are linkages for protected areas with many other development sectors. So significant are these linkages that it is worth-while to examine some variations on the basic theme.

LINKED LINKAGES

There are often linkages between linkages. These occur in circumstances where economic linkages reflect or reinforce environmental linkages and vice versa. These linked linkages are significant in part because one set of linkages often exerts a compounding impact on the other, which is far more important than a merely additive impact, and in part because the amplifying effect is more likely than a single linkage to carry sizable consequences for lands far removed from the original site. As the global economy becomes more integrated and as the global ecosystem becomes more stressed, the repercussions of these linked linkages will surely become more marked and widespread. Indeed, they could soon become a pervasive aspect of one-Earth living.

The situation is exemplified by a number of well-known instances, which have been widely documented and so are not dealt with in detail here. They include the hamburger connection, the cassava connection, the songbird connection, the cash-crop/desertification connection, and the debt/development connection (for extensive analysis, see Myers 1986). These linked linkages highlight three dimensions of the environmental prospect in general and the wildlands prospect in particular for the foreseeable future. First, it is all too easy for a nation to export its problems, whether economic in cause and environmental in consequence, or the other way around. The United States has sought to counter inflationary trends in its fast food markets by buying artificially cheap beef from Central America, thus supplying the principal pressure for deforestation in Central America and thereby exerting additional pressure on protected areas. Much the same applies with respect to Western Europeans and their search for foreign-aid–subsidized beef from Kenya, Botswana, and other countries of savanna Africa. The United States, Great Britain, and several other industrialized nations are dumping their fossil fuel pollution onto neighboring nations through acid rain and all communities that burn fossil fuels are enjoying a "free ride" on everybody's climate, with grossly adverse repercussions for protected areas and other environmental assets.

Second, and by extension from the first factor, it is not always sufficient for a nation to seek to safeguard its own environmental endowment through its national policies (just as it becomes increasingly difficult for a nation to safeguard its economic interests through unilateral action). The best-protected parks cannot be kept immune from the impacts of acid rain, global warming, and other climatic dislocations, and from atmospheric degradation such as enhanced UV-B radiation. Rather, the entire community of nations, reflecting an emergent global constituency, must consider its joint needs as an indivisible objective of collective global well-being. All too often we shall find that, "national" parks have an international dimension by virtue of their environmental and economic interconnections with lands far away. Plainly the health of the planetary ecosystem, and that of the totality of nations with their national parks, is greater than the sum of the well-being of the components of the ecosystem.

Third, and most important of all, the situation illustrates a new version of the tragedy of the commons (Hardin 1968), this time on a global scale. The actions of one nation, as perceived by that nation within the context of its own sovereign interests, may appear rational and productive, as illustrated by the search on the part of several leading nations for cheap beef, cassava, peanuts, and hardwood timber from overseas, with their spillover effects for wildlands. It is only when these actions are perceived within the context of the global community that they turn out to be irrational and destructive, not only to the overall interests of the global community but also to the separate interests of individual nations.

Moreover, one linkage can often work in conjunction with another linkage to generate a synergized (or mutually amplified) impact. The impact is not doubled—it is increased several times over because the consequence is not additive but multiplicative. To put it another way, one problem interacting with another problem does not produce a double problem, it generates a superproblem. To cite a simple illustration from the realm of ecology, a biota's tolerance of one stress tends to be lower when other stresses are at work. If low sunlight reduces a plant's photosynthetic activity, the plant becomes more susceptible to cold weather. Conversely, cold weather increases a plant's vulnerability to low sunlight. The result of compounded interactions can be an order of magnitude greater than the sum of the component effects.

We know all too little about environmental synergisms. Ecologists cannot even identify many of their natural manifestations, let alone document their impacts. If we can discern potential synergisms in the environmental upheavals ahead, we shall be better able to anticipate, and even prevent, some of their adverse repercussions (Myers 1995b). A simple illustration of synergistic reinforcing of impacts occurs with respect to acid rain in the humid tropics. This is already a recognizable problem in the forests of southern China and it will soon affect several other sectors of tropical forests, including central Indonesia, central-southern Thailand, southwestern India, West Africa, southern Brazil, and northern Colombia. Present stresses on tropical forests, such as overlogging and agricultural settlement, plus (possibly) excess UV-B radiation from ozone layer depletion, are grossly increasing the forests' susceptibility to injury from acid rain. Conversely, acid rain on undisturbed forests greatly increases their vulnerability to UV-B radiation (Rodhe and Herrera 1988). The human encroachment factor is especially pertinent to parks and reserves in tropical forests that suffer illegal logging and agriculture; and UV-B radiation, like acid rain, can affect all parks and reserves, no matter how well protected.

As an example of a compounded impact linkage at a higher level, consider the linkage between agriculture and global warming. As we have seen, agriculture will surely be a key to the survival of protected areas and other wildlands. The higher temperatures and reduced soil moisture expected in a greenhouse-affected world will not prove appropriate for most agricultural crops that are finely attuned to current climatic regimes. Therefore, the need to expand the genetic underpinnings of our crops places a premium on germplasm variability to increase resistance to drought in the wake of global warming. Yet the gene reservoirs of many crop plants are being more rapidly depleted than ever before because of large-scale elimination of biodiversity. To reiterate a basic point, this is especially important for protected areas because the greatest threat to these areas already stems from pressures on the part of land-hungry farmers.

Global warming will also interact synergistically with population growth. Consider the case of Bangladesh, already one of the most overcrowded nations on Earth. Within another three or four decades the country may well lose a sizable portion of its territory to sea-level rise (Huq et al. 1995). By that time, Bangladesh's population is projected to nearly double from today's 120 million people. (The situation will be further exacerbated by troubles associated with global warming such as disruption of the monsoon system.) Each of the two basic problems will make the other much more severe in its impact. Their synergized impacts will grossly reduce the survival prospect for Bangladesh's protected areas, much more than if we consider the two in isolation from each other. What is the outlook, then, for the famous Sunderbans tiger area on the coastline of the Bay of Bengal?

When we consider all environmental disruptions together, there are a multitude of synergistic interactions that could have pronounced adverse impact. For this reason, we should anticipate a greater environmental debacle overall, overtaking us more rapidly than is usually anticipated. At the same time, let us bear in mind that there can be constructive synergistic interactions, too, particularly as concerns management interventions. For example, grand-scale tree planting in the humid tropics, undertaken to generate a sink for atmospheric carbon dioxide to counter the greenhouse effect, could supply many spinoff benefits, for example, through commercial forestry plantations that will relieve excessive logging pressure on remaining natural forests (Marland 1988; Trexler et al. 1989). In turn, reduced deforestation will help safeguard the uniquely abundant stocks of species and genetic resources in tropical forests (with sometimes large agricultural benefits; a wild rice in India's forests helped save much of the Asian rice crop from a blight disaster). Both tree plantations and surviving natural forests supply many hydrologic functions, such as their capacity in upland catchments to regulate water flow and reduce downstream flooding, among other spinoff advantages for wildlands.

As should be apparent from this short account of linked linkages, the phenomenon is now a pervasive and profoundly important feature of everyday life. Linkages are becoming more numerous and more significant as more people engage in more activities of more sorts. Far from being isolated and occasional aspects of everyday life, linkages are a built-in feature that impinges on all aspects of human enterprise. Indeed, they are endemic to most if not all spheres of human activity, especially in a world where that activity is becoming more complex and integrated (if only by virtue of interdependency relationships, whether environmental or economic). It is virtually axiomatic that linkages will become an even more salient feature of our world as our economic systems interact more closely with the earth's environmental systems, generating feedback responses in both directions. The consequences for protected areas could hardly be more profound.

ECOLOGICAL DISCONTINUITIES

The linkages phenomenon, especially as concerns its reinforcing effects, will often lead to an overshoot outcome. In turn, this outcome can precipitate a downturn in the capacity of environmental resources to sustain human communities at their previous level. Designated as jump effects of ecological discontinuity, or threshold effects of irreversible injury, these occur when ecosystems have absorbed stresses over long periods without much outward sign of damage, then eventually reach a disruption level at which the cumulative consequences of stress finally reveal themselves in critical proportions. We can well anticipate that as human communities continue to expand in their numbers and demands, they will exert increasing pressures on ecosystems and natural resource stocks, whereupon ecological discontinuities will surely become common (Myers 1995b), with all that entails for protected areas.

An example has arisen in the Philippines, where the agricultural frontier closed in the lowlands during the 1970s. As a result, multitudes of landless people started to migrate into the uplands, leading to a buildup of human numbers at a rate far greater than that of national population growth. The uplands contain the country's main remaining stocks of forests and they feature much sloping land. The result has been an exceptional increase in deforestation and a rapid spread of soil erosion, with decline in agricultural productivity (Cruz et al. 1992; Myers 1988; Myers and Kent 1995). In other words, there has occurred a breakpoint in patterns of human settlement and environmental degradation. As long as the lowlands were less than fully occupied, it made little difference to the uplands whether there was 50 percent or 10 percent space left. Only when hardly any space at all was left did the situation change radically. What had seemed acceptable became critical, and the profound shift occurred in a very short space of time. As a consequence, there are now greatly expanded pressures on protected areas in the country's uplands.

Similarly, in Costa Rica agricultural expansion finally reached both oceans and both frontiers during the 1980s. For the first time in 400 years of their history, Costa Ricans (currently increasing in numbers at 2 percent per year) have no ready access to new land. Their predominantly agrarian society is having to adjust to a sudden change from land abundance to land scarcity (Cruz et al. 1992). This is all the more regrettable for a country with an exemplary record concerning protected areas. Whether these areas will survive as well as they have done to date is now an open question.

This problem of land shortages is becoming widespread in many if not most developing countries, where land provides the livelihood for almost 60 percent of populations and where most of the most fertile and most accessible land has already been taken. During the 1970s, arable areas were expanding at roughly 0.5 percent per year. But during the 1980s the rate dropped to only half as much; primarily because of population growth, the amount of per capita arable land declined by 1.9 percent per year (United Nations Population Fund 1994). Moreover, as far back as 1975 some 25 million km^2 of land already supported 1.2 billion people, yet only 563 million could be sustainably fed with the low-technology farming methods generally practiced. Most of this land was in semiarid or montane zones, unusually susceptible to soil erosion, decline of soil fertility, and loss of agricultural productivity. The population overloading aggravated the pace of land degradation (Food and Agriculture Organization 1984). The implications of a potential quantum increase in pressure on protected areas are plain.

Consider an instance in which a potentially renewable resource suddenly becomes overwhelmed by rapid population growth. Most people in the developing world derive their energy from fuelwood. As long as the number of wood collectors does not exceed the capacity of the tree stock to replenish itself through regrowth, the local community can exploit the resource indefinitely. They may keep on increasing in numbers for decades, indeed centuries, and all is well if they do not surpass a critical level of exploitation. But what if the number of collectors grows until they finally exceed the self-renewing capacity of the trees—perhaps exceeding it by only a small amount? Quite suddenly a point is reached at which the tree stock starts to decline. Season by season the self-renewing capacity becomes ever more depleted. The exploitation load remains the same, so the resource keeps dwindling more and more—meaning, in turn, an ever-increasing overloading of the resource. The vicious circle is set up and it tightens when the level of exploitation becomes nonlinear. Note that this scenario applies even if the number of collectors stops growing. The damage is done. But if the number of collectors continues to expand through population growth, the double degree of overloading (derived from a dwindling stock exploited by more collectors) becomes compounded. There ensues a positive feedback process that leads to fuelwood scarcity, and all too quickly the stock is depleted to zero. It is a process that occurs all the more rapidly as the stock is depleted.

The essence of the situation is that the pace of critical change can be rapid. As soon as a factor of absolute scale comes into play, the self-sustaining equilibrium becomes disrupted. A situation that seemed as if it could persist into the indefinite future suddenly moves to an altogether different status. It is as if two lines on a graph approach each other with seeming indifference to each other; once they cross, the situation is radically transformed.

We encounter this nonlinear relationship between resource exploitation and population growth with respect to many other natural resource stocks, notably forests, soil cover, fisheries, water supplies, and pollution-absorbing services of the atmosphere. Whereas resource exploitation may have been growing gradually for very long periods without any great harm, the switch in scale of exploitation induced through a phase of unusually rapid population growth can readily result in a slight initial exceeding of the sustainable yield, whereupon the debacle of resource depletion is precipitated with surprising rapidity. The lessons for protected areas are apparent. In many instances we shall be prudent to anticipate a sharp increase—indeed, a qualitatively different increase—in land shortage pressures on protected areas.

CLIMATE LINKAGES

We have already noted that the best-protected areas cannot be safeguarded from atmospheric degradation such as acid rain and enhanced UV-B radiation—the winds recognize no boundaries. But the greatest change of all will surely arise as a result of global warming. For many of the earth's species, a few degrees' change in temperature makes the difference between survival and extinction. As the planet warms up, temperature banks will move away from the equator and toward the poles. Vegetation will try to adapt by following the temperature bands, though with limited success. For one thing, climatologists project that changes will not only be large, they will also arrive suddenly, almost overnight (Intergovernmental Panel on Climate Change 1994). At the end of the last ice age, when the glaciers covering much of North America retreated, trees and other plants followed the ebbing ice northward. But they moved slowly, at a rate of only 50 km or so per century. The sudden arrival of the greenhouse effect will require communities of plants and animals to migrate at a rate 10 times faster. Many if not most species will find it impossible to adapt. Unless special dispensations can be made through heroic management measures, large numbers of species will die out.

Species that can make the quick transition will encounter another problem. In the past, they enjoyed a free run with only geographic obstacles—mountains, rivers, and the like—blocking their path. Following the last ice age, wild papayas and oranges migrated as far north as Toronto and tapirs and peccaries reached Pennsylvania. This time, wild species will find their way blocked by farmlands, cities, and other paraphernalia of human communities, which are "development deserts" for wildlife (Peters 1991).

This raises profound questions about the future of protected areas: their role, their effectiveness, and their survival. The very rationale for protected areas is threatened in a world that will not be an extension of the past but will be marked by radical shifts in its makeup and workings. Can the protected areas movement achieve a shift in approach that is equally radical?

INSTITUTIONAL INDIFFERENCE

Despite their great and growing importance, linkages are increasingly tuned out by our institutional limitations. We often remain indifferent to linkages on an individual level because an effort to recognize that everything is interconnected (so we can never do only one thing) runs counter to our professional backgrounds, single-disciplinary as they are likely to be.

Similarly, it is contrary to the linear thinking of the Western tradition for us to "think sideways." Despite the persistent emphasis on linkages in the Brundtland Report, there has been little effort to confront the challenge through systematic institutional adaptations.

As a measure of how much we balk at the prospect, consider some organizations that are engaged in activities extending far into the future and that ostensibly suppose that the long-term future will feature climate conditions unchanged from today's, even though it is increasingly apparent that that will not be the case. In the main, foresters worldwide continue to plant tree species that reflect past conditions of warmth and moisture, even though the trees will surely experience radically different conditions by the time they reach maturity.

POLICY RESPONSES

Fortunately, there is a good deal we can do to counter our multiple environmental problems, especially through policy initiatives leading to changes in programming and planning.

Measures to Tackle Linkages

As we have seen, the question of linkages lies at the heart of many of our problems. There is much that policymakers can do to address this challenge, and they must adopt a linkage mindset. Traditionally policymakers have been inclined to look upon the two great policy spheres of environment and development as two separate arenas of activity. Instead, they should perceive them as two sides of the one coin. As long as the two are viewed as separate entities, policymakers will continue to view environment as a constraint on development, to be articulated as little more than an add-on factor instead of the built-in dimension that it has demonstrably become. They should appreciate that this momentous step is no longer an option, it is an imperative—a solidly entrenched fact of life. Yet despite its detailed documentation in the Brundtland Report, there has been next to no attempt on the part of governments and international agencies, strategists and planners, and other parties involved to adapt their policy approaches accordingly. This is not surprising because the challenge is no less than a seismic shift in the entire outlook of policymakers and a basic departure from the day-to-day workings of governments and agencies.

Natural Resource Accounting

To break the established mold, a sound start could be made through the introduction of natural resource accounting (NRA). This would entail a radical revision of the way governments assess the state of their economies and their natural resource endowments (including wildlands, biotic communities, and other environmental assets). To date, our economic accountancy procedures fail to reflect our use, or rather misuse and overuse, of natural resources. National budgets and other annual audits rarely consider the depletion of natural resources, nor do they portray the declining goods and services that these resources generate. Whereas we hear much about rates of economic growth, we hear all too little (in equally systematized fashion, at least) about rates of natural resource depletion. This discrepancy between the way we appraise our economic activities and the way we evaluate the state of the natural resource base that ultimately sustains economic activity leads to a grossly distorted view of our economic health - (Jansson et al. 1994).

When we use human-made assets such as equipment and buildings, we write off our use as depreciation. But we do not view our environment as productive capital, even though we use it as such. When we excessively exploit forests, overwork croplands until the soil erodes, use our skies as a free garbage can and our rivers as free sewers, commercially exploit wildlands, and generally abuse our environments, our income as measured by GNP almost always registers an increase. When we engage in efforts to reduce, neutralize, or repair environmental damage, these economic activities are also registered as additions to GNP, but they should more realistically be deducted from GNP (Daly 1994).

Some preliminary analysis is revealing. In the case of Indonesia, for example, during the period 1971–1984 the nation's GNP was conventionally reckoned to be growing at an average rate of 7.1 percent per year. But if account had been taken of resource depletion, notably overlogging and other degradation of natural forests, the rate would have been reduced to around 4 percent per year, meaning that almost half the growth rate was unsustainable (Repetto 1986).

In short, NRA would correct the partial accounting system used at present, with its misleading signals for use of the natural resource stocks and environmental services that underpin our economies. As a policy intervention of comprehensive scope, it would soon generate a plethora of accurate messages about our legitimate use of our environmental resource base including our use of wildlands, particularly as exemplified by protected areas. In turn, policymakers would soon find they had little alternative but to engage in a whole series of adjustments when evaluating their policies, programs, projects, and other activities. The immediate effects would be significant and pervasive and would quickly generate their own multiplier effects. As a high-leverage intervention, NRA would engender linkage thinking in development and environment circles.

Pricing Policies

Although pricing policies may seem an esoteric factor in the future of protected areas, policymakers can take measures that would have many beneficial spillover effects for protected areas and other wildlands. Many current pricing policies send out perverse messages about sustainable use of environmental resources because they do not reflect all costs of production.

A well-known adjustment strategy is known as full-cost pricing. Suppose we were to internalize the many externalities of fossil fuel burning, perhaps through a carbon tax to reflect acid rain, carbon dioxide emissions, and other forms of pollution (all of which, as we have seen, are exceptionally significant to the future welfare of protected areas). We would then be obliged to pay a price for fossil fuels that reflects all costs involved. The price of a gallon of gasoline might double in European countries and increase much more in North America and certain other developed regions, which account for three-quarters of fossil fuel emissions of greenhouse gases.

The result would be a massive shift in our perception of the role of (formerly cheap) gasoline in our economies and to bring an end to the free ride we currently enjoy. Car transportation would swiftly decline, with a concomitant shift to public transport systems that, reflecting suddenly expanded demand, would be enhanced by both official and private bodies. Many other fuel-saving measures would ensue. The ramifications would ripple throughout our economies, our societies, and our understanding of our status vis à vis our environmental support base, including the role of protected areas and other wildlands. Indeed, the process would foster linkage thinking in many ways on the part of policymakers, business leaders, and private citizens. In other words, we would find ourselves having to think twice not only when we visit the gas station but when we visit the supermarket, and at dozens of other points in our daily lives. The overall effect would be to incorporate environmental values across the economic system, to do it in systematic fashion, and to make us all undertake the revised thinking that is at the heart of linkage-derived adjustments.

These three sets of policy responses, together with a number of related initiatives such as implementation of the polluter-pays principle and precautionary principle (Pearce et al. 1991) must be undertaken quickly. They will require a radical reorientation to shift policymakers' outlook from their former mode of linear thinking, which can also be seen as nonlinkage thinking. Attitudes and perceptions are deeply entrenched in myriad ways, inducing an institutional inertia in virtually all spheres of human activity. This inertia factor is a powerful force (although fortunately a new form of inertia can work to our benefit once the corrective measures are in place, generating a constructive form of momentum). The longer we delay in adopting corrective measures, the more deeply the present patterns of environment-depletive activities will become entrenched. Time is at a premium.

CONCLUSION

Linkages are not something to be tackled when the time is finally ripe. They are a central fact of the world we live in right now, and they are growing more numerous and significant all the time. As this chapter has demonstrated, the gray areas of dynamic interactions between sectors are sometimes as important in strictly economic terms (let alone other development considerations) as are the sectors themselves–including the sector of protected areas. The more we overlook the linkages, the more we shall find that the sectors fail to function efficiently and productively. In many cases, in fact, we shall increasingly find that linkages are all.

Yet linkages tend to be ignored because our view of the world is traditionally grounded in a practice of splitting it up into manageable components. In a world made more interdependent by virtue of multiple relationships—whether environmental, economic, political, or social—we can no longer afford the luxury of supposing that linkages are an incidental factor that is too complex to be reflected in institutional responses. Yet we still try to direct our daily affairs with institutional systems that reflect a bygone and far less integrative era. These institutional systems are singularly unsystematic in that they disregard what is now a predominant phenomenon of the planetary ecosystem and the world order. Far from adapting in accord with objective reality, they remain static and out of tune with the times. Of all the features of an interdependent world, institutional systems exert a proliferate impact by virtue of their deficiencies.

The bottom line is that we will respond to linkages either by responses of sufficient scope and character or by salvage measures in a world impoverished by our disregard of linkages. Linkages will eventually be addressed, whether by design or by default. Fortunately, there is abundant scope for us to tackle the question of linkages through policy initiatives that are becoming well understood, some of them already being implemented in exploratory fashion. We still have time—but not much. Let us count ourselves lucky that we still have maneuvering room to turn profound problems into profound opportunities. Our fellow species might count themselves lucky, especially the millions that otherwise face the prospect of extinction within our lifetimes.

REFERENCES

Brown, K. and D. W. Pearce, eds. 1994. *The causes of tropical deforestation*. London: University College London Press.
Brown, L. R. et al. 1990. *State of the world 1990*. New York: WW Norton.
Cruz, C., C. A. Mayer, R. Repetto, and R. Woodward. 1992. *Population growth, poverty and environmental stress: Frontier migration in the Philippines and Costa Rica*. Washington, D.C.: World Resources Institute.
Daly, H. 1994. Operationalizing sustainable development by investing in natural capital. In A. M. Jansson, M. Hammer, C. Folke, and R. Costanza, eds., *Investing in natural capital: the ecological economics approach to sustainability*, 22–37. Washington, D.C.: Island Press.

Ehrlich, P. R. and A. H. Ehrlich. 1990. *The population explosion*. New York: Simon & Schuster.

Ehrlich, P. R. and A. H. Ehrlich. 1991. *Healing the planet*. New York: Addison-Wesley.

Food and Agriculture Organization. 1984. *Potential population supporting capacities of lands in the developing world*. Rome: Food and Agriculture Organization.

Groombridge, B., ed. l992. *Global biodiversity: Status of the earth's living resources*. London: Chapman & Hall.

Harden, G. 1968. The tragedy of the commons. *Science* 162: 1243–1248.

Huq, S., S. I. Ali, and A. A. Rahman. 1995. Sea-Level rise and Bangladesh: A preliminary analysis. In R. J. Nicholls and S. P. Leatherman, eds., *Journal of Coastal Research*, special issue 14: 44–53.

Intergovernmental Panel on Climate Change. 1994. *Conference report of the World Coast Conference 1993 : " Preparing to Meet the Coastal Challenges of the 21st Century. "* Rijkswaterstaat, The Hague, November 1–5, 1993.

Jacobson, J. L. 1988. *Environmental refugees: Yardstick of habitability*. Washington, D.C.: Worldwatch Institute.

Jansson, A. M., M. Hammer, C. Folke, and R. Costanza, eds. 1994. *Investing in natural capital: The ecological economics approach to sustainability*. Washington, D.C.: Island Press.

Marland, G. 1988. *The prospect of solving the CO_2 problem through global reforestation*. Washington, D.C.: Carbon Dioxide Research Division, U.S. Department of Energy.

McNeely, J. A. 1990. The future of national parks. *Environment* January/February: 16–20, 36–41.

McNeely, J. A., J. Harrison, and P. Dingwell. 1994. *Protecting nature: Regional review of protected areas*. Gland, Switzerland: IUCN.

Myers, N. 1986. Economics and ecology in the international arena: The phenomenon of "linked linkages." *Ambio* 15(5): 296–300.

Myers, N. 1988. Environmental degradation and some economic consequences in the Philippines. *Environmental Conservation* 15: 205–214.

Myers, N. 1990. *Future worlds: Challenge and opportunity in an age of change*. New York: Doubleday.

Myers, N. 1991. *Environment and development: The question of linkages*. Geneva, Switzerland: UNCED.

Myers, N. 1995a. Tropical deforestation: Population, poverty and biodiversity. In T. Swanson, ed., *The economics and ecology of biodiversity decline*. Cambridge, U.K.: Cambridge University Press.

Myers, N. 1995b. Environmental unknowns. *Science* 269: 358–360.

Myers, N. and J. Kent. 1995. *Environmental exodus: An emergent crisis in the global arena*. Washington, D.C.: Climate Institute.

Pearce, D., E. Barbier, A. Markandya, S. Barrett, R. K. Turner, and T. Swanson. 1991. *Blueprint 2: Greening the world economy*. London: Earthscan.

Peters, R., ed. 1991. *Proceedings of the conference on the consequences of the greenhouse effect for biodiversity*. New Haven, Conn.: Yale University Press.

Peters, W. J. and L. F. Neunschwander. 1988. *Slash and burn farming in Third World forests*. Moscow: University of Idaho Press.

Ramphal, S. 1992. *Our country the planet*. London: Lime Tree Press.

Repetto, R. 1986. *Natural resource accounting for countries with natural resource–based economies*. Washington, D.C.: World Resources Institute.

Rodhe, H. and R. Herrera. 1988. *Acidification in tropical countries*. Chichester, U.K.: Wiley.

Schumann, D. A. and W. L. Partridge, eds. 1989. *The human ecology of tropical land settlement in Latin America*. Boulder, Colo.: Westview Press.

Trexler, M. C., P. E. Faeth, and J. M. Kramer. 1989. *Forestry as a response to global warming: An analysis of the Guatemala Agroforestry and Carbon Sequestration Project*. Washington, D.C.: World Resources Institute.

United Nations Population Fund. 1994. *The state of world population 1994*. New York: United Nations Population Fund.

World Commission on Environment and Development. 1987. *Our common future*. Oxford, U.K.: Oxford University Press.

CHAPTER 11

THE FACTS OF LIFE (ON EARTH)

Thomas E. Lovejoy

The biodiversity crisis poses an enormous challenge to both science and society. Although some may deny its existence, there is essentially total agreement among biologists that extinction rates are elevated and accelerating. The scale of the problem is large both because of direct effects such as habitat destruction and because, through indirect effects, virtually all environmental problems have an impact on biodiversity. In a sense, biodiversity is the bottom line of the planet's environmental ledger.

Although there is serious need for major efforts in science to address the problem, it can also be argued that science and society have not done enough with the knowledge that does exist. José María Figueres, visionary president of Costa Rica, speaks of the need for bioliteracy. Too many people in the United States and elsewhere do not know the basic facts of life in the sense of the benefits we derive in goods and ecological services from biological diversity. An important policy conclusion is that real effort must be put into education, including adult education. As scientists and environmentalists, we should be spending less time talking to one another and more time talking to others and illuminating our place in nature.

IMPORTANCE OF BIODIVERSITY

There is a serious need to help people understand the many benefits biodiversity already confers upon us, including food, fiber, shelter, and medicine.

Similarly, people need to appreciate the values of ecological services, including watershed function, air and water purification, pollination, and pest control. These services, which are usually treated as free (and therefore tend to be degraded), derive from the biodiversity that provides ecosystem structure as well as function; that is, we would not have forests and all they provide without the species of trees of which they are composed.

There is also a need to recognize the indirect values, which include aesthetic and ethical, as well as the extraordinary indirect benefits biodiversity provides as the basic library from which the life sciences are built. The history of medicine, agriculture, and other sciences is littered with examples of improvements and solutions that arose from the study of some esoteric species. This is how the life sciences are built; we owe many of the advantages in our daily lives to some obscure life form. Rather than isolating us from nature, the life sciences highlight our dependence on it.

Biodiversity presents a magnificent opportunity as science and technology find new ways to generate wealth at the level of the molecule. As much as the twentieth century will be thought of as the century of the information revolution, so will the twenty-first century be thought of as the century of biology, as biotechnology reveals almost undreamable human benefit from bioindustry, bioremediation, bioconcentration, and even nanotechnology (the ultimate in miniaturization, using molecules from nature).

Science is more candid about uncertainty than any other form of human endeavor. This is often thrown back in our faces when our conclusions or recommendations are unpopular. Nevertheless, as scientists we have an important obligation to explain the pervasive nature of uncertainty in all areas of human activity while maintaining our objectivity about our own science and its implications for society.

SUSTAINABLE DEVELOPMENT

We have to learn to live within nature rather than think of nature as something that survives here and there in a sea of human activity. This is the essence of ecosystem management, where large ecologically cohesive units of landscape are managed to conserve their processes and characteristic biodiversity. The latter could be thought of simple-mindedly as maintaining the particular landscape's characteristic species list. Started early enough, this approach provides a great deal of flexibility and multiple options for human activity.

Ecosystem management of necessity involves a considerable degree of local participation in decision-making. It also has a large element of voluntary participation. It works because in return for giving up (voluntarily) a small amount of control over a particular piece of the landscape mosaic, the participant gains the opportunity to influence decisions relating to many other portions of the landscape that might affect the particular land owner's individual piece. An important experiment is taking place with the southern California coastal sage scrub (home of the California gnatcatcher and less well-known endemics). There, the combined efforts of country, state, and federal agencies working with development interests and environmentalists seem to be succeeding.

Ecosystem management must take all factors, including those intrinsic and extrinsic to the ecosystem, into account, including population growth and climate change. Were the world to be divided up into a set of ecosystems (admittedly arbitrary because there is no clearly defined single set), the sum of their sustainable management would equal sustainable development:

Sustainable development = $\sum$ Sustainable ecosystem management

Achieving this understanding will require going way beyond concerns of the biological sciences to confront the complex Gordian knot of socioeconomics. There is a need to reorder the way economics and markets work and to harness market forces to work for environment and sustainable development more than they currently do. Some environmentalists find it difficult to think about that, but it is abundantly clear that sustainable development will never be achieved without the wholehearted participation of the private sector. Costa Rica, under the leadership of president Figueres and building on an already strong historical base, is providing important inspiration in sustainable development in which different elements support one another synergistically.

We need to become more sophisticated in distinguishing sustainable development from raw, untrammeled economic growth. Interesting contributions have been made by the private sector, including by the World Business Council for Sustainable Development (in two stimulating books, *Changing Course* and *Financing Change*) and recently by the President's Council on Sustainable Development in the United States. The two forms of biological growth provide a suggestive contrast: simply getting large (growth based on consuming more resources) and getting more complex (so-called growth by intussusception), perhaps analogous to service and information industries.

The economics of sustainable development are more complex than is currently recognized. Discount rates undervalue, or value not at all, anything of long-term consequence (Chichilnisky 1996). With a tyranny of discount rates, is it surprising to find overemphasis on quarterly results? Currently a forest or a tree is usually valued only for its timber or carbon sequestration, which is rather like valuing a computer chip for its silicon.

In contrast, if biodiversity conservation remains largely endangered-species–driven, in the end the game will be lost. This is in part because at some point the list becomes so long as to be unmanageable. It is also because in the United States there has been a tendency to wait until there is a problem, by which time it is harder to do something about it, and almost by definition there is a set of vested economic interests squared off against some organism with an esoteric and easily caricatured name. Ecosystem management, properly practiced, will take species into account, but should act in advance to avoid problems and provide solutions. The Endangered Species Act will still have a vital role to play, but rather than overuse of this safety net, there is an urgent need for comprehensive preemptive conservation of biodiversity in ecosystems. Some might say that this is what habitat conservation is about, but I believe it has fallen short of the ecosystem conservation goal.

One important initiative would be for both the private and public sectors to undertake adaptive management, wherein natural resource management is designed and conducted as a scientific experiment, essentially learning by doing. That way, the results of a management approach can be clearly evaluated and contribute to further improvements in resource management. Usually natural resource management is ad hoc (although sometimes brilliantly so), such that a plethora of opportunities to learn how to do better are wasted.

RESEARCH NEEDS

Although an ecosystem management approach should be undertaken on the basis of available knowledge, it clearly requires continuing scientific research to provide ongoing refinement of management practices. It would clearly be wrong to assume that the world of nature is static even if we did not have to contend with our own interventions. Among important research needs are the following:

- *Biological survey:* This volume, by highlighting what we do know about life on Earth, has pinpointed enormous lacunae in our knowledge. Biological survey (*sensu strictu*), which I view as just good biological housekeeping, can be usefully thought of as involving a series of approximations about what occurs where, as well as status and trends. Rather than being merely a collection of individual, strictly curiosity-driven scientific endeavors, biological survey must be conducted with at least partial reference to a conservation, bioprospecting, and pure research framework. Given the dynamics of the natural world, biological survey is essentially a task that can never be completed.
- *Systematics:* Systematics is a vital underpinning to biological survey, conservation, and realization of the magnificent practical potential of biological diversity. There is a vital need to train more systematists for the gargantuan task of categorizing and inventorying the diversity of life on Earth. This responsibility falls in large part on natural history institutions (museums and botanical gardens) of the industrialized nations, for they house the bulk of all natural history collections and still a great many of the professional scientists. This is inherently an international exercise, for no nation has or ever will have the full array of systematics expertise they will need to conduct biological survey. The innovation of parataxonomists, as seen in Costa Rica, is an important element of meeting this challenge, which together with the new interactive electronic media identification aides (e.g., compact disks) enable the amateur to do more and to conserve the time of specialists for tasks that only they can do.

Australia's Environmental Resource Information Network (ERIN), which is essentially a distributed network of databases, brings new power to systematics and related exercises by calling up and analyzing computerized databases related to natural history collections. In the United States, with such data housed in hundreds of institutions, the challenge of building such a network is considerably greater. The benefits of computerized data are so great, however, that computerization is happening spontaneously in a number of places. The Global Biodiversity Information Facility (GBIF) under consideration by the OECD Megascience Forum would be just such a distributed network on a global basis. These efforts all can tie into Geographic Information Systems (GIS).

Systematics Agenda 2000 lays out a challenge and vision of real importance to society (Systematics Agenda 2000 1994). I believe it is important to think of it as a new age of discovery, as thrilling as the age of exploration launched by Prince Henry the Navigator in the fifteenth century.

• *Conservation biology:* Conservation biology is essential to the success of ecosystem management. One of the important and interesting questions is how biological diversity relates to ecosystem function. When this question is addressed normally, it is examined as a snapshot of time and there seem to be a large number of rare or marginally relevant species, and indeed apparently "redundant" species with respect to function. It is probably more appropriate to think of them as functional equivalents. It is likely that many of these will be of greater functional significance if the ecosystem is studied over significant periods of time and environmental fluctuation. Of particular interest in this regard is the work of David Tilman and colleagues (Tilman et al. 1996) that demonstrates resistance to stress and greater retention of nutrients and productivity in more diverse grassland ecosystems.

There is a necessary array of research needs that range from basic natural history, to the genetics of small populations, to population biology, community biology, ecosystems, and ultimately biogeochemical cycles. Lack of knowledge should not impede some action, but new knowledge allows continual refinement.

• *Socioeconomic research:* In the end, people often will not make the right decisions about natural resources even when well informed. Socioeconomic research can identify factors that drive in that direction and help design systems that harness market forces in favor of the environment. A powerful example of the latter is how the market for SO_2 emission rights (to counter acid rain problems) drove the price from $2,000/ton to essentially 5 percent of that figure.

• *Remote sensing:* Remote sensing has a great deal to contribute to natural resource management, even if at a somewhat coarse level. A periodically updated real image-based vegetation map of the world could help greatly in habitat and therefore biodiversity conservation.

NEEDED ACTIONS

Many of the needed elements to support biodiversity conservation and sustainable development are to be found in the Convention on Biological Diversity and the associated Agenda 21. Some nations, including the United States, have yet to ratify the convention. In the United States some mistakenly see the convention, and indeed the National Biological Survey (now the Biological Service Division of the U.S. Geological Survey), as a threat to development. To the contrary, the new knowledge should help solve problems and provide economic opportunity. Although it is ultimately science based, it is quite possible for other, sometimes perverse agendas to drive meetings of the Conference of the Parties, including the Subsidiary Body on Scientific, Technical and Technological Advice (SBSTTA). Any nation not a full member forfeits some of its ability to influence the debate and outcome, including such critical and complex issues as intellectual property rights. Its nonmember status is a serious handicap for the United States and also for the convention, which is therefore denied some of the scientific contributions that this nation could make.

The promise of massive new financial support for developing nations to help them implement both the biodiversity and climate conventions has not materialized. Worse, there has generally been retrenchment in overseas development assistance, with little apparent prospect for any improvement in the near future. It is to be hoped that other forms such as joint implementation under the climate convention or the bioprospecting contracts with institutions such as Costa Rica's INBio can facilitate research and technology transfer.

All the above presupposes attention to human population dynamics, to consumption patterns and equity (Ehrlich et al. 1995). Even though we are very good as a species at augmenting the carrying capacity for humans (Cohen 1995), the biological impoverishment of the planet (Wilson 1992) is pushing in the opposite direction.

REFERENCES

Chichilnisky, G. 1996. The economic value of the earth's resources. *Trends in Ecology and Evolution* 11: 135–140.
Cohen, J. 1995. *How many people can the Earth support?* New York: WW Norton.
Ehrlich, P. R., A. H. Ehrlich, and G. C. Daily. 1995. *The stork and the plow*. New York: Putnam.
Systematics Agenda 2000. 1994. *Systematics Agenda 2000: Charting the biosphere*. New York: American Museum of Natural History.
Tilman, D., D. Wedin, and J. Knops. 1996. Productivity and sustainability influenced by biodiversity in grassland ecosystems. *Nature* 379: 718–720.
Wilson, E. O. 1992. *The diversity of life*. Cambridge, Mass.: Harvard University Press.

CHAPTER 12

CONVENTION ON BIOLOGICAL DIVERSITY: PROGRAM PRIORITIES IN THE EARLY STAGE OF IMPLEMENTATION

Kalemani J. Mulongoy, Susan Bragdon, and Antonella Ingrassia

Alarmed by the rapid erosion of the earth's biological diversity and its components, the world's nations gathered to negotiate a treaty to address the problem. After almost 4 years of discussions and negotiations, the Convention on Biological Diversity was adopted in May 1992 in Nairobi, Kenya. Affirming that the conservation of biological diversity is a common concern of humankind, the treaty's objectives are the conservation of biological diversity, the sustainable use of its components, and the fair and equitable sharing of the benefits arising from the use of genetic resources, including appropriate access to genetic resources, transfer of relevant technologies, and funding. The convention defines biological diversity as the variation between and within species of microorganisms, plants, and animals, and between ecosystems, both terrestrial and aquatic, including marine ecosystems. Components of biological diversity are considered at three levels: genes and genomes, species and communities, and ecosystems and habitats.

The convention opened for signature in June 1992 at the United Nations Conference on Environment and Development in Rio de Janeiro, Brazil, and entered into force on December 29, 1993, less than 19 months after its opening for signature. As of April 5, 1995, 117 nations and the European Union have become parties to the convention. The rapid entry into force of the convention and the large and growing number of parties reflect the commitment of the world to the objectives of the convention.

The entry into force of the convention is just the starting point. The convention envisions that parties will continue to work together through its institutions, particularly the Conference of the Parties (COP), the Secretariat, and the Subsidiary Body on Scientific, Technical and Technological Advice (SBSTTA), to give practical meaning to its broad provisions. The first meeting of the COP held in Nassau, the Commonwealth of the Bahamas, from November 28 to December 9, 1994, took the important first steps to give life to the convention's institutions with the selection of the United Nations Environment Programme (UNEP) to provide the Convention's Secretariat and the selection of the Global Environment Facility (GEF) as the interim structure to operate the financial mechanism. The COP agreed on a medium-term program of work and identified 13 program priorities to guide GEF in making funding decisions.

This chapter elaborates on views and actions of the parties and other signatories to the convention to identify and briefly describe the priorities in the area of conservation of biological diversity and sustainable use of its components.

PROGRAM PRIORITIES UNDER THE CONVENTION

Noting that biological diversity is seriously threatened today by its main beneficiary, humankind, nations decided in recent years to identify some urgent actions that would minimize the threats and reverse the pattern of biological diversity loss and to affirm the unwavering commitment they made by signing and ratifying the convention.

Country Studies

In accordance with recommendations formulated in Chapter 15 of Agenda 21,[1] a number of countries carried out country studies. The UNEP is coordinating a set of country studies in Latin America and the Caribbean, Africa, Asia, and Eastern Europe. The first phase of this project, funded partly by GEF, is currently being evaluated.

Major outcomes of country studies include the following:

• Realization that new activities should be initiated in accordance with the objectives and provisions of the convention and founded on harmonization and integration of these objectives and provisions with national socioeconomic and cultural decision-making. The principle of sector integration has been a cornerstone of environmental policies of many countries since the late 1980s, with realization on the global level of the importance of the environment in the development of nations.

• Identification of areas in which priority action is needed to protect genes and genomes, species, and communities of economic, social, scientific, and cultural importance, and those that are under threat, as well as habitats and ecosystems that have social, economic, cultural, and scientific importance or contain high diversity or large numbers of endemic or threatened species.

• Realization or confirmation that major threats to biological diversity are caused by human activities[2] such as intensified farming practices, city and road construction, and industrial pollution.

The next phase is the development of national strategies and action plans (see "National Strategies and Action Plans" later in this chapter).

The UNEP Expert Panel I on the Follow-Up of the Convention on Biological Diversity

Recognizing that the successful implementation of the provisions of the Convention on Biological Diversity depends on adequate scientific and technical input, UNEP established four expert panels to prepare background documents on relevant matters for consideration by the Intergovernmental Committee on the Convention on Biological Diversity (ICCBD)[3] (see the next section). The UNEP Expert Panel I was mandated to consider action priorities for the conservation and sustainable use of biological diversity and an agenda for scientific and technical research.

The report of the panel provides guidance on the implementation of articles 6–14 and 17–18 of the convention, which are directly relevant to conservation and sustainable use of biological diversity. Tables 1 to 3 of the report translate the provisions of the convention into a sequence of actions that countries could use in setting their priorities. Table 4 of the same report, particularly its part A, which provides a preliminary research agenda, served the Open-Ended Intergovernmental Meeting of Scientific Experts on Biological Diversity as a framework for the identification of gaps and as a source of elements for a scientific and technical agenda for the implementation of the convention.

The Open-Ended Intergovernmental Meeting of Scientific Experts on Biological Diversity

Conventions such as the Framework Convention on Climate Change and the International Convention to Combat Desertification in Those Countries Experiencing Serious Drought and/or Desertification, particularly in Africa, benefited much from a number of scientific meetings before the first meeting of the COP. In the framework of the Convention on Biological Diversity, the ICCBD established by the UNEP Governing Council to consider critical issues in preparation of the first meeting of the COP recommended,[4] at its first session, that an open-ended intergovernment meeting of scientific experts on biological diversity be convened to consider the following items:

- Identification of scientific programs and international cooperation in research and development related to the conservation and sustainable use of biological diversity
- Organization of the preparation of an agenda for scientific and technological research on the conservation of biological diversity and the sustainable use of its components, including possible institutional arrangements ad interim for sci-

entific cooperation among governments for the early implementation of the provisions of the Convention on Biological Diversity

• Identification of innovative, efficient, state-of-the-art technologies and know-how relating to the conservation and sustainable use of biological diversity and the ways of promoting development and transferring such technologies

In relation to the survey of scientific programs and international cooperation in research and identification of gaps and areas to be strengthened, the committee did the following:

• Examined the main processes or mechanisms for identifying scientific programs and international cooperation and for strengthening national capacity to develop and carry out scientific programs.
• Identified a number of gaps in scientific programs in research and development related to the conservation and sustainable use of biological resources. These gaps served as a basis for developing elements for an agenda for scientific and technological research[5]covering articles 5 ("Cooperation"), 6 ("General Measures for Conservation and Sustainable Use"), 7 ("Identification and Monitoring"), 8 ("*In-Situ* Conservation"), 9 ("*Ex-Situ* Conservation"), 10 ("Sustainable Use of Components of Biological Diversity"), 11 ("Incentive Measures"), and 19 ("Handling of Biotechnology and Distribution of its Benefits") of the Convention on Biological Diversity.

Of particular interest to the international scientific community, the committee identified scientific programs and research and development areas that would benefit from international cooperation. These areas include in-situ and ex-situ conservation, strengthening of taxonomy, systematics and conservation biology, biological diversity prospecting, conservation and sustainable use of all types of forests, understanding the role of biological diversity in maintaining ecosystem functions, and sustainable use of marine and coastal zones.

The committee also listed technologies relevant to identification, characterization, monitoring, and conservation of components of biological diversity; methods to measure sustainability; ways to integrate, knowledge, innovations, and practices of indigenous and local communities into modern management practices; and training programs.

The Conference of the Parties

At its first meeting the COP gave a list of program priorities in the conservation of biological diversity and sustainable use of its components.[6] Although the purpose of the list is to provide direction to the interim structure operating the financial mechanism, these priorities may be applicable to all the parties.

In accordance with the country-driven nature of the convention, national priorities that fulfill the obligations of the convention are recognized as the first priorities. Articles 6 ("General Measures for Conservation and Sustainable Use"), 7 ("Identification and Monitoring"), 8 ("*In-Situ* Conservation"), 16 ("Access to and Transfer of Technology"), and 18 ("Technical and Scientific Cooperation") constitute the basis for the other program priorities (see the list of program priorities in appendix 12.1).

The COP also defined a medium-term program of work for 1995–1997 and provided SBSTTA with a list of matters on which advice is required. The medium-term program of work[7] and the agenda of SBSTTA[8] represent the major activities to be carried out in the coming years at the international level to fulfill the objectives of the Convention on Biological Diversity.

NATIONAL STRATEGIES AND ACTION PLANS

Development of national strategies, plans, or programs for the conservation and sustainable use of biological diversity or adaptation of existing ones and the integration of the objectives of the convention into relevant sectoral or cross-sectoral plans, programs, and policies (article 6 of the convention) are useful first steps in the implementation of the convention. The need for these activities is also stressed in Agenda 21, chapter 15, paragraphs 4 and 5.

In order to facilitate the development of national strategies and action plans, the World Resources Institute (WRI) in collaboration with the International Union for the Conservation of Nature (IUCN) and the UNEP prepared some guidelines for biodiversity planning based on the experience of a few developing and developed countries. Basically, the guidelines consists of seven steps: (1) establishment of a focal point in government; (2) assessment of the status and trend of the nations' biological diversity and related programs, with the identification of gaps between desired and actual situation; (3) formulation of a strategy to close the gaps; (4) formulation of a plan of action; (5) implementation involving the whole society; (6) monitoring of the impact of the plan; and (7) reporting. The authors of the guidelines noted that biodiversity planning is a continuous, cyclical, interactive, adaptive, multistakeholder and multisectoral process. The process should use a participatory approach and would be greatly facilitated if it were supported by strong political will and commitment.

Experience shows that formulation of national strategies and action plans catalyzes and improves coordination of actions being implemented for sustainable development and eradication of poverty. In addition, these strategies and plans provide an operational framework that may guide allocation of national funds and the screening of foreign investments toward conservation of biological diversity and sustainable use of its components.

A few countries, including Australia, Canada, China, India, Sweden, and the United Kingdom, have already drafted their national biodiversity strategies and action plans. About 25 other countries are currently developing national biodiversity strategies. As suggested in chapter 15 of Agenda 21, this task may involve governments at their appropriate levels, United Nations bodies, and, as appropriate, international organizations, indigenous people and local communities, nongovernment organizations, and other groups including the business and scientific communities.

The process may begin, as in Canada, with an analysis of current gaps and potential opportunities in existing policies, programs, and legislation relevant to the conservation of biological diversity and the sustainable use of biological resources. This analysis can be carried out as part of a coordinated country study to record the status of biological diversity and its components, particularly in relevant economic sectors including agriculture, forestry, fisheries, and pharmaceutical and biotreatment industries.

The COP recognized at its first meeting that the building of capacity, including training of human resources and establishment of appropriate institutions for conservation of biological diversity and sustainable use of its components, is a prerequisite to the development and implementation of national strategies and action plans. Development of national strategies and action plans involves costs that may constrain development. For example, the Philippines had to rely on external funds to carry out this activity.

In March 1995, the GEF Secretariat selected for funding the following three projects out of a first batch of projects submitted by developing country parties:

- Development of a national biodiversity strategy with specific emphasis on the conservation and sustainable use of plant genetic resources (Djibouti)
- Nature reserve management project (China)
- Conservation of the Lake Titicaca ecosystem (Peru and Bolivia)

The first of these projects deals with development of national strategies and action plans. The other two projects are related to priority activities described in the next section.

PRIORITY ACTIVITIES RELATING TO IDENTIFICATION AND MONITORING (ARTICLE 7) AND CONSERVATION (ARTICLES 8 AND 9) OF BIOLOGICAL DIVERSITY

Priority activities relating to identification and monitoring (article 7) and conservation (articles 8 and 9) of biological diversity are specified in the medium-term program of work of the COP as follows:

- Identification and monitoring of wild and domesticated biodiversity components, particularly those under threat, and implementation of measures for their consideration and sustainable use
- The strengthening of conservation, management, and sustainable use of ecosystems and habitats, with particular emphasis on coastal and marine resources

under threat and other environmentally vulnerable areas such as arid, semiarid, and mountainous areas

With regard to the identification and monitoring of wild and domesticated biodiversity components, the approach used by countries includes the following:

- The use of new technologies, such as computer-aided taxonomy, biotechnology, and remote sensing (India, China)
- The following measures to minimize the loss of threatened species[9]: maintaining wild flora and fauna populations, diverse ecosystems, landscapes, and waterscapes; restoring degraded ecosystems where practical and where restoration will make a significant contribution to the conservation and sustainable use of biological diversity; developing and implementing policies and programs aimed at preventing or reducing human-caused atmospheric changes that threaten biological diversity; ensuring that development and use of nonrenewable resources does not result in the decline of biological diversity; elimination of policies and programs that unintentionally act as disincentives to conservation.

The Open-Ended Intergovernmental Meeting of Scientific Experts on Biological Diversity proposed elements of an agenda for scientific and technological research relating to articles 7–9, thus including the two priority activities mentioned earlier, under the following headings:

- Identification, inventory, and documentation of status and distribution of biological diversity (article 7 (a) and (d))
- Monitoring and evaluation of changes in biological diversity caused by natural fluctuations or human impacts (article 7 (b) and (c))
- Understanding the role of biological diversity in maintaining ecosystem structures and functions (article 8 (a)–(f))
- Assessment of threats to and adverse impacts on biological diversity (articles 7 (c), 8 (g), (h), and (l), and 14)
- Conservation and restoration of biological diversity (articles 8 (a)–(f) and (k)–(l), 9, and 10 (d))
- Development of criteria and methods for sustainable use of components of biological diversity (articles 8 (c) and (i) and 10)
- Screening of biological resources for potential use (articles 5, 7 (a), and 10 (e))
- Studies on ethnobiology and adaptation of traditional knowledge and skills (articles 8 (j) and 10 (c))

These elements of an agenda for scientific and technological research as well as the examples of facilitating research activities provided in the report of the Open-

Ended Intergovernmental Meeting of Scientific Experts on Biological Diversity may guide nations in developing their action plans and programs.

Considering the importance and urgency of program priorities relating to articles 7–9 of the convention, the COP asked the SBSTTA to advise on alternatives ways in which the COP could start the process of considering the threatened components of biological diversity, particularly in coastal and marine zones and in vulnerable areas, and the identification of action that could be taken under the convention.

Desirable Characteristics of Program Priorities

The list of program priorities developed by the COP (see appendix 12.1) also contains elements that increase chances of obtaining funds through the financial mechanism of the convention, including projects that do the following:

- Promote access to, transfer of, and cooperation for joint development of technology, including biotechnology
- Use innovative measures including, in the field of economic incentives, measures that help developing countries to address situations in which opportunity costs are incurred by local communities and to identify ways by which these can be compensated
- Provide access to other international, national, or private sector funds and scientific and technical cooperation and promote the sustainability of project benefits
- Offer experience applicable elsewhere and encourage scientific excellence
- Strengthen the involvement of local and indigenous people and integrate social dimensions, including those related to poverty

Many of these elements can be found in existing national biodiversity strategies and action plans. Two examples of international cooperation in program priority areas can be cited to illustrate one of these desirable elements:

- Biodiversity planning in the Philippines began in the 1980s with the development of National Integrated Protected Areas System (NIPAS), consolidated by the NIPAS Act of 1992. The design, feasibility studies, management planning, legislation studies, and other support studies such as training and consensus-building were supported by a $5-million Japan–World Bank Technical assistance grant. Japan is currently helping the Philippine government to address conservation of various threatened animal species. Both in situ and ex situ conservation are being used by government and nongovernment organizations in cooperation with local or foreign institutions. A follow-up project will involve actual management of 10 priority sites representative of 15 biogeographic zones of the Philippines and will be submitted to GEF for funding.
- The International Coral Reef Initiative is one of the most emerging partnerships in the area of conservation and sustainable use of coral reefs. Coral reefs con-

tribute to national economies through recreation and tourism. They have also been cited as possible indicators of climate change and as natural indicators of the health of coastal zones. The International Coral Reef Initiative has been developed to protect, manage, and monitor coral reef resources and related ecosystems, such as mangroves and sea grass beds. This initiative has attracted participation from Australia, France, Jamaica, Japan, the Philippines, the United Kingdom, and the United States. The United States is providing first-year funds for a Global Coral Reef Monitoring Programme under the auspices of UNEP, Intergovernmental Oceanographic Commission, IUCN, and World Meteorological Organisation.

CONCLUSION

The Convention on Biological Diversity is an adequate framework to reverse the current pattern of loss of biological diversity and degradation of its components. It contains general provisions and states some specific actions for the conservation and sustainable use of biological diversity at the local, national, and international levels. Its strength and uniqueness lie in its holistic approach and in the integration of its objectives into sectoral and cross-sectoral programs for sustainable development.

At its first meeting, the COP prepared a medium-term program of work that offers opportunities for the development of biological diversity science and policy. The COP also gave a list of program priorities to guide allocation of funds through the convention's financial mechanisms. It is now for the nations to consider these priorities in their action plans and projects. The few projects selected early in 1995 by GEF for funding concern mainly priorities 1–3 and 5 (see appendix 12.1); they include development of national biodiversity strategies, development of protected areas, and sustainable use of an aquatic ecosystem. It is expected that new projects will also take into account the other priority areas encompassing social and economic considerations. Biological diversity science should address socioeconomic aspects on which decision-makers can base their management policies.

NOTES

1. See UNCED. 1992. *Agenda 21: Programme of action for sustainable development.* Chapter 15: "Conservation of biological diversity."

2. See Department of Environment. 1994. *Biodiversity: The UK action plan* . Cm 2428. London: HMSO.

3. The ICCBD was established pursuant to resolution 2 of the Conference for the Adoption of the Agreed Text of the Convention.

4. The Open-Ended Intergovernmental Meeting of Scientific Experts on Biological Diversity was held in Mexico City April 11–15, 1994. Its report is contained in UNEP. 1994. *Report of the Intergovernmental Committee on the Convention on Biological Diversity. Report of the Open-Ended Intergovernmental Meeting of Scientific Experts on Biological Diversity, including the*

agenda for scientific and technical research. Document UNEP/CBD/COP/1/16, prepared for the first meeting of the Conference of the Parties to the Convention on Biological Diversity held in Nassau, The Bahamas, November 28–December 9, 1994.

5. See annex X to document UNEP. 1994. *Report of the Intergovernmental Committee on the Convention on Biological Diversity. Report of the Open-Ended Intergovernmental Meeting of Scientific Experts on Biological Diversity, including the agenda for scientific and technical research.* Document UNEP/CBD/COP/1/16, prepared for the first meeting of the Conference of the Parties to the Convention on Biological Diversity held in Nassau, The Bahamas, November 28–December 9, 1994.

6. See UNEP. 1995. *Report of the first meeting of the Conference of the Parties to the Convention on Biological Diversity .* Document UNEP/CBD/COP/1/17.

7. See the annex to *Decision I/9 medium-term programme of work of the Conference of the Parties.* In UNEP. 1995. *Report of the first meeting of the Conference of the Parties to the Convention on Biological Diversity .* Document UNEP/CBD/COP/1/17.

8. See *Decision I/7 Subsidiary Body on Scientific, Technical and Technological Advice* (SBSTA). In UNEP. 1995. *Report of the first meeting of the Conference of the Parties to the Convention on Biological Diversity .* Document UNEP/CBD/COP/1/17.

9. Drawn from Environment Canada. 1995. *Canadian biodiversity strategy. Canada's response to the Convention on Biological Diversity.* Similar measures are found in other national biodiversity action plans.

APPENDIX 12.1

Programme Priorities for Access to and Utilization of Financial Resources Under the Convention on Biological Diversity*

1. Projects and programmes that have national priority status and that fulfill the obligations of the Convention;
2. Development of integrated national strategies, plans or programmes for the conservation of biological diversity and sustainable use of its components in accordance with Article 6 of the Convention;
3. Strengthening conservation, management and sustainable use of ecosystems and habitats identified by national governments in accordance with Article 7 of the Convention;
4. Identification and monitoring of wild and domesticated biodiversity components, in particular those under threat, and implementation of measures for their conservation and sustainable use;
5. Capacity-building, including human resources development and institutional development and/or strengthening, to facilitate the preparation and/or implementation of national strategies, plans for

* See UNEP. 1995. *Report of the first meeting of the Conference of the Parties to the Convention on Biological Diversity .* Document UNEP/CBD/COP/1/17.

priority programmes and activities for conservation of biological diversity and sustainable use of its components;

6. In accordance with Article 16 of the Convention, and to meet the objectives of conservation of biological diversity and sustainable use of its components, projects which promote access to, transfer of and cooperation for joint development of technology;

7. Projects that promote the sustainability of project benefits; that offer a potential contribution to experience in the conservation of biological diversity and sustainable use of its components which may have application elsewhere; and that encourage scientific excellence;

8. Activities that provide access to other international, national and/or private sector funds and scientific and technical cooperation;

9. Innovative measures, including in the field of economic incentives, aiming at conservation of biological diversity and/or sustainable use of its components, including those which assist developing countries to address situations where opportunity costs are incurred by local communities and to identify ways and means by which these can be compensated, in accordance with Article 11 of the Convention;

10. Projects that strengthen the involvement of local and indigenous people in the conservation of biological diversity and sustainable use of its components;

11. Projects that promote the conservation and sustainable use of biological diversity of coastal and marine resources under threat. Also, projects which promote the conservation of biological diversity and sustainable use of its components in other environmentally vulnerable areas such as arid and semi-arid and mountainous areas;

12. Projects that promote the conservation and/or sustainable use of endemic species;

13. Projects aimed at the conservation of biological diversity and sustainable use of its components which integrate social dimensions including those related to poverty.

PART IV

WHAT NEEDS TO BE DONE

STRANGE BEDFELLOWS: WHY SCIENCE AND POLICY DON'T MESH AND WHAT CAN BE DONE ABOUT IT

Jeffrey A. McNeely

SCIENCE, POLITICS, AND BIODIVERSITY

Scientists have much to contribute to public policy, especially in fields such as biodiversity, where so much of the factual basis of the issue depends on scientific observations. Scientists therefore have called for biodiversity to be given greater attention by policymakers, which indeed has happened. The Convention on Biological Diversity was signed by the leaders of 157 countries and the European Community at the Earth Summit in Rio de Janeiro in June 1992 and entered into force at the end of 1993; 172 countries had ratified the convention by the spring of 1998. The First Conference of the Parties, held in Nassau, the Commonwealth of the Bahamas, in late 1994, had seemed an important opportunity for scientists to influence policy. However, what actually happened was that politicians took over the debate, leading it down familiar and sometimes unproductive pathways that polarized North and South; the scientific content was modest, even inconsequential. What went wrong?

Ironically, perhaps part of the problem is that scientists have been too convincing about the importance of biodiversity to human welfare. When biodiversity was seen as a rather narrow issue of a few endangered species or protected areas where few interests were at stake, the issue could safely be left with scientists and specialized agencies. But science has been mobilized so successfully to increase public awareness about the issues of biodiversity loss and unsustainable use of biological resources (such as forests, fisheries, and agricultural soils) that political parties, industrialists, religious leaders, farmers, economists, diplomats, indigenous peoples, conservation organizations, and even the legal profession hopped on the biodiversity bandwagon. As biodiversity became a concern of central significance to the larger society and serious money was involved in the decisions to be made, policymakers were put in the position of mediating among competing interest groups to seek the course of development that best serves the larger public. Thus issues such as equitable sharing of benefits, intellectual property rights, sustainable development, and national sovereignty have taken over the modern conservation scene, and the scientific issues of extinction rates, biogeography, and ecosystem function have been pushed to the sidelines.

It is clear, therefore, that science and policy provide two different approaches to reality. Scientists tend to view research as an end in itself, driven by ideas or techniques. The scientific method often forces research to be reductionist, isolating factors to be manipulated by experiments. Results are often presented with statistical degrees of certainty or reliability and typically lead to more questions that must be answered. Scientists tend to view themselves as an intellectually elite segment of society, believing that science deserves support simply because it is scientific research; accountability is largely restricted to the cultural process of peer review. It often seems that the relevance of the research to the needs of society is overwhelmed by the sanctity of the individual researcher's curiosity. Ideally suited to carrying out research, scientists are seldom suitably placed to understand the pressures under which policymakers work.

Public policy, on the other hand, addresses problems, and many policy-making activities seem to be little more than attempts to contain crises with inadequate resources. By definition, these policies are in the public eye and subject to public scrutiny. The primary aim of the top policymakers in the line agencies is to acquire resources and control information so that public comment is minimized, or to ensure that such comment is favorable. Resource management policies are often determined by committees that may or may not have the necessary competence to make meaningful value judgments. They must deal with the larger picture and are not able to indulge in the luxury of manipulating a single variable in an experiment; research results are only one of a multitude of factors that must be considered. Making policies about biodiversity certainly requires a basic understanding of science and a familiarity with the scientific community, but policymakers are seldom scientists and do not have time to digest the detailed information that would enable them to make full use of the scientific advice. They are especially nervous about statistically reasoned analysis that underlines uncertainty (Warren 1993), preferring clear-cut guidance in black and white. Policy-making in biodiversity has not been accorded high status in the academic community, and managers have typically not been accorded the intellectual status given to scientists engaged in research. Though usually poorly equipped to carry out scientific research, policymakers often have very clear ideas about the kind of information they require for developing resource management polices.

Thus the priests of the scientific and policy-making subcultures of modern society have never communicated very well as both have sought to protect their own power bases and tended to ignore how they might be able to enhance that power base by reference to the other subculture. The situation is made more difficult because the concerns of society and the actions of the numerous other actors on the biodiversity stage are in constant and unpredictable flux, as is science. This dynamism helps explain why controversy is so pervasive in politically sensitive fields such as forestry, fisheries, water resources, and grazing, where science often is used by official environmental agencies to underwrite inaction or to provide political reassurance. Science has a virtually infinite scope to redefine biodiversity issues scientifically, to embrace constantly increasing numbers of real-world variables, or to refocus the significance attached to those already acknowledged (Grove-White 1993). It follows that the same is true also of the scope for criticism and disagreement over much more fundamental matters than extinction rates or energy flows through ecosystems.

In seeking to improve the quality of policies affecting biodiversity, this chapter assesses how science is used by policymakers, reviews some of the limitations of science, considers how to use science to build public support for conserving biodiversity, and recommends steps for bringing science and public policy closer together.

HOW SCIENCE IS USED BY POLICYMAKERS

The need for science pervades the policy-making arena. Senior civil servants need scientific advice in preparing ideas for regulations, legislation, programs, projects, and budgets; legislators need scientific advice to translate these ideas into draft legislation; the various interest groups need science to help ensure that their concerns are built into the legislation and to support its passage through Congress; and everyone needs science to assess how policies, programs, and projects are affecting biodiversity. But how science is actually used in these various contexts depends very much on the users.

Generally speaking, the concepts of environmental law and biodiversity policy are based on science as interpreted by the nonscientist. Because policymakers typically lack scientific expertise, their reaction to scientific information is often either highly critical, questioning the validity of the science as the basis for action, or overly accepting, adopting the scientific data uncritically. The quality of the scientific information is likely to suffer in either case, but the policymaker still requires the science to be presented in simple, easily digested morsels even when such oversimplification will weaken the basis for decision-making. When it comes to complex principles such as the conservation of biodiversity and sustainable use of biological resources, where scientific and social factors are closely related, it is not surprising that a scientifically based consensus is so difficult to reach.

When the resulting law is tested in the courts, scientific information may be an essential part of the evidence; science is often marshaled by both sides of a case, requiring juries to make judgments on highly complex issues. And of course, the managers on the ground need advice on how the legislation is to be implemented. Thus the views and practices of nonscientists exercise a considerable influence over the way scientific observations are used. Institutions such as the mass media and the law inevitably act as filters and mediators, affecting the public prominence of scientific aspects of biodiversity. Scientists dealing with such a complex topic as biodiversity therefore are mere contributors to a dialogue in which many other disciplines participate.

In a World Bank study of decision-making about biodiversity, Metrick and Weitzman (1994) found that both scientific and emotional elements play important roles in determining whether a species is put on the U.S. Endangered Species List. However, the scientific characteristics appear to have little influence on the way funds are spent by the federal and state governments to address the problems of these species, as the emotional characteristics seem to dominate.

THE LIMITATIONS OF SCIENCE

When applied to biodiversity, science has limits that stretch beyond the straightforward issue of lack of knowledge. Biologists tend to view the loss of biodiversity as a problem arising from the relationship between people and the environment, when the problem is perhaps more accurately seen as an outcome of the economic relationships among people that determine how any piece of land or any set of resources is to be used. But because a given development or conservation activity will inevitably advance the interests of some while prejudicing others, it is a mistake to assume that conservation objectives may be achieved through some combination of improved scientific information, education, and technical remedies. In cases where these fail, the tendency is to fall back on state regulation supported by strengthened law enforcement (Painter 1988), but this approach is often a failure when it ignores the distribution of costs and benefits, who gains and who loses, who has power and who is vulnerable.

It is perhaps worth recalling that science is not really a fixed, objectively verifiable body of knowledge of nature's workings available only to highly qualified scientists; rather, science exists as a social phenomenon in which doctrines of "objective" practice rest on a web of conventions, practices, understandings, peer pressures, and negotiated ambiguities (Grove-White 1993). In discussing the science of field biology, Schaller (1993) pointed out that

> A fact is not a fact until someone has posed the question, and slowly the world of an animal emerges from the questions raised and facts collected. But if someone else asks a different question, a different creature, a different reality comes into existence. The animal is an illusion created out of the animal's interaction with an observer who decides what to measure and record and what to ignore. We constantly infer the unseen, we confuse ideas with facts. Furthermore, animals bound, prowl, slither, and flap through our subconscious in the form of myths about lions, bats, foxes, owls, snakes, and doves, each culture with its own fantasies. Victor Hugo

wrote, "Animals are nothing but the forms of our virtues and vices, wandering before our eyes, the visible phantoms of our souls." In such a way is science fashioned. Any biologist who observes a tiger, gorilla, panda or other creature and says he or she has done so with total objectivity is ignorant, dishonest, or foolish.

The Convention on Biological Diversity perhaps inadvertently undermined science by noting that "Where there is a threat of significant reduction or loss of biological diversity, lack of full scientific certainty should not be used as a reason for postponing measures to avoid or minimize such a threat." This is the precautionary principle, which appears to be an acknowledgment that biological systems are so complex that it may be impossible to obtain statistical and scientifically valid proof of cause and effect, at least within a meaningful timescale (Warren 1993). Some interpret this principle to imply that the burden of proof should be placed on those who propose a project to prove that it will not significantly reduce biological diversity, which arguably is good for biodiversity; but the precautionary principle also can be used to take the objectivity out of scientific judgment by giving undue weight to results that are not statistically significant. Part of the problem is that the precautionary principle seems to be a juxtaposition of scientific reasoning and administrative policy, a sure recipe for confusion.

Biodiversity itself may be part of the problem. As Dudley (1992) pointed out, biodiversity does not fit very well into government legal and regulatory contexts. Translating biodiversity into regulatory language comprehensible to nonbiologists may be impossible. It may be more important to demonstrate the applicability of research results to the actual management decisions faced by government decision-makers. Therefore, Dudley argues, the Endangered Species Act may be more useful as a surrogate for biodiversity than would be misguided attempts to use biodiversity as a regulatory criterion.

Furthermore, although science has played the dominant role in developing our understanding of biodiversity, it is value neutral (or at least claims to be); some may charge that it has also contributed to the loss of biodiversity, especially when science is married to technology. For example, scientific research produced pesticides, ozone-eating chlorofluorocarbons, and uniform varieties of plants, all of which may threaten biodiversity (or help save it). Science thus has been an essential collaborator in much of the ecological destruction associated with modern industrial society while also helping to define problems and solutions.

Science can thus be mobilized in support of both sides in conflicts affecting biodiversity. For example, the 1989 Exxon Valdez oilspill seems to have sullied more than the waters and wilderness of Prince William Sound. Six years later, no scientific consensus has been reached on what really happened to the sound after the 35,000-ton oilspill (only the 34th largest in history), for which Exxon agreed to pay $1.1 billion to the state and federal governments. Many of the scientific investigations were kept secret, possibly because scientists paid by the oil companies came up with conclusions the opposite of those of scientists paid by the government. Not too surprisingly, the long and detailed studies of the Exxon scientists concluded that the spill in the cold and highly productive Alaskan waters had little or no impact. Exxon now faces $2.6 billion in additional lawsuits brought by Native Americans and fishermen and is therefore very cautious about releasing any of its own data. With at least 18 government agencies on the scene, jurisdictional disputes were frequent, complex, and often bitter. The first priority of the government agencies was to collect evidence to help win the case against Exxon, and the strategy was to keep evidence confidential until it was introduced in court, lest the opposition find out what cards were in hand. Censorship by lawyers was so frustrating that some charged that there was less secrecy after Chernobyl.

The broad participation of many actors in developing public policy on biodiversity is appropriate because the critical assessment of information and how it should be used must involve all interests, not merely those with narrowly defined technical expertise. On the other hand, biodiversity issues often can be critically assessed only through knowledge and understanding, rare commodities that often seem to be cornered by scientists. Thus science remains uniquely authoritative, especially if scientists can marshal their evidence in ways that will be useful to policymakers anxious to ensure better management of biological resources. Public support will be essential to encourage them to do so.

KEEPING BIODIVERSITY ON THE PUBLIC AGENDA

Concern about biodiversity was first crystallized not by governments responding to or using science, but by poor and powerless nongovernment organizations and academics who mobilized their own science and communicated it effectively through the mass media and direct mail campaigns. But "official science" has been the effective measure of whether issues are "real" and are given attention by government policymakers. Because biodiversity is unable to speak up for itself, it needs a stand-in, and in modern industrialized societies science often seems to be the only stand-in capable of commanding widespread legitimacy. However, scientists do not control how scientific evidence is used to influence public opinion. The mass media seem to demand a rough balance between competing viewpoints and give a premium to controversy and the mediagenic, often leading to polarization and distrust. Thus information provided to the public by the mass media is often misleading; for example, although most of the global focus is on biodiversity loss in tropical forests, the developed countries accounted for over 80 percent of global forest products in 1992, with the U.S. alone earning three times more than the highest-exporting developing country (Indonesia). Whereas the mass media focus on tigers, rhinos, and pandas when extinction is mentioned, by far the greatest numbers of recorded extinctions have taken place in developed countries, namely the United States and Australia (Groombridge 1992).

Keeping biodiversity on the public agenda requires overcoming at least three formidable problems (Tobin 1990). First, current practices that are depleting biodiversity often are extremely popular. The fact that the desire for consumption is far more powerful than the conservation-oriented advice of scientists should come as no surprise, as incentives to consume far outweigh incentives to conserve. A typical American meal travels 1,300 miles from farm field to dinner plate; in much of Africa it is generally just a few hundred feet. The resource requirements of microwave-ready foods are about 10 times higher than those for preparing meals from scratch. It takes 94 times the amount of energy to obtain an out-of-season piece of fruit or vegetable from a foreign locale and 30 times more from a local greenhouse, than if obtained in season and locally produced. Yet most of us act as if December strawberries are part of our birthright. As another example, a 1994 report of the Commission on Sustainable Development found that worldwide the amount of money governments spent to support environmentally destructive behavior amounted to $1 trillion per year. Another indicator is the amount of money spent on advertising, basically encouraging people to consume more than they might otherwise consume; globally, advertising budgets rose 7.4 percent in 1995 to $364 billion, more than the annual GNP of Australia or the Netherlands. McDonald's restaurants, for example, spent more than $425 million for advertising in 1990, and while the United States was cutting its support to the World Food Program designed to feed the starving masses in developing countries, Ultra Slim-Fast Diet Food was spending more than $77 million in advertising. And of course, the public likes to receive benefits without paying the costs. According to former Senator Warren B. Rudman, R–New Hampshire, adults today will receive $14 trillion more in benefits from the government than they pay in taxes. This translates into a $150,000 gap for every household.

Second, no easily identified opponent is available against which conservation forces can be rallied; unlike such headline-makers as Bhopal, the Exxon Valdez, and Chernobyl, no newsworthy disasters have yet linked human welfare with the loss of biodiversity. On the contrary, many people are making substantial profits from overexploiting biological resources, and those with the highest political profiles tend to be among those making the largest profits through overexploitation. It is apparent that the larger and the more immediate the prospects for gain, the greater the political power that is used to facilitate unlimited exploitation, often through mobilizing significant economic incentives provided by government.

Third, the loss of biodiversity has no immediately observable impact on lifestyles, especially those of people living in cities far removed from the biological resources that support their consumption. If we are losing dozens or hundreds of species per day, as many experts assert, then we are already living with the consequences of extinction without any discernible effects on our daily lives. And when scientists argue that efforts to conserve endangered species deserve especially high priority, it is difficult to link this argument directly with the human welfare issues of concern to policymakers because these species have already been reduced to such low population levels that they usually can be used only as symbols.

Not everybody even believes that we are facing an extinction crisis. Economist Julian Simon (Myers and Simon 1994:43) contends that the high rate of extinctions widely supported by scientists is pure guesswork. Selecting his evidence very carefully, Simon claims that "recent scientific and technical advances—especially seed banks and genetic engineering—have diminished the importance of maintaining species in their natural habitat." Unfortunately, cornucopians such as Simon are selling a message that is very popular to policymakers but roundly denounced by ecologists: resources are plentiful and perpetual growth is feasible.

The modern public has little appetite for complex arguments. In the Lincoln–Douglas debates in the last century, the two political candidates debated each other for 7 hours at a stretch, as points were discussed, debated, questioned, and challenged. Audiences sat through the debates from beginning to end, following the often complex political and historical arguments being offered. Ronald Reagan summed up his presidential campaign with an 18-minute music video. Today's TV generation simply does not have the patience for information, truth, or facts, preferring the mental massage of images to provide them with the illusion of knowledge.

Wright (1995) contends that modern information technology has not only failed to cure America's ills but actually seems to have made them worse, as intensively felt public opinion leads to the impulsive passage of dubious laws and fosters a gridlock that keeps the nation from balancing its budget as various interest groups clamor to protect their benefits. The increasingly democratic face of interest groups, he suggests, means that the American government is asked to provide more benefits, which means finally that Americans of all classes are also paying more and the rates of resource consumption continue to climb.

Biodiversity also has become a popular theme for the private sector, which tends to put its own spin on the issues. In 1990 the number of eco-ads more than quadrupled in the United States, according to an audit conducted by the J. Walter Thompson advertising agency. Although some may well have indicated environmental concern, many others gave new luster to the word *greenwash* as clever advertisers took credit for complying with regulations under the Endangered Species Act. For example, *Newsweek* reported that Chevron was obliged to build artificial dens on its California oil fields to replace the natural dens of the endangered kit fox that were destroyed by their drilling operations. Number of dens built to date: 11. Cost per den: around $1,000. Cost per TV and print ad campaign running the ad, "The Little Fox and the Coyote": over $1 million.

A crucial factor in sustaining policy interest in biodiversity is the perception of reality by the general public. A 1989 Harris Poll asked, "Do you think this country should be doing more or less than it does now to protect the environment and curb pollution?" An incredible 97 percent felt that the country should be doing more and only 1 percent that it should be doing less. Clearly, a large majority of the American public still favors increased government efforts to protect the environment. Because being against environmental values appears to have no legitimacy in the public debate, those promoting perpetual growth or a weakening of conservation laws have taken a different approach: they attack the very role of government itself as being interventionist, overcentralized, and out of touch with the needs of common people. Thus biodiversity, an issue with profound impacts for all people, is instead being painted as a government issue of little relevance to the general public.

This disparity between the abstract principle of conserving biodiversity or species that live in distant wilderness and the real-world practice of conserving individual sites or species is at the root of the problem, but this disparity has very little to do with science. In most cases, preserving endangered species or protected areas provides primarily abstract benefits to individual members of the public, whereas the people who are expected to make economic sacrifices by restricting their activities in the habitat of these species, sacrifice some of their livestock, or give up their land for expanding protected areas tend to be developers, miners, large ranchers, forestry interests, and others who are very effective in conveying their wishes to politicians (Tobin 1990). It should be no surprise that gold, silver, and other valuable metals are mined from federal land at no cost and livestock is grazed on the 280 million acres of public rangeland at rates far below the market price.

Wolves provide a good example of this rural–urban split. Before their successful introduction into Yellowstone National Park, conserving wolves was an important issue for urban conservationists, but meetings in Wyoming that argued for returning the wolf to Yellowstone were picketed by people carrying signs proclaiming the wolf "the Saddam Hussein of the animal world" and bumper stickers read: "Predators: nature's criminals." Record numbers of people in Montana, Idaho, and Wyoming were wearing baseball caps with wolves on them, which might sound like good news. Wrong: the wolves have rifle scopes superimposed over them and the hat's logo reads, "Wolf management team."

Although the largely urban American public may support conserving biodiversity in the abstract, the support by rural people for specific action tends to be much weaker because they pay more of the costs and perceive fewer of the benefits. This is to be expected because the kind of public policies that are most popular to voters are those that call for modest changes in current practices to address immediate problems, rather than policies that call for comprehensive changes in deeply embedded social behavior (Tobin 1990). The major problems of conserving biological diversity lie not in the biology of the species concerned, or even in the functioning of ecosystems; rather, they involve the social, economic, and political arenas within which people operate. This is the spawning ground for the "wise use" movement, which threatens to overturn some of America's most important conservation legislation.

Part of the reason environmental groups have become more politically influential, at least for the time being, is that they have helped to symbolize widely shared and deeply rooted concerns of society about an uncertain future and the welfare of subsequent human generations. These groups have helped to give definition to the risks and dangers of current human behavior. The public may be increasingly concerned that our complex modern society is now being shaped and dominated by technological and corporate bureaucratic systems that operate outside the influence of formal political institutions in which there is no genuine sense of shared public control. This feeling of powerlessness may also help explain the rise of fundamentalist movements, both in the United States and abroad.

CONCLUSIONS

This chapter has explored some of the reasons why science is not having more of an impact on the biodiversity debate. Because policymakers want to deliver benefits, not constraints, scientists who advocate policy changes need to build on science to demonstrate the real benefits of conserving biodiversity to farmers, ranchers, and foresters, balance the attention given to loss of biodiversity with concern for sustainable use of harvestable species, and build a broader constituency among business, the public, and academics.

Scientists providing expert advice to policymakers on biodiversity need to ask questions such as

- Why is the advice wanted? Is it to support a decision already made, to provide deep background, or to provide real options?
- Is this the sort of advice that is actually required in the circumstances? For example, detailed species inventories may have little influence on whether a road should be built through a forest and may be little more than displacement behavior.
- Will the users of the advice understand the terms being used? If not, the scientist will need to modify the language being used.
- If detailed advice is provided, will it be summarized by a nonscientist? Would the summary be better prepared by the scientist? Or will the scientist have a chance to review the summary before it is released?

A policymaker assessing the relevance of the advice given by scientists may ask questions such as

- How credible is the advice? Is it based on objective assessment of evidence? Is it consistent with other evidence? Can it be replicated by other scientists?
- How independent is the advisor? Does he or she owe allegiance to any special interests?
- What are the gaps in the information provided? Should alternative scientific views be sought?
- How can the advice be applied to the specific case in question?

Warren (1993:110) suggested a rather simple principle to cope with the interface between scientist and nonscientist: "In communications between scientists and nonscientists, the scientist should assume from the outset that the other party has no understanding of scientific principles and that all terms need to be explained. At the same time, the other party should assume that the scientist is narrow-minded and, in particular, has no knowledge of other specialties."

Modern environmental policy is founded on scientific observations, requiring a basic understanding of scientific principles if workable environmental policy is to be developed, implemented, and enforced. Although expert advice can fill in gaps in knowledge, the final decision must be made by the appropriate authority, requiring that key personnel possess the necessary scientific skills to evaluate the advice given and incorporate it into practical results. This also requires scientists to have sufficient grasp of the legal implications of policy to determine what kind of advice is needed.

In short, research generally must continue to be science driven if scientists are to be true to their training and motivations. At the same time, at least some scientists need to work more actively to ensure that their findings are applied to real problems. The application of science to policy need not be done by the same scientists who are doing the research. But if present trends of exploitation and destruction of biological resources continue unabated, we may not have the luxury of worrying about whether research should be science or policy driven.

Helping find solutions to the biodiversity crisis in the political quagmire of modern society is the challenge facing scientists today. In the times of ecological, social, and economic instability that are just around the corner, if not upon us already, the renewable and locally available biological resources—and the knowledge of how to use these resources sustainably—will be more important than ever. Scientists therefore need to be much more effective in communicating the problems and solutions to policymakers and building the public support that will be necessary for supporting the difficult choices ahead.

ACKNOWLEDGMENTS

My thanks to Dan Walton for his insights on scientists and policymakers, Martha Rojas, Carleton Ray, and Caroline Martinet for their comments on an earlier draft, and Sue Rallo for her tireless work on the various incarnations through which this chapter has gone.

REFERENCES

Dudley, J. P. 1992. Rejoinder to Rohlf and O'Connell: Biodiversity as a regulatory criterion. *Conservation Biology* 6(4): 587–589.

Groombridge, B., ed. 1992. *Global biodiversity: Status of the Earth's living resources*. London: Chapman & Hall.

Grove-White, R. 1993. Environmentalism: A new moral discourse for technological society? In K. Milton, ed., *Environmentalism: The view from anthropology*, 18–30. London: Routledge.

Metrick, A. and M. L. Weitzman. 1994. Patterns of behavior in biodiversity preservation. In *World Bank policy research working paper* 1358: 1–28. Washington, D.C.: World Bank.

Myers, N. and J. L. Simon. 1994. *Scarcity or abundance: A debate on the environment*. New York: W. W. Norton.

Painter, M. 1988. *Co-management with whom? Conservation and development in Latin America*. Paper presented to the symposium "Culture: The Missing Component in Conservation and Development," Washington, DC, April 8–9, 1988.

Schaller, G. B. 1993. *The last panda*. Chicago: University of Chicago Press.

Tobin, R. 1990. *The expendable future: US politics and the protection of biodiversity*. Durham, N.C.: Duke University Press.

Warren, L. M. 1993. The precautionary principle: Use with caution. In K. Milton, ed., *Environmentalism: The view from anthropology*, 97–111. London: Routledge.

Wright, R. 1995. Hyperdemocracy. *Time* 23 (January): 53–58.

CHAPTER 14

SEEING THE WORLD AS IT REALLY IS:
GLOBAL STABILITY AND ENVIRONMENTAL CHANGE

Peter H. Raven and Joel Cracraft

THE STATE OF THE WORLD

Over the course of the 4.5-billion-year history of this planet, including the last 500,000 years when *Homo sapiens* appeared and established itself as the dominant species, the world has been transformed in astonishing ways (Turner et al. 1990; Tolba and El-Kholy 1992; McMichael 1993; Simmons 1996). But only after the introduction of agriculture 8,000 to 10,000 years ago, and the development of increasingly sophisticated technologies, did human-driven global change and environmental impact accelerate to the point that a major proportion of the earth's biodiversity is on the cusp of an extinction event fully congruent with the five major extinction events of the geological past. And it is the only event of its kind created by the activities of a single species.

When agriculture arose, there were far fewer people in the entire world than there are in metropolitan New York today. When our ancestors built Stonehenge, the great monolithic monuments of Europe, and the great pyramids of Egypt, there were fewer people in the world than there are in New York State now. At the time of Christ, there were only 130 million or so people in the entire world, about the same number that inhabited the United States during World War II. But global numbers have since exploded to 2.5 billion in 1950 and to nearly 6 billion people now, with 80 million people being added each year. This, along with unprecedented technological change that enables more efficient and expansive exploitation of the biosphere, has created a unique situation: no time in world history has even remotely resembled our present capacity to assault the natural resources of this planet.

It is sometimes said that human societies have arrived at a good point in world history because more people are better off than they ever were in the past. The latter part of the statement is true; the former part of the statement is clearly untrue because it fails to take into account what we call global stability. Consider several examples:

- The most telling indicator of the sheer impact we have had on the world is to point out that human beings right now, only one of 10 million or so species of eukaryotic organisms, are either using directly, wasting, or diverting more than 40 percent of the total net photosynthetic productivity on land (Vitousek et al. 1986).
- Over a billion hectares of land have been degraded because of the loss of topsoil and deformation due to erosion and abuse (Tolba and El-Kholy 1992). One result is that about 15 percent less land is available to feed a growing population. Another is that as agricultural land is taken out of production because of loss of topsoil, more and more wildlands are converted to farmland to compensate.
- Despite improvements in the livelihoods of people living in poor nations over the past several decades, the human condition in these countries is still appalling and getting worse in some instances: life expectancy in the least developed nations is 35 years less than in the most developed, more children less than 5 years old die in poor countries each year of respiratory infections and diarrhea than do all people in the world from malignant neoplasms, and in Europe and the United States adult female literacy is 97 to 99 percent, whereas in Africa it is 40 percent and Asia 60 percent (World Resources Institute 1996).
- All people aspire to a better life, yet growing inequities in wealth are likely to prevent many from attaining their goals and dreams. Half the world's people account for less than 15 percent of the world's gross domestic product, but the 15 percent wealthiest people account for over 50 percent of the world's wealth (World Resources Institute 1996). Even though economic growth in the developing world has improved dramatically, the gap between the rich and poor is widening everywhere.
- Over 80 percent of the world's forest ecosystems that existed at the dawn of agriculture have already been lost and 39 percent of what remains is under threat (World Resources Institute 1997). The vast majority of Earth's biodiversity is housed in these last remaining forest ecosystems.
- About 16 percent of the total animal protein consumed by people comes from marine ecosystems and 950 million people, largely in developing countries, depend on fish as their primary source of protein (World Resources Institute 1996). Yet the world's oceans and large marine ecosystems are being overfished and some of those ecosystems are collapsing.
- Most of the world's river systems and freshwater ecosystems have already been damned, diverted, polluted, or somehow modified by human activities, particularly by the introduction of exotic species. This has led to the widespread loss of biodiversity and ecological services.

The consequence of these actions is that we are increasingly losing life-sustaining biological diversity and we are threatening biodiversity at such a rate that perhaps as much as 20 percent of it may disappear over the next 30 years. This amounts to 200 species being lost every day, if you take a conservative estimate of the number of plants, animals, fungi, and microorganisms existing now (Wilson 1992). By the end of the next century, three-quarters of the earth's species may be extinct or on the way to extinction.

Furthermore, we have increased the carbon dioxide in the atmosphere since the end of World War II by nearly 20 percent. We have injured the ozone layer—which was so critical for life's beginnings several billion years ago—to the point that malignant skin cancer at a latitude such as New York's has increased about 25 percent, a clear manifestation of the protective function of the ozone layer in maintaining life on Earth.

Far from everything being all right at present, and far from our having established a level of development from which we could spring to further greatness and prosperity, we are, in the words of Herman Daly (see Daly and Cobb 1989), using the world as if it were a business in the process of liquidation. In effect, we are in the position of someone who has torn a porch from the front of his house to burn for heat and says, "Isn't it nice? Let's go for the front walls next and then the side walls." The subsidized loss of biodiversity that many have written about and the enormous price in instability that we are paying give us the illusion that the world is going to get better, yet if current trends continue, it cannot. It will not.

As a species, we humans are operating globally as if our ancestral memory of unrestrained exploitation of resources actually was good, a strategy in which it was important to horde resources when they were available and to garner as many of them as we could around our persons, our families, our groups. We continue to operate this way, but in a world that does not even remotely resemble the times when that kind of behavior might have been adaptive.

THE STATE OF THE DEVELOPED AND DEVELOPING WORLDS

In a very real sense, the report of the World Commission on Environment and Development, the famous Bruntland Report, which was notable in that it was adopted by the United Nations and was the first coordinated world statement on the environment, is grossly misleading. It said, in effect, that the industrialized nations are well off—although they all want to be better off—and that by some magical alchemy, the poor nations of the world will develop until they too share in the global prosperity enjoyed in a country such as the United States.

The sustaining forces in the global ecosystem that would allow this to happen are simply not there. If we continue to use resources as we are, the only way that the world will develop to a higher state is not by pretending that the industrialized nations' standards will eventually be met by everyone else. The only way that the world will improve is by an acute realization of what the world is really like: one that requires a stabilization of population levels, a more rational and intelligent level of consumption everywhere, and the use of more appropriate technologies to support ourselves. Social justice is a necessary ingredient of sustainable development.

What of the developing world? The groups of people that make up the developing world have grown from two-thirds of the world population in 1950 to four-fifths at present and they will be 85 percent of the world's population by 2020. In other words, for every person living in the industrial countries in 1950 there were two people living elsewhere, but by 2020—just 70 years later and within a single human lifetime—there will be five people living elsewhere.

The 4.8 billion people who live in developing countries include 1.4 billion who are living in absolute poverty, unable to find adequate food, shelter, or clothing for themselves or their families on a day-to-day basis. There are 400 million malnourished people whose bodies are literally wasting away and whose brains cannot develop properly in their formative years of childhood. Four-fifths of the people in the developing countries live at a standard one-twentieth or one-thirtieth of an average citizen in the United States. They have access only to 15 percent of the world's economy, 15 percent of the its industrial energy, 15 percent of its iron and steel, 6 percent of its aluminum, and comparable percentages of any other ingredient that you could think of as contributing to one's standard of life.

Women who live in these countries—having to gather firewood, which is their only source of fuel for cooking, having to go out to find clean drinking water, which is rarely directly available to them, and having little opportunity to get an education or contribute to the welfare of the communities in which they live—are therefore unable to contribute effectively to the world's vision of sustainable development. By the same token, children who live under those conditions obtain marginal education and must engage in the same pastime, joining their mothers in the search for firewood, which in turn is typically burnt in poorly ventilated housing, thus making them susceptible to respiratory disease and other health problems.

What the developing countries do have that is of value is about 80 percent of the world's biodiversity. That 80 percent will be protected only if we begin to address ourselves seriously to some of the questions and relationships discussed earlier in this chapter.

THE STATE OF THE UNITED STATES

If the world's environmental problems are to be addressed effectively and sincerely, then the United States must have a more realistic and honest view of the world and a renewed sense of stewardship of the earth's natural resources, not their opportunistic exploitation. Citizens of the United States must attain a deeper understanding of their actual and potential contribution to global stability. We must ask,

What are our belief system and values toward the remainder of the world's countries, their environments, and their people?

Some basic observations are in order. The United States has about 4.5 percent of the world's population, a proportion that has remained steady for over a century. During that period, we have captured about 25 percent of the world's economic activity—in other words, a mere 4.5 percent of the world's people, represented by the United States, have for the past 125 years or so been able to support their standard of living by using about 25 percent of the world economy. At the same time, more and more acutely with each passing year, the United States has been producing about 25 to 30 percent of the world's pollution.

The fact that the citizens of the United States are an amalgam of cultures from all over the world and the fact that this country has used the lion's share of the world's resources for its sole benefit ought to make the United States the most internationally oriented country that has ever existed. But just the opposite seems to be the case. Indeed, the United States may be one of the least internationally oriented countries that has ever existed. Since the end of the nineteenth century, we have believed and acted as if this global economic hegemony were our birthright: it supports us, so never mind that it results in the persistent erosion of resources around the world that might be better used to contribute to a condition of overall global stability.

What expectations do political and economic institutions appear to have in the United States? To listen to many of today's policymakers, they expect to revisit an era similar to the 1950s, when, because of dislocation caused by World War II, the United States temporarily controlled about 40 percent of the world's economy. But what is the reality of the world today? The United States will never again control 40 percent of the world's economy, as it did in the 1950s. The world economy has a different dynamic now and cannot be expected to work as in the past, despite what politicians might promise.

In the United States, a very basic fact is forgotten, or conveniently disregarded: we are the richest nation that has ever existed on the face of the earth. We are not merely the richest nation that exists on the earth now; we are the richest that has ever existed. Our standard of living in the United States is 20 to 30 times the standard of living of most people in the world, yet we indulge ourselves by pretending that we are constantly suffering economic hardship or do not have the monetary resources to address important problems in effective ways.

Again, what is the reality? Citizens of the United States think—largely because political leaders keep promoting the view—that we are grossly overtaxed, when in fact we pay the lowest rate of taxes per capita of any industrialized country. The economy of the United States is organized in such a way as to allow an enormous amount of individual and corporate initiative, more than in any other industrialized country. Yet at this low level of taxation, we still fool ourselves into thinking we are so highly taxed that if that burden were only reduced further, the economy would spring forward to a new higher level of productivity and prosperity. There is little reality to this expectation, and it is often forgotten that this experiment was undertaken in the 1980s, in the process running up another trillion and a half dollars worth of deficit.

Although we are the richest nation in the world and therefore have the most to gain by promoting global stability and sustainability, our national actions do not acknowledge that fact. The United States is the lowest donor per capita of foreign development assistance of any industrialized country. Other than the special cases of Israel and Egypt, generated by the Camp David accords, and Russia, we gave $6.2 billion annually in foreign development assistance at the beginning of the Clinton Administration. This amount was then cut to $4 billion, and now we find sympathy among our policymakers to cut our foreign development assistance still further. Because we have the most to lose if the world is not stable and sustainable, why do we allow politicians to hoodwink us into supposing that, with our $1.7-trillion budget and the $200- to $300-billion annual deficit, if we adjust our foreign development assistance to $4 billion, it will bring us back into register?

Why cannot we acknowledge that our fate is inextricably tied with the fate of Mexico? Or with any other country? Why did there have to be a national debate about bailing out Mexico? Whether there was a NAFTA or not, this continent is a partnership of the United States, Canada, and Mexico. These countries are so intertwined in terms of stability, whether it be ecological, financial, employment, or scientific, that it ought to have been a matter of national shame when Congress was unwilling to face the necessity of aiding Mexico, when so much here depends on it, when our futures are bound so closely together? What do we think we are—an island floating isolated in the sky, totally independent of all of these influences? Are the tens of billions of dollars in exports to Mexico and the 500,000 to 2 million people who enter the United States illegally every year negligible with respect to regional and global stability? Is it all right to pretend that we have no economic or other relationships with them at all?

We also belie our dependence on global stability by the moral and ethical choices we sometimes make in our economic activities with our global partners. For example, we blithely export substances banned in the United States, such as DDT and other chemicals, which cause untold environmental and health damage; we attempt to expand markets for items such as cigarettes, thereby creating exorbitant social costs for those countries in the future; and most of all, we have the dubious status of the largest arms exporter in the world, with sales of about $40 billion a year (compared to $4 billion in foreign assistance). Can the United States assume a stewardship role for the global environment and for global stability without a change in our ethics toward the global economy?

Most Americans do not know how the majority of people in the world live, and most do not appear to want to know. They are very content to worry about problems at home. At the same time, citizens of the United States consume resources avidly by being, so to speak, too kind to themselves. Consider two examples. Although most would agree that true health care also includes environmental health, sustainability, and ecological stability, the United States has an inadequate health system and spends one-sixth of its gross national product on health care. A sizable portion of that is spent on health care in the last year of life. Few people question whether prolonging life by 3 months is more important than prolonging the productive life of this planet so that it could be a place where our children and grandchildren will live with something like the privileges and opportunities that are found here.

Second, in the United States gasoline is sold at a price that is one-third less than was charged for gasoline in 1945, using constant dollars. In constant dollars, gasoline in 1945 was 21 cents a gallon, but now, using the same measure, it is about 14 cents a gallon. Still, politicians and citizens alike posture as if the world were coming to an end at the thought of a 5-cent tax increase on a gallon of gasoline. This strategy will not work in the long run. If the world were a balloon filled with petroleum and if that were being used at its present rate, the entire balloon would be deflated in 600 years. Of course, the world is not like this imaginary balloon, so petroleum reserves will have a much shorter life span. A $250-billion subsidy on gasoline prices is just another way of lowering still further what are already the lowest tax rates of any industrialized country.

While we use twice as much energy per capita as many other industrialized nations use, the growth in United States population from about 135 million at the time of World War II to about 270 million at present has caused us to look abroad for sources of petroleum and to be deeply preoccupied with the Middle East, Mexico, and Venezuela. It causes us to drill around our shores, to threaten our wildlife refuges, and to turn to nuclear energy. The United States gains little, if anything, compared with Sweden, Switzerland, or Germany by using twice as much energy per capita because these countries live about as well as we do, if not better. Furthermore, because our living standards are some 30 times those of many people in developing countries, the ecological impact of the 135 million of us added since World War II is equivalent to 4 *billion* people in certain parts of the developing world.

SOLUTIONS

What are the solutions? Saving biodiversity is not an academic exercise, nor can it be seen in isolation from the social contexts that underlie the use of biodiversity or the values human societies place on the natural world. As Norman Myers pointed out so eloquently in the past and in this volume, saving biodiversity can be accomplished only by arriving at a condition of global stability (Myers 1993). If a condition of global stability cannot be achieved, biodiversity will not be saved. It does not matter how elaborate our schemes are; it does not matter how well-thought-out they might be; it does not matter how much we understand about them. We still will not be able to save biodiversity without achieving global stability.

At the same time—and here is the tragedy and irony of the situation—if a substantial portion of biodiversity is not saved, we will not have the organisms around that will enable us to restore the earth and build sustainable communities.

We need a new way of thinking. The old ways of thinking will not do because the questions and problems are too profound. If we, in our utter fascination with our own welfare and the welfare of our communities and our nations, continue to operate under the delusion that being extremely self-serving and self-sufficient will save the world's biological diversity, then we will probably lose it.

Because the basic conditions for biological sustainability will not change, we must change. We must define a new set of values. If we are to live in peace, tranquility, and stability in the world, and if this condition is to be passed on to our children, then we need a new covenant with the world and all its peoples—a new commitment to understanding and a new commitment to action.

For the United States: Seeing the World as it Really Is

When we talk about a confrontation between the environment and economics or development, we forget that the environment is the context in which all economics must take place. It is not a simple tradeoff between a set of environmental laws and a set of economic laws. The environment is all that we have on this planet, and within that environment, all hopes, dreams, and aspirations, including our economic and financial hopes, dreams, and aspirations, must be forged.

Although economists, working mainly in Europe and the United States over the past 200 years, have invented a series of financial equivalencies and so-called laws, which are believed to govern the way the world operates, they do not, in fact, govern the way the world operates. No single economic law or principle will make this planet increase in size by 1 centimeter or will make it more resilient or more able to operate into the future. The environment is all that we have; we must understand and cherish it because it is the only thing that supports us.

The point at which the world population finally stabilizes will depend on whether the community of nations will act on the excellent recommendations of the Population Conference in Cairo in September 1994 and, as a matter of urgent priority for all of our security, supply modern contraceptives to the estimated 300 million women in the world who would like to have them but do not and address the social conditions that underlie fertility rates. If that were to happen, the world population might stabilize between 7.8 and 8.5 billion people. If that does not happen, the world population will not stabilize until it reaches perhaps 14 billion people at the middle of the next century, 20 billion people at the end of the next century, or something really unspeakable. But even those figures do not tell the full story.

For the world to stabilize at the level of 14 billion, 20 billion, or any number, we will have to be devoted to the cause of population stabilization around the world as well as to the problems associated with it. In the United States and the rest of the industrialized world, no future figure of world population will just happen by chance. That future depends on our choices and our actions.

Social Justice, Global Discrimination, and Sustainability

One of the ways in which developed countries must change is in the attention they give to social justice. Social justice is now considered by some to be an old-fashioned concept. Congress often does not seem to worry much about social justice, especially in the developing world and especially as it relates here and abroad to issues of the environment. What this form of social justice means is that all people have the opportunity, capacity, and freedom to express their humanity by caring for one another and caring for their environment. Others have pointed out that there is no possibility of caring for nature, either in an ethical sense, a spiritual sense, or a factual sense, if we do not first pause to care for one another. As Richard Leaky has noted, "Saving the environment is not possible without one square meal a day."

Anyone who professes to gain strength from religious teachings ought to be deeply troubled by the fact that the developed world draws a standard of living from the world environment that is 20 to 30 times higher than the condition that most people find themselves in, unless at the same time we accept some obligation to help to improve the world and to use our energies and institutions to enhance social justice for others.

If we want to embrace all peoples and benefit from their philosophical and cultural diversity for the betterment of the world and the development of a sustainable global society, the industrial world must explicitly discard the idea of getting as much as possible as soon as possible, regardless of how adaptive that notion might have been in the past. People everywhere must be taught to recognize that the earth is our single planetary home and must embrace human diversity and empower its potential. But people must learn that this is possible only in the context of a healthy environment.

It has already been mentioned that internationalism is not popular in the United States, although internationalism is fundamental for creating the conditions for global sustainability. If each of us does not find ways for ourselves, our children, and our fellow citizens to understand the conditions in which a great majority of the people in the world are living, we will not find the wherewithal—emotional, financial, intellectual, or any other kind—to truly contribute to a common view of the world and to the attainment of global stability. This point is critical. Building a blueprint for global sustainability based on the maintenance of biological diversity will not emerge from decisionmakers in Gland, New York, or Washington about how the world ought to make choices about conservation. Ultimately it must derive from social empowerment set within a broad consensus and value system of societies that recognizes the mutual interdependence of the global community (Raven 1990).

Science as a Form and Mechanism of Empowerment

Six percent of the world's scientists and engineers live in the developing world. In other words, 80 percent of the people in the world, with 80 percent of its biodiversity, have to make do with just 6 percent of the world's scientists. One of the worthiest aims of U.S. international assistance programs is to provide training and opportunities for people in developing countries. INBio in Costa Rica and the analogous bodies in Mexico, Taiwan, and elsewhere are organizations in which economic use, proper management, education, conservation, and academic study come together in such a way that people will pursue the preservation of biodiversity because they understand it to be in their own self-interest. This is another way of saying they are pursuing regional stability.

Going a step further, not only do just 6 percent of the world's scientists and engineers live in developing countries, but most of that small number are concentrated in a few countries such as China, India, Mexico, Brazil, Colombia, and Venezuela. Immediately one can see that for over half the countries in the world, there is virtually no scientific or technical expertise.

Imagine what it might mean to live in a country without any scientific or technical expertise and then imagine one had to decide whether to agree to an international convention or to try to get together with other nations to achieve common objectives. There would be no way to assess those objectives, based on a critical evaluation. Nothing could be more in the interests of the United States and other industrial countries than to build this kind of self-confident, informed, and empowered desire for sustainability in every country of the world. The industrialized world cannot afford isolationist thinking that involves reducing foreign development assistance. On the contrary, these countries should be expanding programs that serve the aims of people everywhere and empower them to become partners in building sustainable and stable societies.

A major part of this effort is fostering institutions dedicated to expanding knowledge of biodiversity: museums, botanical gardens, research stations, and universities (Cracraft 1995). One cannot save intelligently or efficiently what one does not know. All knowledge about biodiversity that can be shared globally—whether through science, poetry, art, or something else—will be essential if we are to preserve biodiversity for its productivity, its restoration, its collective economic use, and its beauty.

One of the most cynical points of view commonly driving policy decisions in the United States is the belief we cannot do anything abroad until we settle matters domestically. Yet this desire itself is misguided, not only because all societies are intricately linked economically, culturally, and ethically to one another, but also because of its irony: to any observer it would appear as if we in the United States have little political will even to deal with our own domestic environmental agenda effectively.

One of the most telling examples of this is polls taken at election time that survey citizens' priorities for our domestic policy agenda. The environment regularly appears far down the list. Instead, Americans tell their candidates something quite different: we want more money (make the economy grow), we want lower taxes, or we want crime eliminated.

There are many dedicated and intelligent people at all levels of government who would love to hear that preserving the environment is a high priority. But if all we can tell them is that we want more economic expansion, lower taxes, more jails, and better military preparedness, rather than a better environment, better education, more economic and social justice, no poverty, or the promotion of human development around the world, why is the general indifference of politicians toward these goals so surprising or disappointing? Never in U.S. history has the president or any politician been free to devote large amounts of money and effort to environmentalism or any other thing that he or she might have wanted to devote money and effort to, unless it was backed up strongly and politically by the people.

The simple fact is that we are not going to have long-term positive results on the environment by demonizing anyone; instead, it will happen only by saying what we want, and saying it often, to those who represent us. Citizens must tell their elected representatives that the long-term health of our country and of the world, and of those who live in it, depends not on the voracious consumption of everything we can imagine, but on forging a stable relationship with the earth.

Human beings have always been able to change for the better, given the right kinds of values, and they can change very rapidly. If we truly control our own destiny, then we cannot continue business as usual. We not only can change, we must change. The world remains finite; that fact does not disappear, even in the face of our individual preoccupations and agendas. Because the earth is finite, we must forge a peaceful relationship with it. To succeed in this goal, each individual must make a personal commitment and not excuses. Everyone is confused, at some level, about what ought to be done in the face of the current environmental crisis. However, it is time to stop using confusion as an excuse for inaction and begin devoting ourselves to building a world in which future generations can be healthy and prosperous.

REFERENCES

Cracraft, J. 1995. The urgency of building global capacity for biodiversity science. *Biodiversity and Conservation* 4: 463–475.

Daly, H. E. and J. B. Cobb, Jr. 1989. *For the common good*. Boston: Beacon.

McMichael, A. J. 1993. *Planetary overload: Global environmental change and the health of the human species.* London: Cambridge University Press.

Myers, N. 1993. *Ultimate security: The environmental basis of political stability.* New York: Norton.

Raven, P. H. 1990. The politics of preserving biodiversity. *Bioscience* 40: 769–774.

Simmons, I. G. 1996 (2d ed.). *Changing the face of the earth: Culture, environment, history.* London: Blackwell.

Tolba, M. K. and O. A. El-Kholy. 1992. *The world environment 1972–1992.* London: Chapman & Hall.

Turner, B. L. III, W. C. Clark, R. W. Kates, J. F. Richards, J. T. Mathews, and W. B. Meyer. 1990. *The earth as transformed by human action.* New York: Cambridge University Press.

Vitousek, P., P. R. Ehrlich, A. H. Ehrlich, and P. A. Matson. 1986. Human appropriation of the products of photosynthesis. *Bioscience* 36: 368–373.

Wilson, E. O. 1992. *The diversity of life.* Cambridge, Mass.: Harvard University Press.

World Resources Institute. 1996. *World resources 1996–1997.* New York: Oxford University Press.

World Resources Institute. 1997. *The last frontier forests: Ecosystems & economies on the edge.* Washington, D.C.: World Resources Institute.

PERSPECTIVE

SCIENTISTS' PUBLIC RESPONSIBILITIES

Strachan Donnelley

For someone with a professional life in philosophy and practical ethics, the *Living Planet in Crisis* conference, with its impressive array of scientific experts and hundreds of attendees, is an extraordinary event. We have come together to discuss what we know and do not know about the present state of the earth's living nature and what we humans ought to do about the threats to its and our long-term viability. We are explicitly recognizing public or citizen responsibilities for the present and future of nature and our human communities. This is truly significant, at least as a first step.

What can I usefully add? As a philosopher in a sea of accomplished scientists, I am unsure whether I feel more like a spotted owl or a zebra mussel, an endangered species or an exotic species, resigned to insignificance or to causing havoc. Whatever, I will play the gadfly and concentrate on the broad theme of ignorance, which we philosophers are professionally bound to profess.

I have been struck by the philosophic tone of the conference: scientists' confessions of ignorance about what they still do not know about the diversity of biological species and their roles in evolutionary and ecological processes. There have been recurrent calls for more and better science, aimed to help us in deciding and

acting on crucial public policy issues. Who could quarrel with the practical importance of such pleas?

Yet I am equally struck by a silent indictment of scientists: their failure to dis-
pel the ignorance of the public, to educate us citizens about nature and science and
how the two ought to figure into meeting public responsibilities to ourselves and
nature. Granted, this is a difficult and complex problem, but so far the scientific
community has failed us. Too many scientists have considered us citizens unedu-
cable, not worth the effort, that we need only recognize and accept the authority of
scientists' expert findings. This is a fatal mistake. It leaves even publicly minded
and concerned citizens dangerously unmoored and in a moral darkness.

For example, scientists fail to educate us about the nature of science—that science is an ongoing exploration of nature and human life, with empirical findings
and concepts always incomplete, tentative, or open to revision. We tend to take the
latest science as definitive, complete, or ready to plug into our technological and
progressivist dreams, ignoring possible distortions of human or natural life that
might await us. Specifically, scientists have failed to help us face human ignorance
with respect to the effects of large-scale corporate, economic, and public policy initiatives. In the main, the scientific community has fed our economic and technological boosterism and left us bulls in the China shop of nature.

Here evolutionary biologists and ecologists should particularly feel the moral
sting. They have failed to grab us citizens by the throat and make us understand
and take to heart that human communities and their activities, economic and otherwise, are nestled within wider and vulnerable living systems. We citizens still do
not get it and blunder about with an all too clear conscience. We remain ignorant
of basic principles of evolutionary biology and ecology, a general ignorance that really matters. Our global economies bulldoze rain forests, gulp down our fresh waters, erode our soils, pollute our air, and plunder our fisheries and ocean depths.
More than human necessity and moral indifference are here at work. The ignorance of citizens and their communities plays a decisive role.

More and better scientific knowledge will not suffice. The conference's White
Paper and these proceedings will not be enough. Who among us ignorant citizens
will read them, take them to heart, and actively heed their recommendations? Scientists will have done their public duty and fulfilled their own citizen responsibilities when conferences such as *The Living Planet in Crisis* are held not only at museums of natural history, but in the atrium of the World Financial Center, the
auditoriums of our teaching hospitals, and the halls of Congress and other legisla-
tive bodies—with portfolio managers and leveraged buyout artists, surgical cow-
boys and medical biotechnologists, and contractors on America and our natural
habitats as much in attendance as scientists and the environmental chorus.

At bottom, whatever our professions, we are all ordinary citizens. Now and
henceforth, we must assume extraordinary responsibilities. Can we do it? Are we
scientifically and morally educable? That is the question that challenges scientists,

ethicists, and all of us alike. No doubt it will take all of us working together to pull it off. Yet if significant numbers of citizens can fathom the *Wall Street Journal* and the *New York Times* business section, laser technologies and transgenic knockout strategies, and arcane bureaucracies and regulations, surely they can grasp the principles and practical implications of evolutionary biology and ecology.

There is no inherent reason why scientists should leave the rest of us in the dark as dangerous bumblers. Scientists do have a professional responsibility to press for more and better science to help inform practical policy. But equally, they have a professional and human responsibility to educate us citizens, to help us get our heads screwed on right and meet our long-term responsibilities to humans and nature. The latter task in public education may prove as formidable, challenging, and practically important as scientific exploration itself.

The gauntlet of public education is thrown down and ought not be ignored. Our citizen ignorance is an integral part of the living planet in crisis.

Susan Bragdon
Genetic Resources Science and Technology Group
International Plant Genetic Resources Institute
Via delle Sette Chiese 142
00145 Rome, Italy
Tel.: 39–06–51892400
Fax: 39–06–5750309
E-mail: s.bragdon@cgnet.com

John Burnett
International Organisation for Plant Information
13 Field House Drive
Oxford OX2 7NT, UK
Tel . : 44–1–865–513–324
Fax: 44–1–865–513–324
E-mail: jhb@easynet.co.uk

Eric Chivian
Center for Health and the Global Environment
Harvard Medical School
260 Longwood Avenue
Boston, MA 02115
Tel . : 617–432–0493
Fax: 617–432–2595
E-mail: chivian@mit.edu

Joel Cracraft
Department of Ornithology
American Museum of Natural History
Central Park West at 79th Street
New York, NY 10024
Tel . : 212–769–5633
Fax: 212–769–5759
E-mail: jlc@amnh.org

Strachan Donnelley
The Hastings Center
Garrison, NY 10524
Tel . : 914–424–4040
Fax: 914–424–4545
E-mail: mail@thehastingscenter.org

Clare Flemming
Department of Mammalogy
American Museum of Natural History
Central Park West at 79th Street
New York, NY 10024
Tel . : 212–769–5481
Fax: 212–769–5239
E-mail: flemming@amnh.org

Francesca T. Grifo
Center for Biodiversity and Conservation
American Museum of Natural History
Central Park West at 79th Street
New York, NY 10024
Tel . : 212–769–5742
Fax: 212–769–5292
E-mail: grifo@amnh.org

Antonella Ingrassia
Via Pacini 93
95129 Catania, Italy
Tel.: 39–347–665–6073
E-mail: media094@telmedia.it

Thomas E. Lovejoy
The Smithsonian Institution
1000 Jefferson Drive S.W.
Washington, DC 20560
Tel . : 202–357–4282
Fax: 202–786–2304
E-mail: siwp01.si.tlovejoy@ic.si.edu

Ross D. E. MacPhee
Department of Mammalogy
American Museum of Natural History
Central Park West at 79th Street
New York, NY 10024
Tel . : 212–769–5480
Fax: 212–769–5495
E-mail: macphee@amnh.org

Jeffrey A. McNeely
IUCN: The World Conservation Union
Rue Mauverney 28
CH-1196 Gland, Switzerland
Tel . : 41–22–999–0001
Fax: 41–22–999–0002
E-mail: jam@hq.iucn.ch

Dominic Moran
Centre for Social and Economic Research on the
    Global Environment
University College London
Gower Street
London WC1E 6BT, UK
Tel . : 44–171–387–7050
Fax: 44–171–916–2772
E-mail: d.moran@ucl.ac.uk

Kalemani J. Mulongoy
International Academy of the Environment
4 Chemin de Conches
Conches, Geneva, Switzerland
Tel . : 41–22–702–1867
Fax : 41–22–702–1899
E-mail: kalemani.mulongoy@iae.org

Norman Myers
Upper Meadow, Old Road
Headington
Oxford OX3 8SZ, UK
Tel . : 44–865–750–387
Fax: 44–865–741–538
E-mail: normanmyers@gn.apc.org

David Pearce
Centre for Social and Economic Research on the
    Global Environment
University College London
Gower Street
London WC1E 6BT, UK
Tel . : 44–171–504–5874
Fax: 44–171–916–2772
E-mail: dpearce@ucl.ac.uk

David Pimentel
College of Agriculture and Life Sciences
Cornell University
Ithaca, NY 14853
Tel.: 607–255–2212
Fax: 607–255–0939
E-mail: dp18@cornell.edu

Norman I. Platnick
Department of Entomology
American Museum of Natural History
Central Park West at 79th Street
New York, NY 10024
Tel . : 212–769–5612
Fax: 212–769–5277
E-mail: platnick@amnh.org

Peter H. Raven
Missouri Botanical Garden
P.O. Box 299
St. Louis, MO 63166
Tel . : 314–577–5110
Fax: 314–577–9595
E-mail: praven@nas.edu

G. Carleton Ray
Department of Environmental Sciences
University of Virginia
Charlottesville, VA 22903
Tel . : 804–924–0551
Fax: 804–982–2137
E-mail: cr@virginia.edu

Melanie L. J. Stiassny
Department of Ichthyology
American Museum of Natural History
Central Park West at 79th Street
New York, NY 10024
Tel .: 212–769–5796
Fax: 212–769–5495
E-mail: mljs@amnh.org

Nigel E. Stork
Cooperative Research Centre for Tropical Rainforest
    Ecology and Management
James Cook University
Smithfield Campus, P.O. Box 6811
Cairns, Queensland 4870, Australia
Tel .: 6170–42–1246
Fax: 6170–42–1247
E-mail: nigel.stork@jcu.edu.au

Diana H. Wall
Natural Resource Ecology Laboratory
Colorado State University
Fort Collins, CO 80523
Tel .: 970–491–1982
Fax: 970–491–1965
E-mail: diana@nrel.colostate.edu

Arthur H. Westing
Westing Associates in Environment, Security,
    and Education
134 Fred Houghton Rd.
Putney, VT 05346
Tel .: 802–387–2152
Fax: 802–387–4001
E-mail: westing@together.net